Lecture Notes in Networks and Systems

Volume 35

Series editor

Janusz Kacprzyk, Polish Academy of Sciences, Warsaw, Poland
e-mail: kacprzyk@ibspan.waw.pl

The series "Lecture Notes in Networks and Systems" publishes the latest developments in Networks and Systems—quickly, informally and with high quality. Original research reported in proceedings and post-proceedings represents the core of LNNS.

Volumes published in LNNS embrace all aspects and subfields of, as well as new challenges in, Networks and Systems.

The series contains proceedings and edited volumes in systems and networks, spanning the areas of Cyber-Physical Systems, Autonomous Systems, Sensor Networks, Control Systems, Energy Systems, Automotive Systems, Biological Systems, Vehicular Networking and Connected Vehicles, Aerospace Systems, Automation, Manufacturing, Smart Grids, Nonlinear Systems, Power Systems, Robotics, Social Systems, Economic Systems and other. Of particular value to both the contributors and the readership are the short publication timeframe and the world-wide distribution and exposure which enable both a wide and rapid dissemination of research output.

The series covers the theory, applications, and perspectives on the state of the art and future developments relevant to systems and networks, decision making, control, complex processes and related areas, as embedded in the fields of interdisciplinary and applied sciences, engineering, computer science, physics, economics, social, and life sciences, as well as the paradigms and methodologies behind them.

Advisory Board

More information about this series at http://www.springer.com/series/15179

Mustapha Hatti

Editor

Artificial Intelligence in Renewable Energetic Systems

Smart Sustainable Energy Systems

 Springer

Editor
Mustapha Hatti
EPST-CDER
UDES, Unité de Développement des
  Equipements Solaires
Bou Ismaïl
Algeria

ISSN 2367-3370   ISSN 2367-3389 (electronic)
Lecture Notes in Networks and Systems
ISBN 978-3-319-73191-9   ISBN 978-3-319-73192-6 (eBook)
https://doi.org/10.1007/978-3-319-73192-6

Library of Congress Control Number: 2017963750

Printed on acid-free paper

This Springer imprint is published by Springer Nature
The registered company is Springer International Publishing AG
The registered company address is: Gewerbestrasse 11, 6330 Cham, Switzerland

Preface

The development of renewable energy at low cost must necessarily involve the intelligent optimization of energy flows and the intelligent balancing of production, consumption, and energy storage. Intelligence is distributed at all levels and allows information to be processed to optimize energy flows according to constraints. It is around this thematic that is shaping the outlines of the economy of the future, and behind these topics is the possibility of transforming society. Taking advantage of the growing power of the microprocessor makes the complexity of renewable energy systems accessible. Especially, since the algorithms of artificial intelligence are gourmand in data making, it possible to take relevant decisions or even reveal unsuspected trends in the management and optimization of renewable energy flows. This book is a good prospect for those who work on energy systems and those who deal with models of artificial intelligence to combine their knowledge and their intellectual potential for the benefit of the scientific community and humanity.

Short Biography

 **Dr. Mustapha Hatti** was born in El-Asnam (Chlef); he was educated at El Khaldounia school, then at El Wancharissi high school, and obtained his electronics engineering diplomat at USTHB, Algiers, and his post-graduation studies at USTO-Oran. He worked at CDSE, Ain oussera, Djelfa, CRD, Sonatrach, Hassi messaoud, CRNB, Birine, Djelfa, UDES/EPST-CDER, Bou Ismail, Tipasa. He is an IEEE senior member, he is the author of several scientific papers, he is a senior researcher, and his areas of interest are sustainable energy systems, innovative, fuel cell, photovoltaic, optimization, intelligent embedded systems.

Contents

Renewable Resources

Intelligent Maximum Power Point Tracking

Smart Home in Smart Cities

Smart Buildings and Occupants Satisfaction: The Case of Cyber Park of Sidi Abdallâh and Some Residential Buildings in Algeria

Tizouiar Ouahiba[1(✉)], Belkadi Fatima[2], and Hamel Thafath[2]

[1] LAE, Laboratoire Architecture et Environnement,
Ecole Polytechnique d'Architecture et d'Urbanisme EPAU, Algiers, Algeria
otizouiar@yahoo.com
[2] Ecole Polytechnique d'Architecture et d'Urbanisme, Algiers, Algeria

Abstract. We have contributed to explain how the project managers and promoters proceed to realize smart buildings, how to optimize their energy efficiency by allowing an effective management of the resources with a minimum of costs. This by analysing some examples of smart residential buildings in Algeria. Then, we tried to find the satisfaction degree that these buildings offer to their occupants, but for this part, it concerns a tertiary building, by doing a diagnosis at the cyber park site of Sidi Abdallah and using also a satisfaction survey.

This study demonstrate that for succeeding such as building and its good energy management, also its techniques related to home automation; integration of smart systems, a less energy consuming. It is necessary to ensure that these facilities meet the needs of the users, ensure the simplicity of use to suit the equipment whether in terms of operation or programming, the reliability and durability of equipment too that does not slow down consumers in their acceptance.

Keywords: Smart buildings · Technology · Behavior · Energy efficiency Comfort

1 Introduction

The building sector is the largest energy consumer among the various economic sectors in the world; its consumption represents 40% of the world's energy, and therefore a carbon dioxide emitter. Faced to this situation, and in order to respond to environmental occupations of today, building must adapt. Taking into account regulatory requirements, technical developments (the advent of technology and renewable energy sources), and the new societal aspects (user behavior). This will give rise to an ecological and smart management building. The smart building is today an element of major response to the different modes of energy consumption as well as in terms of comfort and simplification of inhabitant life. This can done only through development of new information and communication technologies (ICT); and therefore the use of home automation equipment; from heating to ventilation, lighting, and shutter and blind management, which is supposed to bring the best solutions to the performance of building.

© Springer International Publishing AG 2018
M. Hatti (ed.), *Artificial Intelligence in Renewable Energetic Systems*, Lecture Notes in Networks and Systems 35, https://doi.org/10.1007/978-3-319-73192-6_1

First, our interest has focused on the subject of reasoned introduction of domotic and multimedia technologies in architectural work, in order to have a house that simplifies life to its inhabitants, a house where everything is comfortable, communicating, scalable, autonomous, safe and economical. Our goal is to draw on the domain of home automation for energy efficiency, which is a very important aspect in a smart residential building. This is through integration of innovative constructive solutions based on the smart management of produced and consumed energy. The research also reveals the emergence of this concept in Algeria and it deepens about various difficulties and deficiencies, which caused this flagrant lack of this concept in our society.

The second part of our research focuses on attractiveness and energy efficiency in the scale of building, using technologies, which seems to be an inescapable approach all over the world. Algeria has taken note of these innovations by creating a technological park, namely the Cyber Park of Sidi Abdellah in Algiers. In this sense, the present research consists in understanding the different concepts and notions related to smart buildings, the study of the interaction between occupant and his physical environment with satisfaction level offered by this specific building. As well, as evaluate the experience success of smart building as an actor of technological attractiveness and energy saving.

2   Problematic

The main problems that arise for residential or tertiary smart buildings are the following:

"What are the architectural features and techniques that allow good energy management in a smart residential building?" This is part of a current debate (fight against global warming) and emanates from a basic fact problematic related to the Algerian context, trying to verify whether smart residential buildings exist in Algeria, and understand how the masters of works and promoters do for their designs and how satisfied occupants with these smart buildings?

"Is the development of attractive smart building subject to constraints social acceptability of technologies?" This concerns the lived experience of technological buildings, specifically for the administrative seat block and multi-building -locator of the Cyber Park of Sidi Abdellah, trying to verify the impact of automation degree on the appropriation of their spaces by users.

3   Cases Study

3.1   Case of Residential Buildings Integrating Some Smart Solutions (Individual and Collective Housing)

We first addressed a presentation of some attempts to build this type of building in Algeria; The individual and the collective, while trying to emphasize different devices and techniques applied in this type of building, in particular in the field of energy efficiency. It concerns smart individual housing in "Paradou, Algiers", smart housing in Oran and "Residence des pins" of the Lebanese promotion in Cheraga, Algiers [7].

3.2   Case of the Tertiary Smart Buildings in Algeria

Algeria, like other countries in the world, has invested in this domain by creating a technology park of world-class that will provide state-of-the-art infrastructure and services to businesses and institutions. It is a real reference in the field of buildings called "smart".

Our study in this case focuses specifically on two blocks: the administration building and the multi-building locator, workspaces distribution of the first one is of two types; Landscaping, and other individual offices. For the second, distribution of workspaces in the business center is open space [1].

4   Methodological Approach

The research begins with a state of knowledge on the different principles of smart buildings and their implementation mechanisms and approaches throughout the world and over time [3, 4, 6, 8, 9].

In the case of residential buildings, this theoretical step has been combined with another phase, which is based on a fieldwork carried out in a qualitative dimension focusing masters of works or companies realizing this type of construction in Algeria. This, in order to situate themselves in relation to this type of architecture and make a comparison with international references. The work is accompanied by two interviews for the residents of this type of building as well as for the project managers; In order to allow a better understanding of the subject.

Also in the case of the Cyber Park, the state of knowledge on so-called "smart or technological" buildings and their attractive role in terms of comfort, telecommunications, security and management has been combined with another phase, the analysis of places in order to list the technical management system in the chosen buildings, and the comfort facilities. Then a survey is carried out on a representative sample in order to qualify the quality of use. Use of the post-occupational qualitative method (POE) is adopted for this case as a tool for assessing performance of smart buildings and assessing user satisfaction based essentially on impressions and perception of users on their workspace [5].

The survey in this part of the study takes place at the first two levels "indicator" and "investigator", without the "diagnostic" level. The pre-survey concerns the collection of data on the state of the premises and this through guided tours. A documentary and iconographic collection was compiled for each building. The second step is "maintenance". It was from this database that the analysis was carried out. We then came up with a grid of analysis by establishing the thematic categories on which the survey is constructed (Fig. 1).

Fig. 1. Technology method used for qualitative analysis and description of the representative sample

5   Results and Discussion

5.1   Analysis of Results for Residential Buildings

For this analysis, we used most of criteria about smart building, taking care of home automation in two domains: home automation for energy efficiency and home automation for comfort [4, 7–9].

Smart Individual Housing in "Paradou, Algiers": The site constraints have conditioned some decisions by designers, which has influenced the passive energy efficiency of the house (implantation and orientation [10], Shape, openings, materials); After which a few criteria of home automation are applied, either to the service of energy or comfort. The Fig. 2 shows the percentage of the criteria used for home automation in the field of energy efficiency and home automation for comfort. These criteria are deduced from the theoretical part of our research (management of sunshades/blinds, control and management of heating, dual flow ventilation plant, access control, and technical detection, sound.

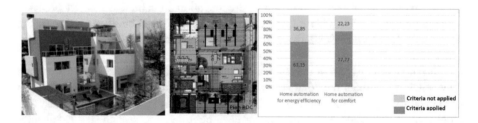

Fig. 2. Passif and technological criteria applied in Smart individual housing in "Paradou, Algiers"

Smart Housing in Oran: The influence of northern orientation of rooms (less of natural light) [10] has increased relatively the use of home automation for energy efficiency (Fig. 3).

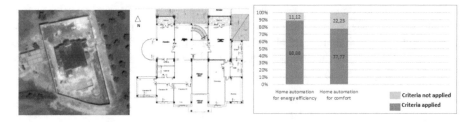

Fig. 3. Passif and technological criteria applied in Smart individual housing in "Oran, Algeria"

"Résidence des pins" of the Lebanese Promotion in "Cheraga, Algiers": It incorporates some smart solutions (Energy management, building security, Building management by a specialized administration). Various parameters of comfort and safety are also programmed during the design of the structure, as the opening of the car park, motion detectors, and the use of heating floor. Home automation is used in a more usual way for comfort service, especially in the field of security and lighting (Fig. 4).

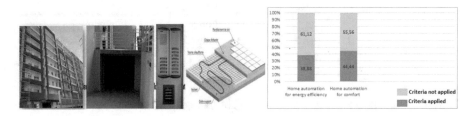

Fig. 4. The criteria applied in smart residential building "Residence des pins" of the Lebanese promotion in Chéraga

Smart buildings are beginning to appear in our company, especially in individual villas, although few or rare design offices that opt for this innovative option. The latter has begun to appear timidly in the residential building sector, and for the moment there are only a few construction attempts that have begun to integrate home automation in a few areas, particularly in the field of building security. On the other hand, inventions in this field are not yet widespread because of their cost and complexity. Only people who live in well-off financial conditions can benefit. Indeed, to be able to speak of smart building in Algeria, there must be a mastery of four basic pillars namely materials, protection, air conditioning and electricity supply but also an essential element which is management in terms of energy, monitoring and maintenance.

Indeed, we have been able to understand also the steps and the stages of realization of an smart building; and that the success of the integration of new technologies requires a good design upstream, the choice of adapted solutions, therefore a match of the solutions to the expectations of the users and an accompaniment in the handling of the systems.

The two interviews [7] made it possible to answer our question about the state of satisfaction of the inhabitants and the level of comfort offered by this kind of concept, and which is still a concept unknown and poorly mastered by the inhabitants. And also to know the various deficiencies and difficulties caused by this flagrant lack of this type of concept in Algeria, among them there is the high cost of the home automation installations and these techniques and also the housing crisis of which our country suffers nowadays. However, for the time being there is no regulation in this sense, for a sustainable and smart construction in its own sense.

5.2 Analysis of Results for Tertiary Buildings

Regarding cases Cyber Park, after the analysis of documents and interviews, we came up with an analysis grid on which we built our survey. The main objective is to discover the living quality environment studied through the levels use-satisfaction, comfort and behavior by evaluating physical factors such as lighting, cooling, heating, air quality and safety [1] (Table 1).

Table 1. Analysis grid

Comfort installation	Comfort		Behavior
	Advantages	Disadvantages	
	Lighting	Lighting	Lighting
	Heating/Cooling	Heating/Cooling	Heating/Cooling
	Air quality	Air quality	Air quality
	Security	Security	Security

First, for the Administration Building: The category that expressed its satisfaction consists of engineers who are actors in the management of these buildings who are optimistic for improvement of this system. Some occupants do not appreciate this type of building in reference to their offices, it feels "bad", this state is mainly related to physical and psychological needs. This state of dissatisfaction mainly concerns the feeling of confinement and the discomfort expressed by the occupants. On the other hand most of the occupants are more or less satisfied with the comfort devices put in place

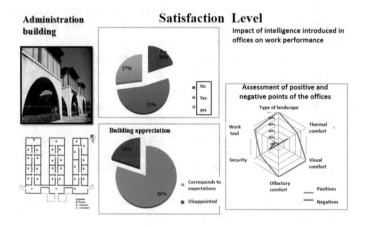

Fig. 5. Satisfaction level for the administration building

because according to the latter they ensure a better deliverability and a speed in the exchanges. At this stage of analysis we can observe that dissatisfaction is linked to factors concerning the envelope and the choice of the technical management system (Fig. 5).

Result of the Level of Satisfaction of the Cooling and Heating Factor: The majority of respondents were satisfied (54%), the rest were unsatisfactory (20%), moderately satisfied (13%) and acceptable (13%). They consider that offices are not suitable or moderately suitable, because of the centralized system used. First, that is generalized for all offices, it is the cold witch be regulated for all offices heating also, the occupant can not intervene to satisfy his own need because the regularization for his office will be the same for the neighboring office whereas the orientations of the offices are different. The period of sunshine of offices in the day is different too. Adding to constant interruption of electricity cause failures in the system whose absence of alternatives planned for this kind of problem [1, 2].

The cooling control is very frequent, 53% of the respondents often intervene in the regulation of the air-conditioning, while 47% are more satisfied with the centralized control, but the control would always do it in their offices.

Same for heating 40% responded with thermal comfort, while 27% said it was acceptable, 20% unsatisfactory and 13% moderately satisfactory. This dissatisfaction is always argued by the constraints of centralized system control. The heating control frequency is high with 87%, 47% "always" and 40% "often" (Fig. 6).

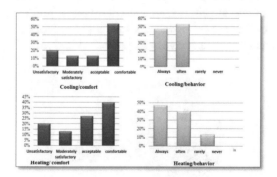

Fig. 6. Satisfaction level of the cooling and heating factor for the administration building

Level of Satisfaction of the Lighting Factor [1, 3]: Overall, the lighting was assessed as "comfortable" and we noticed respondents' appreciation of introduced intelligence (presence detection). Indeed 60% often regulates the intensity, 27% never and 13% rarely. Whereas other (40%) prefer to turn off the light, (47%) rarely and (13%) justify what is satisfactory. These choices are justified by the good sunshine of the offices and their orientations.

Level of Satisfaction of the Air Quality Factor: Air conditioning was designed as the only source in offices, two-thirds of the occupants did not perceive it (73%) because of unsatisfactory new air intake and low and non-automatic extraction, respondents pointed out that the extraction of air requires a great power especially maintenance of the installations what absent in these buildings. For this reason, the occupants seem to perceive another source of ordinary aeration by the opening of the doors while the windows are conceived to be closed.

Satisfaction Level of the Safety Factor: Almost all occupants expressed satisfaction at being safe. However, 13% consider this supervision as a limitation of the freedoms of the workers who have as objectives in these buildings to give their best for a better deliverability. All occupants are sensitized to the subject of fire and they prefer to follow the instructions of the services concerned by the intervention in case of danger, instead of acting individually. We found that 40% of the respondents tried the manual fire system arguing that this is the best way to test the operation of these. But sometimes this is meant to stimulate a fire and make evacuation tests.

Level of Satisfaction of the Internet Network: Overall satisfaction expressed by the occupants, they argue that the speed of the Internet allows a speed in the exchange of data and a reactive flexibility.

Assessment of Awareness of Energy Consumption: The group of respondents are aware of the increases in consumption for this reason they say that they seldom open the doors while 20% justifies by the absence of alternative aeration with low air extraction and the openings designed closed. 13% of respondents returned to the main objective of smart building design, the one aimed at reducing consumption and more than producing energy, which is not the case for cyber park buildings. Reduce consumption by introducing the CTM (centralized technical management) (Fig. 7).

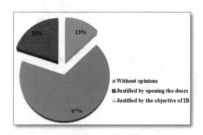

Fig. 7. Occupants trend level facing the energy consumption for the administration building

Second, for the Analysis of Survey Data on Multi-Locator Building:

Level of Satisfaction: The category that says the type of building has no impact on work performance is totally influenced by the physical urban setting of their workplace. The urban context of these buildings is unpleasant, an unsatisfactory of the service and accompanying equipment, and the difficulty of access to the site.

The overall satisfaction concerns the expectations linked mainly to the well being lived in a framework of professional life offered by this type of architecture. The expressed dissatisfaction with power cuts seems to be accentuated among the people who occupy the offices every hour of work, this disturbance results in delays and inconveniences in performing the occupant's spots.

The open space organization seems unfavorable in some because it limits privatization (Fig. 8).

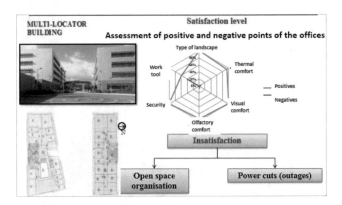

Fig. 8. General satisfaction level formulti-locator building

Level of Satisfaction of the Heating and Cooling Factor: Heating and air conditioning are acceptable. Complaints from occupants reveal the influence of the factors of office volume, number of occupants and control over the satisfaction of the management system and desired comfort (Fig. 9).

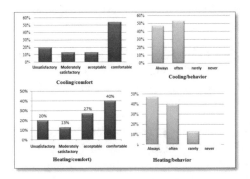

Fig. 9. Satisfaction level of cooling and heating factor for the multi-locator building

Lighting Factor Satisfaction Level [1, 3]: All the occupants are satisfied with the lighting of the offices either natural or artificial. The need for lighting is high expressed, on average 60% of respondents who turn off the light, 20% "never" while 20% expressed their willingness to turn off the light. The active control question showed that 47% (rarely) regulates intensity while 53% are satisfied with the intensity already programmed.

Level of Satisfaction of the Air Quality Factor: The air quality in the surveyed office is rated as acceptable (53%), (27%) feel comfortable, (13%) unsatisfactory and (7%) moderately satisfactory.

Air conditioning is not the only source of ventilation in the offices, unlike offices of the first building. As the air extraction is low so the return to the windows opening is maintained to renew indoor air. Indeed, there was satisfaction with the occupants (80%) nevertheless (20%) of people open the windows for no specific reason while the heating or air conditioning is working (Fig. 10).

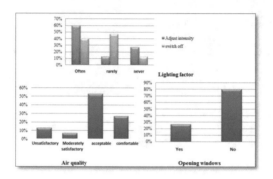

Fig. 10. Satisfaction level of lighting factor and air quality for multi tenant building

Safety Factor Satisfaction: All respondents are satisfied with the security system, they pointed out that the safety factor is one of the criteria for which tenants chose to occupy these offices. About fire, (20%) of the respondents are satisfied with the safety devices against fire while (80%) of which (20%) prefers to go towards the rescue and 60% did not answer the question.

Level of Satisfaction of the Internet Network: Overall satisfaction with the speed of the Internet network introduced in the offices facilitated exchanges and communication and minimized travel. However, electricity cuts are a major problem in the normal course of work.

Assessment of Awareness of Energy Consumption: (13%) of respondents say that the energy consumption in these offices is very high during the months of the year throughout the summer season, because demand for cooling is very important for all the volume of offices space) and the number of occupants is important. They add that most of the time certain overtures lets open to renew the indoor air in addition the office heats up quickly.

6   General Conclusion

Despite the difficulty of accessing examples of smart houses that rarely appear on Algiers. Given the difficulty of having a large percentage of the sample, we were able to conclude and understand that:

- Successful integration of new technologies requires good upstream design. This is through the choice of appropriate solutions, the suitability of solutions to the expectations of the users and the accompaniment in handling of the systems.
- Before thinking smart building it is necessary to begin by respecting the natural environment, thus restoring its nobility letters to constructive science, before moving to the veal of technology and has to home automation.
- In Algeria, inventions in this field of new information and communication technologies are not yet widespread because of their costs and complexity. It begins to appear in the individual villas and timidly in the residential buildings sector. For the moment there are only a few attempts of constructions that began to integrate this concept in some areas including the field of the security and of lighting.

The effect on the employees attitudes to their work and their performance was checked on the administrative block and the multi-locator building of Cyber Parc of Sidi Abdallah. The result is a better understanding of the relationship between the physical environment and the behavior of the people who live there, and which affects the architecture and technical management systems used:

- The results of this work show that the appreciation of this type of architecture seems rather acceptable for most of the respondents for the two buildings. The main results show trend reactions related to the programming and operation of comfort facilities.
- Technical quality of the building questions is ability to satisfy the requirements of technical and environmental performance, via compatible installations or systems and adapted to users. The intelligibility of these systems and their ease of handling must contribute to this technical quality as well as their intrinsic performances. The aim is to ensure that interfaces and commands are adapted to the skills and availability of those who will be required to manipulate them. Indeed, active control contributes positively to the lived quality of an environment, the most satisfactory comfort devices are those which offer the greatest margin of control such as heating and cooling. Also think about maintenance and guarantee the supply of electrical energy.
- The results of this study led us to determine that the search for quality of use is to conceive beyond techniques and norms (without to get rid of them), Needs of the users who will attend it daily, make it work and maintain it.

As part of future research, complementary research could be considered in order to refine the understanding in this vast field of intelligent buildings and to encourage people to this kind of concept especially in the field of energy efficiency.

References

1. Belkadi, F.: L'attractivité des bâtiments technologiques, l'expérience algérienne de conception au vécu: Le cyber parc de Sidi Abdallâh, 72 p. Master memory in architecture at Polytechnic School of Architecture and Urbanism of Algiers, directed by Tizouiar, O., October 2016
2. Chabane, I.J.: Evaluation de la qualité vécue des environnements hermétiques en mur-rideau de verre, 160 p. Magister memory in sciences in architecture at the Polytechnic School of Architecture and Urbanism of Algiers, directed by Pr Bensalem Rafik (2006)

3. Chenailler, H.: L'efficacité d'usage énergétique : pour une meilleure gestion de l'énergie électrique intégrant les occupants dans les bâtiment, 280 p. Ph.D. thesis, electronic department, National Polytechnic Institute of Grenoble (INPG), Grenoble (2012)
4. Dutreix, A.: Bioclimatisme et performances énergétiques des bâtiments, 239 p. Edition Eyrolles, Bd Saint Germain, Paris (2010)
5. Fischer, G.N., Vischer, J.C.: Evaluation des environnements de travail: la méthode diagnostique», 264 p. Montréal: Presse de l'Université de Montréal, and Brussels, DeBoek (1998)
6. Gonzalo, R., Habermann, K.: Architecture et efficacité énergétique: principes de conception et de construction, 215 p. BIRKHAUSER Edition, Boston, Berlin (2006)
7. Hamel, T.: Investigation sur les bâtiments résidentiels smarts en Algérie: L'étude de l'efficacité énergétique de ces bâtiments, 72 p. Master memory in architecture at the Polytechnic School of Architecture and Urbanism of Algiers, directed by Tizouiar, O, November 2016
8. Jeuland, F-X.: La maison communicante. 306 p. 2nd edn. Eyrolles, Saint Germain, Paris (2004)
9. Jeuland, F-X.: Réussir son installation domotique et multimédia. 381 p. 2nd edn. Eyrolles, Saint Germain, Paris (2008)
10. Tizouiar, O.: Disponibilité de l'éclairage naturel en milieu urbain dense - investigation sur les performances de puits de lumière naturelle, 185 p. Memory of magister in sciences in architecture at the Polytechnic School of Architecture and Urbanism of Algiers, directed by Pr Bensalem Rafik (2012)

Human Activity Recognition in Smart Home Using Prior Knowledge Based KNN-WSVM Model

M'hamed B. Abidine$^{(\boxtimes)}$ and Belkacem Fergani

Laboratoire d'Ingénierie des Systèmes Intelligents et Communicants
(LISIC Laboratory), Faculty of Electronics and Computer Sciences,
University of Science and Technology Houari Boumediene (USTHB),
32, El Alia, Bab Ezzouar 16111, Algiers, Algeria
abidineb@hotmail.com

Abstract. The ability to recognize human activities from sensed information become more attractive to computer science researchers due to a demand on a high quality and low cost of health care services at anytime and anywhere. In this work, we proposed a new hybrid classification model to perform automatic recognition of activities in a smart home using KNN-WSVM combined the K-Nearest Neighbors (KNN) with Weighted Support Vector Machines (WSVM) learning algorithm, allowing to better discrimination between the classes of activity. We also added the temporal features (TF) in the used method KNN-WSVM. Experiments show our proposed approach outperforms the KNN, WSVM used alone in terms of recognition performance, highlighting the advantages of this method.

Keywords: Smart homes · Activity recognition · KNN
Support vector machines · Weighted SVM

1 Introduction

The growing population of elders in our society calls for a new approach in care giving [1] to ensure the comfort of old people and because the healthcare infrastructures are unlikely to accommodate the drastic growth of elderly population. Smart systems are equipped with sensor networks able to automatically recognize activities about the occupants and assist humans. They must be able to recognize the ongoing activities of the users in order to suggest or take actions in an intelligent manner. Activity recognition can be used to automatically monitor the activities of daily living (ADLs) of old people such as cooking, brushing, dressing, cleaning, bathing and so on substantially increased as well.

Feature extraction and classification are two key steps for activity recognition in a smart home environment [2]. In this paper, we proposed a new hybrid classification method KNN-WSVM using *the K-Nearest Neighbors (KNN) [3] and Weighted Support Vector Machines (WSVM)* [4]. The last is used to deal the class imbalance in the training data due to the fact that people do not spend the same amount of time on the

© Springer International Publishing AG 2018
M. Hatti (ed.), *Artificial Intelligence in Renewable Energetic Systems*, Lecture Notes in Networks and Systems 35, https://doi.org/10.1007/978-3-319-73192-6_2

different activities. We also integrated the prior knowledge [5] in this method to improving the classification performances of KNN-WSVM. We introduced a new feature: the temporal feature (TF) of activity when the activity is performed. Then we added them to the existing features obtained by the collected wireless sensors data. The 'Prepare breakfast' and 'Prepare dinner' activities share the same model as they involve the same set of object interactions. These two activities are distinguished by time of taking place, i.e. 'Prepare breakfast' takes place in the morning hours and "Prepare dinner" takes place in the afternoon or evening hours of the day.

Our paper addresses these issues and contributes on the following topics. Firstly, we have presented the related works of human activity classification methods. Then, we explained the proposed activity recognition method using the feature insertion using the temporal attribute to recognize activities of daily living from binary sensor data. The next section presents the experimental setup and discusses the results acquired throughout a series of benchmark dataset [6, 7] constituted of highly imbalanced datasets under different metrics. Finally, conclusions and future work are drawn in the last section.

2   Related Works

The machine learning of the observed sensor patterns is usually done in a supervised manner and requires a large annotated data sets recorded in different settings [6, 7]. For this purpose, annotation of data for classification task can be performed in many different ways, e.g. use of cameras [8], self-reporting approaches [7] and monitoring the diary activity [6]. Several classification algorithms have been employed for ADL recognition tasks [6, 8, 9], e.g. Hidden Markov model (HMM) [6], Conditional Random Fields (CRF) [6], Linear Discriminant Analysis (LDA) [9], Bayes approach [8], Support Vector Machine (SVM) [9] and its Soft-margin multiclass SVM extension [10]. In [11], we developed a new classification method named PCA-LDA-WSVM based on a combination of Principal Component Analysis (PCA), Linear Discriminant Analysis (LDA) and Weighted Support Vector Machines (WSVM). We demonstrated the ability of this method to achieve good improvement over the standard used methods as HMM and CRF. This method resolved two problems result in a degradation of the performance of activity recognition; the non-informative sequence features and the class imbalance problem. In the next section, we present the proposed approach using KNN-WSVM.

3   Proposed Approach

3.1   System Overview

As pointed out in the introduction part of this paper, the core of our proposal strategy relies on the use of a temporal feature (TF) with the new hybrid model combining KNN and WSVM. This fusion is performed as follows: we added two new features into original data (binary state vectors). The first corresponds to the predicted Labels generated by KNN classifier. The second is the temporal feature (TF). This new distribution data is therefore used for learning and testing the multi-class weighted SVM classifier using a new automated criterion for weighting the data. The choice of WSVM like classifier [12, 13] in this context is motivated by previous work as shown in [11]. A generic diagram is shown in Fig. 1. Particularly, the initial dataset is divided into training and test datasets.

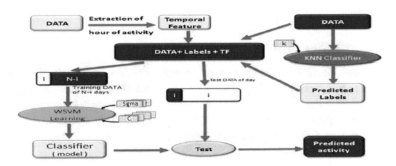

Fig. 1. Multi-class structure of TF-KNN-WSVM

In this work, we improve the classification performances of class activities by introducing the feature insertion stage. We added a new feature to the existent data matrix. This attribute corresponds to the hour of beginning of the activity. We extract this feature directly from the data structure. The sensor activations are collected by the state-change sensors distributed all around the environment. To find out the sensors ID corresponding to the room label of performed activity, we search the different object types he is manipulating in the sensors, see the below figure. ID: is a number representing the sensor ID. Each sensor has its own unique ID (Fig. 2).

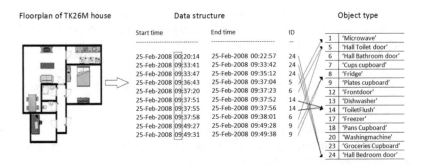

Fig. 2. Temporal feature (TF) for TK26 M dataset. In red, the hour of beginning activity.

3.2   K-Nearest Neighbor (KNN)

The K-nearest neighbor algorithm is amongst the simplest of all machine learning algorithms [3], and therefore easy to implement. The m training instances $x \in R^n$ are vectors in an n-dimensional feature space, each with a class label. In the KNN method, the result of a new query is classified based on the majority of the KNN category. This rule is usually called 'voting KNN rule'. The classifiers do not use any model for fitting and are only based on memory to store the feature vectors and class labels of the training instances. They work based on the minimum distance from an unlabelled vector (a test point) to the training instances to determine the K-nearest neighbors. K (positive integer) is a user-defined constant. Usually Euclidean distance is used as the distance metric.

3.3 Support Vector Machines

Support Vector Machines (SVM) is based on statistical learning theory developed by Vapnik [12]. For a two class problem, we assume that we have a training set $\{(x_i, y_i)\}_{i=1}^{m}$ where the observations are $x_i \in R^n$ and y_i are class labels either 1 or -1. The primal formulation of the margin in SVM maximizes margin $2/K(w, w)$ between two classes and minimizes the amount of total misclassifications (training errors) ξ_i as follows:

$$\min_{w,b,\xi}   1/2.\,K(w, w) + C \sum_{i=1}^{m} \xi_i \tag{1}$$
$$subject \quad to \quad y_i(w^T \phi(x_i) + b) \geq 1 - \xi_i, \; \xi_i \geq 0, \; i = 1, \ldots, m$$

where w is normal to the hyperplane, b is the translation factor of the hyperplane to the origin and $\phi(.)$ is a non-linear function which maps the input space into a feature space defined by $K(x_i, x_j) = \phi(x_i)^T \phi(x_j)$.

We used the Gaussian kernel as *follows*: $K(x, y) = \exp\left(-\|x - y\|^2/2\sigma^2\right)$ where σ is the width parameter. The construction of such functions is described by the Mercer conditions [13]. The regularization parameter C is used to control the trade-off between maximization of the margin width and minimizing the number of training error.

Solving the formulation dual of SVM [13] gives a decision function in the original space for classifying a test point $x \in R^n$

$$f(x) = \text{sgn}\left(\sum_{i=1}^{m_{sv}} \alpha_i y_i K(x, x_i) + b \right) \tag{2}$$

with m_{sv} is the number of support vectors $x_i \in R^n$.

In this study, a software package LIBSVM [14] was used to implement the multiclass classifier algorithm. It uses the one-vs-one method [13].

3.4 Weighted Support Vector Machines

For daily recognition activities applications, especially in daily recognition tasks, the misclassification of minority class members due to large class imbalances is undesirable. An extension of the SVM, weighted SVM, was presented to cope with this problem. Two different penalty constraints were introduced for the minority and majority classes:

$$\min_{w,b,\xi} 1/2. K(w,w) + C_+ \sum_{y_i=1} \xi_i + C_- \sum_{y_i=-1} \xi_i$$

$$\text{subject to} \quad y_i(w^T \phi(x_i) + b) \geq 1 - \xi_i, \ \xi_i \geq 0, \ i = 1, \ldots, m \tag{3}$$

C_+ and C_- are cost parameters for positive and negative classes, respectively, to construct a classifier for multiple classes. They are used to control the trade-off between margin and training error.

Some authors [13, 15] have proposed adjusting different cost parameters to solve the imbalanced problem. Veropoulos et al. in [15] proposed to increase the cost of the minority class (i.e., $C_- > C_+$) to obtain a larger margin on the side of the smaller class. Huang et al. [4] raised a Weighted SVM algorithm. The coefficients are typically chosen as:

$$C_+ = C \times w_+ \tag{4}$$

$$C_- = C \times w_- \tag{5}$$

Where C is the common cost parameter of the WSVM. w^+ and w^- are the weights for +1 and −1 class respectively. They put forward the corresponding solutions to deal with this problem in the SVM algorithm like this:

$$\frac{C_+}{C_-} = \frac{m_-}{m_+} \tag{6}$$

To extend Weighted *SVM* to the *multi-class* scenario in order to deal with N classes (daily activities), we used different **misclassification C_i per class similar to** [11]. By taking $C_- = C_i$ and $C_+ = C$, with m_+ and m_i be the number of samples of majority classes and number of samples in the ith class, the main ratio cost value C_i for each activity can be obtained by:

$$C_i = \text{round}(C \times [m_+/m_i]) \ i = 1, \ldots, N \tag{7}$$

[ ] is integer function. This criterion respects this reasoning that is to say that the tradeoff C_- associated with the smallest class is large in order to improve the low classification accuracy caused by imbalanced samples. It allows the user to set individual weights for individual training examples, which are then used in WSVM training.

4 Simulation Results and Assessment

4.1 Datasets

For the experiments, we use an openly datasets gathered from three houses having different layouts and different number of sensors [6, 7]. Each sensor is attached to a wireless sensor network node. The activities performed with a single man occupant at each house are different from each other. Data are collected using binary sensors such as reed switches to determine open-close states of doors and cupboards; pressure mats to identify sitting on a couch or lying in bed; mercury contacts to detect the movements of objects like drawers; passive infrared (PIR) sensors to detect motion in a specific area; float sensors to measure the toilet being flushed. Binary sensor output is represented in a feature space that used by the model to recognize the activities. We choose the ideal time slice length for discretizing the sensor data $\Delta t = 60$ s. Time slices for which no annotation is available are collected in a separate activity labelled 'Idle'. The data were collected by a Base-Station and labelled using a Wireless Bluetooth headset combined with speech recognition software or a Handwritten diary. Table 1 shows also the number of data per activity in each dataset.

Table 1. List of activities annotated for each dataset and the number of observations of each activity (.).

Dataset	Annotation	Activities
TK26 M	Bluetooth headset	$\text{Idle}_{(4627)}$; $\text{Leaving}_{(22617)}$; $\text{Toileting}_{(380)}$; $\text{Showering}_{(265)}$; $\text{Sleeping}_{(11601)}$; $\text{Breakfast}_{(109)}$; $\text{Dinner}_{(348)}$; $\text{Drink}_{(59)}$
TK57 M	Bluetooth headset	$\text{Idle}_{(2732)}$; $\text{Leaving}_{(11993)}$; $\text{Eating}_{(376)}$; $\text{Toileting}_{(243)}$; $\text{Showering}_{(191)}$; $\text{Brush teeth}_{(102)}$; $\text{Shaving}_{(67)}$; $\text{Sleeping}_{(7738)}$; $\text{Dressing}_{(112)}$; $\text{Medication}_{(16)}$; $\text{Breakfast}_{(73)}$; $\text{Lunch}_{(62)}$; $\text{Dinner}_{(291)}$; $\text{Snack}_{(24)}$, $\text{Drink}_{(34)}$; $\text{Relax}_{(2435)}$
OrdonezA	Handwritten diary	$\text{Idle}_{(1307)}$; $\text{Sleeping}_{(7886)}$; $\text{Toileting}_{(173)}$; $\text{Showering}_{(121)}$; $\text{Breakfast}_{(132)}$; $\text{Grooming}_{(154)}$; $\text{Spare\_Time/TV}_{(8646)}$; $\text{Leaving}_{(1692)}$; $\text{Lunch}_{(331)}$ $\text{Snack}_{(14)}$

4.2 Setup and Performances Measures

We splitted the initial dataset into training and testing subsets using the 'Leave one day out' approach, retaining one full day of sensor readings for testing and using the remaining sub-samples as training data. Sensors outputs are binary either '0' or '1' and represented in a feature space which is used by the model to recognize the activities. As the activity instances were imbalanced between classes. We evaluate the performance of our models using the F-measure, which is calculated from the Precision and Recall scores. On the other, in order to evaluate the sensitivity of the classifiers, the notions of true positive (TP), false negatives (FN) and false positives (FP), have also been implemented. These measures are calculated as follows:

$$F - Measure = \frac{2 \cdot \text{Precision Recall}}{\text{Precision} + \text{Recall}} \text{ with Precision}$$

$$= \frac{1}{N} \sum_{i=1}^{N} \left[\frac{\text{TP}_i}{\text{TP}_i + \text{FP}_i} \right]; \text{ Recall} = \frac{1}{N} \sum_{i=1}^{N} \frac{\text{TP}_i}{\text{TP}_i + \text{FN}_i} \qquad (8)$$

4.3   Results

We optimized the SVM hyper-parameters (σ, C) for all training sets in the range [0.1–2] and {0.001, 0.01, 0.1, 1, 5}, respectively, to maximize the error rate of Leave-one day-out cross-validation technique. Then, for WSVM classification method, we optimized locally the cost parameter C_i adapted to different classes [11]. In Table 2, the results show that the proposed method TF-KNN-WSVM outperforms SVM, KNN, WSVM, KNN-WSVM and the baseline method PCA-LDA-WSVM [11]. TF-KNN-WSVM model using TF contributes to significantly enhance the performance of KNN-WSVM classifier. One also notices that the WSVM is better than SVM and KNN for recognizing activities. For KNN method, we used also cross-validation technique to select the optimal value of K parameter in the range [1–9].

Table 2. Recall, Precision, F-Measure and Accuracy results for all approaches in (%). Bold values are the results for the proposed approaches for each dataset.

Dataset	Approach	Recall	Precision	F-Measure	Accuracy
TK26M	SVM$_{(\sigma=1.3, \ C=1)}$	61.64	74.43	67.43	95.49
	KNN$_{(K=9)}$	67.87	75.84	71.63	94.39
	WSVM$_{(\sigma=0.8, \ C=0.01)}$	75.56	76.53	76.04	92.46
	KNN-WSVM$_{(\sigma=1.4, \ C=0.01)}$	**57.43**	**78.3**	**76.84**	**93.86**
	TF-KNN-WSVM$_{(\sigma=1.5, \ C=0.01)}$	**79.18**	**79.1**	**79.14**	**92.47**
TK57M	SVM$_{(\sigma=1.2, \ C=1)}$	35.83	36.33	36.08	80.89
	KNN$_{(K=7)}$	35.09	34.22	34.65	79.25
	WSVM$_{(\sigma=2, \ C=0.01)}$	37.86	37.86	39.27	77.08
	KNN-WSVM$_{(\sigma=1.3, \ C=0.01)}$	**43.10**	**41.84**	**42.46**	**76.98**
	TF-KNN-WSVM$_{(\sigma=1.3, \ C=0.01)}$	**47.21**	**51.93**	**49.46**	76.80
OrdonezA	SVM$_{(\sigma=1.9, \ C=5)}$	60.15	58.31	59.22	90.16
	KNN$_{(K=9)}$	55.43	56.67	56.04	84.19
	WSVM$_{(\sigma=1.4, \ C=0.01)}$	66.52	68.27	67.38	84.36
	KNN-WSVM$_{(\sigma=1, \ C=0.001)}$	**62.39**	**69.79**	**65.89**	**77.36**
	TF-KNN-WSVM$_{(\sigma=2, \ C=0.1)}$	**69.94**	**71.26**	**70.60**	**76.48**

We report in Fig. 3, the classification results in terms of accuracy measure for each class for TK26M dataset. We show that our proposed combination outperforms the other approaches for 'Toileting', 'Breakfast' and 'Dinner' activities and similar results with other methods for 'Leaving', 'Showering' and 'Sleeping' activities. The majority activities 'Leaving' and Sleeping' are better for all methods while the 'Idle' activity is

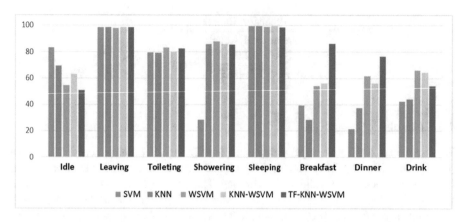

Fig. 3. Accuracy recognition rate for each activity for TK26M dataset.

less accurate for the proposed method compared to other methods. Additionally, the kitchen-related activities as 'Breakfast', 'Dinner' and 'Drink' are in general harder to recognize than other activities.

In order to quantify the extent to which one class is harder to recognize than another one, we analyzed the confusion matrix of TF-KNN-WSVM for TK26M dataset in Table 3. One notices that the activities 'Leaving', 'Toileting', 'Showering', 'Sleeping', 'Breakfast' and 'Dinner' are better recognized comparatively with 'Idle' and 'Drink'. The kitchen activities seem to be more recognized using the proposed method. The high performance obtained in the case of TK26M dataset, which seems to be less vulnerable to class-overlapping than other datasets. This overlapping between the activities is due to the layout of the house. In the TK26M house, there is a separate room for almost every activity. The kitchen activities are food-related tasks, they are worst recognized because most of the instances of these activities were performed in the same location (kitchen) using the same set of sensors. Therefore the location of sensors strongly influences the recognition performance.

Table 3. Confusion matrix (values in %) for TF-KNN-WSVM for the TK26M dataset.

Activity	Id	Le	To	Sh	Sl	Br	Di	Dr
Id	**50.94**	8.21	3.35	8.41	3.39	11.84	9.96	3.89
Le	1.09	**98.42**	0.2	0.26	0.0	0.0	0.02	0.0
To	8.95	2.89	**82.63**	0.79	1.84	2.11	0.26	0.53
Sh	3.4	0.0	4.91	**85.66**	0.0	6.04	0.0	0.0
Sl	0.94	0.0	0.4	0.0	**98.66**	0.0	0.0	0.0
Br	10.09	0.0	0.92	0.0	0.92	**86.24**	0.0	1.83
Di	11.78	0.57	1.15	0.0	0.0	0.0	**76.72**	9.77
Dr	13.56	3.39	3.39	0.0	0.0	5.08	20.34	**54.24**

5 Conclusion

Our experiments on real-world datasets from smart home environment showed that TF-KNN-WSVM strategy can significantly increase the recognition performance to classify multiclass sensory data, and can improve the prediction of the minority activities. It significantly outperforms the results of the typical methods KNN, SVM, WSVM. TF-KNN-WSVM is better than KNN-WSVM. The space features needs a prior knowledge about the smart home using the temporal feature, which makes a model very specific for that environment.

The location attribute can also discriminate between the different activity classes that performed in different locations. In the future, it will be interesting to use both spatial and temporal features to improve the activity classification performance. Also, we are going to test the scalability of our approach by considering datasets containing increased classes and various amounts of sensors.

References

1. Brumitt, B., Meyers, B., Krumm, J., Kern, A., Shafer, S.: Easyliving: technologies for intelligent environments. In: Handheld and Ubiquitous Computing, pp. 97–119. Springer (2000)
2. Abidine, M.B., Fergani, L., Fergani, B., Fleury, A.: Improving human activity recognition in smart homes. Int. J. E-Health Med. Commun. (IJEHMC) 6(3), 19–37 (2015)
3. Darko, F., Denis, S., Mario, Z.: Human movement detection based on acceleration measurements and k-NN classification. In: The International Conference on Computer as a Tool, EUROCON 2007, pp. 589–594. IEEE, September 2007
4. Huang, Y.M., Du, S.X.: Weighted support vector machine for classification with uneven training class sizes. In: Proceedings of the IEEE International Conference on Machine Learning and Cybernetics, vol. 7, pp. 4365–4369 (2005)
5. Fleury, A., Noury, N., Vacher, M.: Improving supervised classification of activities of daily living using prior knowledge. Int. J. E-Health Med. Commun. 2(1), 17–34 (2011)
6. Kasteren, T.V., Noulas, A., Englebienne, G., Krose, B.: Accurate activity recognition in a home setting. In: UbiComp 2008, pp. 1–9. ACM, New York (2008)
7. Ordonez, F.J., de Toledo, P., Sanchis, A.: Activity recognition using hybrid generative/discriminative models on home environments using binary sensors. Sensors 13, 5460–5477 (2013)
8. Logan, B., Healey, J., Philipose, M., Tapia, E.M., Intille, S.: A long-term evaluation of sensing modalities for activity recognition. In: Proceedings of the 9th International Conference on Ubiquitous Computing, pp. 483–500. Springer, Berlin (2007)
9. Abidine, M.B., Fergani, B.: Evaluating C-SVM, CRF and LDA classification for daily activity recognition. In: Proceedings of IEEE ICMCS, Tangier-Morocco, pp. 272–277. IEEE (2012)
10. Chen, D.R., Wu, Q., Ying, Y., Zhou, D.X.: Support vector machine soft margin classifiers: error analysis. J. Mach. Learn. Res. 5(Sep), 1143–1175 (2004)
11. Abidine, M.B., Fergani, L., Fergani, B., Oussalah, M.: The joint use of sequence features combination and modified weighted SVM for improving daily activity recognition. In: Pattern Analysis and Applications. Springer, London, August 2016
12. Cortes, C., Vapnik, V.: Support vector network. Mach. Learn. 20, 1–25 (1995)

13. Osuna, E., Freund, R., Girosi, F.: Support vector machines: training and applications. Technical report. Massachusetts Institute of Technology, Cambridge, MA, USA (1997)
14. Chang, C.C., Lin, C.J.: LIBSVM: a library for support vector machines. ACM Trans. Intell. Syst. Technol. **2**, 1–27 (2011). http://www.csie.ntu.edu.tw/~cjlin/libsvm/
15. Veropoulos, K., Campbell, C., Cristianini, N.: Controlling the sensitivity of support vector machines. In: Proceedings of the International Joint Conference on AI, pp. 55–60 (1999)

IoT-Safety and Security System in Smart Cities

El-Hadi Khoumeri[(⊠)], Rabea Cheggou, and Kamila Farhah

LTI Laboratory, Ecole Nationale Supérieure de Technologie, ENST,
Algiers, Algeria
{elhadi.khoumeri, rabea.cheggou, kamila.ferhah}@enst.dz

Abstract. Today's cities must respond to multiple challenges related to the evolutions of the contemporary world. The safety of people and goods has become a strategic issue for countries and companies. An intelligent system for Smart Cities is a desired technology in the 21st century. The main attraction of any automated system using the Internet of Things (IoT) is to reduce human labour, effort, time and errors due to human negligence. Managing the security of people in a public space is very important in the smart cities. The challenge is to find a system that offers many security solutions at the same time? An affordable and effective tool. In this paper a global system that manages different solutions in a single application using the IoT is proposed. Various sensors based on facial recognition, fire detection, vehicle number plate recognition can be added at this prototype to improve the intelligence and ability to make more accurate decisions.

Keywords: Smart Cities · IoT · Raspberry Pi · Security · Safety
Sensors

1 Introduction

The management of security and surveillance requires considerable attention in the Smart Cities, errors must be very limited. Fires, thefts or intrusions are undesirable events that could lead to a great loss of social wealth and human life. To avoid these losses, various alarm systems have been developed by the industry such as smoke detectors, temperature sensors, intelligent surveillance cameras, and this with the development of technologies at affordable prices. Among the new technologies is the Internet of Things (IoT), which consists of connecting all devices to the Internet in order to communicate with each other. The integration of this technology will create practical and effective means in the area of surveillance and security. The system proposed in this document includes the use of affordable instruments based on the Raspberry Pi card, including a camera module, a motion sensor and an ultrasonic sensor. Three main functions are performed. Firstly, individuals and vehicles flow control, into the entrance of a building or a public institution using a facial recognition and read the license plate. Secondly, the camera is designed to detect fire, it is linked to an alarm system to alert user. Thirdly, detection of movement and intrusion in the building. A web page has been developed to manage the building entrance (real-time information on vehicles and individuals, database manipulation, history of the inputs

© Springer International Publishing AG 2018
M. Hatti (ed.), *Artificial Intelligence in Renewable Energetic Systems*, Lecture Notes
in Networks and Systems 35, https://doi.org/10.1007/978-3-319-73192-6_3

and output). The design and the implementation of a facial recognition, license plate identification and fire detection algorithms are based on image processing. The use of the OpenCV and Python is largely in line with our solution with a high level of reliability.

2 Related Works

Surveillance and security research generally proposes a unique solution for each module. Our solution offers a global solution with several modules in one application. In [1] a novel filters which is unified to detect either the vehicle license or the vehicles from the digital camera imaging sensors of urban surveillance systems in the smart cities. Another example [2] an intelligent system for facial recognition. In [3] they describe the development of visualisation application software used to control operational and technical functions in the Smart Home system or Smart Home care system via the wireless control system. In these articles [4–6] a system with different means of communication is proposed for the smart home. Al-Audah et al. proposed a system that uses the Lab-VIEW software based on a camera vision [7]. Another system called ALPR (Automatic License Plate Recognition) is developed for the detection and recognition of vehicle license plates [8]. A real-time system based on FPGA that provides an optimal solution for facial recognition is presented in [9]. In [10] they propose a specific face recognition system designed around ARM9 platform. Premal and Vinsley presented a mobile camera for real-time detection of forest fires [11].

3 System Description

The solution proposed in our project, security and surveillance with Raspberry Pi consists of different systems: facial recognition, vehicle license plate recognition, fire detection with access control.

3.1 Hardware Design

The proposed solution is based on a real-time processing video sequences, which requires the use of a powerful tool to ensure constraints (speed and real time). We choose the Raspberry Pi 2 card that meets our needs. It is a small single-board computer that was originally developed for computer science education and has since become popularised by digital hobbyists and makers of Internet of Things (IoT) devices. Raspberry P is about the size of a credit card; it has a 64-bit quad-core ARMv8 processor and uses a Raspbian Linux distribution as operating system (OS) [12] (Fig. 1).

The Raspberry Pi is contained a single circuit board and features ports for: HDMI, USB 2.0, composite video, analog audio, power, internet, and SD card. All sensors used in this project shows in the Fig. 2.

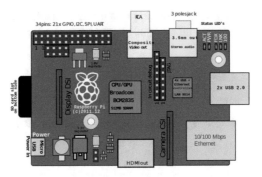

Fig. 1. The raspberry Pi 2 components.

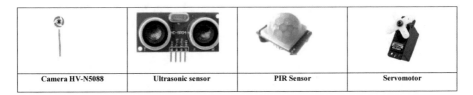

| Camera HV-N5088 | Ultrasonic sensor | PIR Sensor | Servomotor |

Fig. 2. The hardware used.

3.2 Block Diagram and Algorithm

Facial recognition algorithm: Lot of algorithms for extracting face characteristics are proposed in the OpenCV library, such as the LBP (Local binary patterns), Eigenface and Fisherface algorithms. We chose the LBP algorithm [13] as a solution in our project. The facial recognition process can be divided into three main phases shows below (Fig. 3):

Fig. 3. Facial recognition process.

License Plate Identification: The developed algorithm consists in three primary steps: Step 1: Finding the license plate, Step 2: Segmenting each of the individual characters from the license plate. And Step 3: Identifying and recognizing each of the characters. The following diagram in Fig. 4 shows the different phases of recognition [14].

Fig. 4. Plate identification process.

The detection of the plate is the most important and difficult phase, it determines the rapidity and the robustness of the system. Steps represented in Fig. 5.

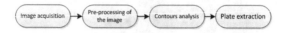

Fig. 5. Block Diagram of Plate identification Module.

The detected characters require an optical character recognition (OCR) system to identify the vehicle's license plate. OCR is a technology that automatically recognizes characters using an optical mechanism. There are several OCR algorithms that are open source and written in different languages (C++, Python, or Java). We choose the "Tesseract" algorithm because of its extensibility and flexibility, that makes the recognition operation simple and reliable [15] (Fig. 6).

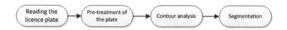

Fig. 6. Block Diagram of Plate identification process

Fire Detection Algorithm: Fire detection is performed in several ways, for example, by using temperature, humidity and smoke analysis. However, these techniques are not reliable because they can not provide other information such as the location of the fire, the size of the fire, and these techniques can lead to false alarms. There are many types of color schemes such as RGB, CMYK, YCbCr, YUV, HSL, HSV. However, each of the color spaces has their advantages and disadvantages. Before you can detect a fire with a camera, you must know the specific properties of a fire that distinguish it from the other objects that the camera sees. The YCbCr color space is used here because of its ability to effectively distinguish luminance information from other color models [16]. The diagram in Fig. 7 shows the diagram of the proposed algorithm.

Fig. 7. Fire detection process.

4   System Implementation

The proposed design of our system is presented in details then we discuss the results of the tests carried out during it, in order to evaluate the functioning and the reliability of system. Our target is to design a system that meets the proposed solution, the general diagram of the prototype design of our system, is presented in the following Fig. 8:

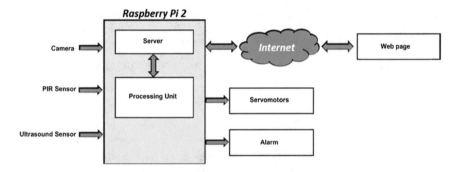

Fig. 8. Architecture of proposed solution.

The Raspberry Pi is the master device of our system, equipped with a camera that presents the video acquisition device and interfaced directly with the Raspberry Pi module via a USB interface, a PIR sensor that is connected directly with the Raspberry Pi ports. The PIR is able to detect movement. The ultrasound sensor is dedicated to managing the camera, acts as a presence detector, so it activates the shot as soon as it detects the presence of a vehicle or person. The processing of the data acquired by the input devices is carried out by the Raspberry Pi, which is responsible to make the decision to actuate the actuator, then trigger an alarm, and to record the data.

Description of the webpage: developed with PHP, JavaScript and HTML. Figure 9 shows the login page. First, every user must identify himself by entering the password and username. After logged on, the user will be redirected to the display webpage as shown in Fig. 10. This webpage allows displaying in real time all functionalities cited before. It is also used to stream video online i.e. we can see the live streaming anywhere through internet. An error message appears if the login is wrong.

Fig. 9. Login page   **Fig. 10.** Home page

5   Experimented Result

Our prototype is illustrated in Fig. 11. Beforehand, all the identification data of the individuals and vehicles must be saved in the data base via the registration page.

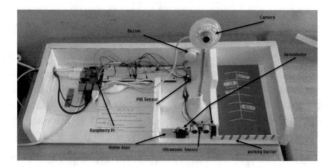

Fig. 11. System design

5.1   Facial Recognition Test

To be able to recognise new face, the system takes 20 photos for the same person and save them in a data base. This person is associated with an identifier previously defined in the database (Fig. 12).

Fig. 12. Creation of the database for facial recognition

Now the system can identify the person, it displays on the screen the identifier and the index of confidence (Fig. 12). This index is calculated by the LBP model. More than the confidence value is minimal, and then recognition is better (Fig. 13).

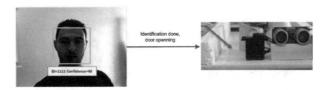

Fig. 13. Door opening with Facial recognition

5.2   Plate License Recognition Test

The raspberry Pi waits for an input from the ultrasonic sensor to activate the camera. The image will be displayed on the webpage and the barrier will be lifted if the license plate is recognised. Otherwise the portal will remain closed (Fig. 14).

Fig. 14. Portal opening with plate license recognition

5.3   Fire Detection

The fire alarm warns the user by sending a sound alarm and submitting a message alert (Fig. 15).

Fig. 15. Fire detection and message sent by Raspberry Pi

5.4   Motion Detection

Motion detection involves the presence of an intruder in the field of view of the PIR sensor. A sound alarm alerts the user and a message is sent by Raspberry Pi is displayed on the webpage (Fig. 16).

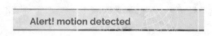

Fig. 16. The screenshot of the message sent by Raspberry Pi

6   Conclusion

In this paper, a design of an advanced security and safety system for Smart Cities has been presented. It reduces the human interactions, by using Internet of Things (IoT). It is absolutely an affordable system that can be used in various areas like supermarket, street, parking, official building, school, etc. This system is dedicated to be used in Smart Cities bringing together three modules in one package. The platform can support other sensors like earthquake, carbon dioxide and radiation.

References

1. Hu, V., Ni, Q.: IoT-driven automated object detection algorithm for urban surveillance systems in Smart Cities. IEEE Internet Things J. **PP**(99), 1 (2017)
2. Aslan, E.S., Özdemir, Ö.F., Hacıoğlu, A., İnce, G.: Smart pass automation system. In: 24th Signal Processing and Communication Application Conference (SIU), Zonguldak, Turkey (2016)
3. Vanus, J., Kucera, P., Martinek, R., Koziorek, J.: Development and testing of a visualization application software, implemented with wireless control system in smart home care. Hum. Centric Comput. Inf. Sci. **4**(1), 1–19 (2014)
4. Li, M., Lin, H.-J.: Design and implementation of smart home control systems based on wireless sensor networks and power line communications. IEEE Trans. Ind. Electron. **62**(7), 4430–4442 (2015)
5. Zuo, F., De With, P.H.: Real-time embedded face recognition for smart home. IEEE Trans. Consum. Electron. **51**(1), 183–190 (2005)
6. Kumar, S.: Ubiquitous smart home system using android application. arXiv preprint arXiv: 1402.2114 (2014)
7. Al-Audah, Y.K., Al-Juraifani, A.K., Deriche, M.A.: A real-time license plate recognition system for Saudi Arabia using LabVIEW. In: 2012 3rd International Conference on Image Processing Theory, Tools and Applications (IPTA), Istanbul, pp. 160–164 (2012)
8. Saleem, N., Muazzam, H., Tahir, H.M., Farooq, U.: Automatic license plate recognition using extracted features. In: 2016 4th International Symposium on Computational and Business Intelligence (ISCBI), Olten, pp. 221–225 (2016)
9. Matai, J., Irturk, A., Kastner, R.: Design and implementation of an FPGA-based real-time face recognition system. In: 2011 IEEE 19th Annual International Symposium on Field-Programmable Custom Computing Machines, Salt Lake City, UT, pp. 97–100 (2011)

10. Ru, F., Peng, X., Hou, L., Wang, J., Geng, S., Song, C.: The design of face recognition system based on ARM9 embedded platform. In: 2015 IEEE 11th International Conference on ASIC (ASICON), Chengdu, pp. 1–4 (2015)
11. Premal, C.E., Vinsley, S.S.: Image processing based forest fire detection using YCbCr colour model. In: 2014 International Conference on Circuits, Power and Computing Technologies (ICCPCT-2014), Nagercoil, pp. 1229–1237 (2014)
12. http://whatis.techtarget.com/definition/Raspberry-Pi-35-computer
13. Soumaya, F.T.: Développement d'un système de reconnaissance faciale à base de la méthode LBP pour le contrôle d'accès. École National Supérieure de Technologie (ENST), Alger, chapitre 2, pp. 19–25 (2016)
14. Chaari, A.: Nouvelle approche d'identification dans les bases de données biométriques basée sur une classification non supervisée. Modélisation et simulation, Université d'Evry-Val d'Essonne, Français (2009). <tel-00549395>
15. Mithe, R., Indalkar, S., Divekar, N.: Optical character recognition. Int. J. Recent Technol. Eng. (IJRTE) **2**, 72–75 (2013)
16. Binti Zaidi, N.I., Binti Lokman, N.A.A., Bin Daud, M.R., Achmad, H., Chia, K.A.: Fire recognition using RGB and YCBCR color space. ARPN J. Eng. Appl. Sci. **10**(21), 9786–9790 (2015)

Smart Home Control System Based on Raspberry Pi and ZigBee

Tahar Dahoumane[✉] and Mourad Haddadi

Department of Electrical Engineering, National Polytechnic School of Algiers, ENP,
16200 El-Harrach, Algeria
`tahar.dahoumane@g.enp.edu.dz`, `mourad.haddadi@enp.edu.dz`

Abstract. The main benefit of smart home system is controlling and monitoring of household appliances remotely and flexibly. Control systems for smart home have been improved with the integration of technologies of communication, digital information and electronics in the home environment. In this paper, a new and flexible smart home control system is presented. The proposed system is based on the Raspberry Pi B+ board, which is programmed as the embedded home server and the standard ZigBee is used to ensure communications within the home network. To accomplish this experiment, XBee ZigBee devices, smart plugs, and Raspberry pi board are used.

Keywords: Smart Home · ZigBee · Embedded server · Web services · Smart plug HTML user interface

1 Introduction

The area of smart home automation is characterized by an infrastructure that enables intelligent networking of devices that use different wireless and wired technologies to ensure seamless integration, making it easier to use the home systems while creating a personalized and secure home space [1]. Smart home is intended to render the life much easier and convenient for its occupants, brings security and also provide some energy efficiency savings [2]. The one important benefit of a smart home is providing an advanced control system, which permits to users to control the household appliances in a simple and flexible manner. Smart home control systems are developed using different technologies of communication, information and electronics. Improving of the smart home control system is based on the integration suitably of new technologies within the home environment and implementing low cost and advanced architectures using smart devices and adequate equipments. The integration of the internet and implementing of server and gateway enables the remote control of devices and appliances within the home environment. In this paper, we proposed a reliable and flexible smart home control system based on different technologies and smart devices.

In the last years, several control system have been proposed based on different technologies and standards. The mobile phone is widely used to control appliances remotely as they presented in [3, 4]. Kaur [5] presents a system of controlling household devices based on a Micro-controller (ATMEL AT89S52).

© Springer International Publishing AG 2018
M. Hatti (ed.), *Artificial Intelligence in Renewable Energetic Systems*, Lecture Notes in Networks and Systems 35, https://doi.org/10.1007/978-3-319-73192-6_4

These systems suffer from limitations such as the lack of flexibility where the appliances are controlled via a one type of device (as mobile phone through GSM). Another inconvenient for systems that proposed SMS messages as solution to remotely control home appliances is that the user has to remember codes or names of all commands to include in the message body. Hence, the development of a dynamic and flexible user interface (UI) for controlling and monitoring applications remain the most promising solution.

In order to overcome the problem of the intrusiveness of the home installation, different works were presented that use wireless technologies. Sriskanthan et al. [6] proposed a system of smart home based on a wireless technology which is a Bluetooth. Zhang et al. [7] present a Wi-Fi smart plug which permits controlling appliances wirelessly based on the WiFi technology, where services are provided using the REST (Representation State Transfer) architecture.

In order to ensure a remote controlling and monitoring services for smart home control systems, the proposed approaches converge to the exploitation of a tiny computers that can be programmed as embedded server and gateway as it is presented in [8] where the control system proposed is designed around the Cortex A8 processor.

In this paper a raspberry pi B+ is used and programmed as the home embedded server. The raspberry pi B+, presented in July 2014, is a tiny computer, which has a size of 85 mm × 56 mm. This board comes with a 512Mo of RAM and an ARM BCM28351 Broadcom SoC (System on Chip) processor running at 700 MHz. The Raspberry Pi has many positive aspects such as its flexibility and the possibility of its utilization in variation of purposes. The Raspberry Pi is suitable for projects that require a graphic interface or the Internet, and this advantage makes Raspberry Pi perfect for usage in Internet of Things vision [9].

2 System Architecture

The Fig. 1 shows the proposed smart home control system. Users can control household appliances locally using a Desktop PC, which is connected to the LAN network via RJ45 interface or via smart devices such as tablet and smart phones that are connected to the Wi-Fi network. The user can also control the appliances remotely via internet. A user interface (UI) is coded in HTML in order to simplify the control task for users and gives them the possibility to use variety of smart devices, and this increase the flexibility of the smart home control system.

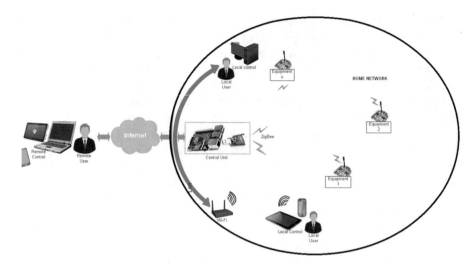

Fig. 1. The control system architecture.

2.1 Description of the Central Unit

The Fig. 2 shows the block diagram of the central unit, which is composed from the Raspberry pi B+ board and an XBee module. This XBee ZigBee module is configured as the coordinator of the network, which is responsible for forming the network and assigning addresses for XBee devices within the home network. The Raspberry pi B+ board is programmed as an embedded server, where the home management system is implemented and data are stored. The implementation details of the home embedded server are discussed further in the Sect. 3.

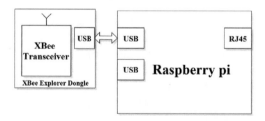

Fig. 2. Diagram block of the central unit

The characteristics of the raspberry pi B+ are given as follow:

- 512 MB SDRAM @400 MHz.
- ARM BCM2835 processor
- Dual step-down (buck) power supply for 3.3 V and 1.8 V
- 5 V supply has polarity protection, 2A fuse and hot-swap protection
- New USB/Ethernet controller chip
- 4 USB ports instead of 2 ports in B model and 1 port in A model

- 40 GPIO pins instead of 26 in A and B model.
- 2 EEPROM Plate identification pins
- Composite (NTSC/PAL) video integrated into 4-pole 3.5 mm 'headphone' jack
- MicroSD card socket instead of full size SD in old models

The Raspberry pi supports different distribution (Operating Systems "OS") as Ubuntu, Arch Linux, Debian and Raspbian as the default one, and this permits a large scale of choice about the OS to use.

Commands for controlling appliances or managing the home network are passed through the central unit. The central unit has the role of a gateway, which permits the execution, in the real home environment, of commands or requests initiated by remote users through internet. The received requests via internet will be interpreted by the gateway and then specifics ZigBee commands will be transmitting by the XBee coordinator to a specific XBee module in the home network.

Exchanging data between the Raspberry pi and the XBee coordinator is done via the serial interface (RS232). Parameters of the serial communications are well defined in a java program and must match the ones entered when configuring the coordinator under the X-CTU software. The FTDI chip enables conversion of the received data on the XBee's pins to the serial protocol, which permits talking suitably with the Raspberry pi board. Using the java XBee library offered by DiGi international, java programs can be programmed to enable the raspberry pi to send commands or handle the received data from the XBee coordinator.

The Raspberry pi is connected to internet via the RJ45 interface, which enables remote controlling of equipments and appliances by remote users.

2.2   Description of the Smart Plug

The smart plug is composed from two important parts: a power and command circuit and sensor board. Controlling Appliances by such smart devices in the smart home is an important task. Smart plug contributes to render smarter the control system and to provide the flexibility and simplicity for users to control their equipments within the home network.

Terminal boards (smart plugs) permit controlling the equipments and turn them ON or OFF and exchange data with a central unit. The data exchanged are the status of the equipment, the value of the electrical consumption and others environmental values such as temperature and humidity. In this paper, a smart plug is developed (see Fig. 3) which is composed from an XBee transceiver module, a sensor board and a power circuit. The smart plug exchanges data wirelessly with the central unit using the ZigBee standard. The ZigBee is performed for a short occupation of the transmission medium and transmitting a low data rate since the commands for controlling equipment in the smart home network are of low size.

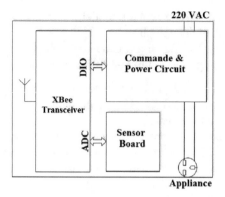

Fig. 3. Diagram block of the smart plug

The XBee device is chosen to satisfy the need of smart plug. The X-CTU software from DiGi International is used to configure XBee devices. A mesh network is configured as the topology of our network. The Coordinator is the responsible of assigning addresses to XBee Routers that are mounted in the developed boards.

Description of the Command and Power Circuit

Upon configuration under C-XTU, one of the pin of the XBee router is configured as an analog input and the other as a digital output. The nineteenth (19) pin is configured as the digital output (DIO1) that serves as a command of the power circuit. As shown in the Fig. 4, the output of the opto-coupler (6N135) is enabled or disabled via a transistor that is connected to the XBee DIO1 pin.

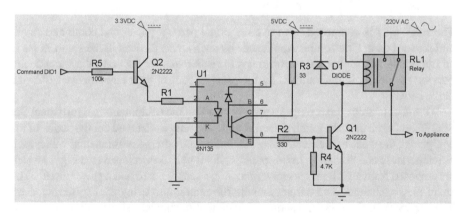

Fig. 4. Smart plug's command and power circuit released under "Proteus"

The power circuit is composed from a relay that receives order from the transistor relied to the output of the opto-coupler. A command (turn ON or OFF the appliance) received on the antenna of the XBee device will activate, through the smart plug's circuit, the coil of the relay, which provides electrical energy to the appliance.

Description of the sensor Board

The Fig. 5 represents the circuit of the sensor node, realized under "Proteus". The XBee module is equipped with four ADC (Analog Digital Converter) which permit to configure four analog input/output on the pins 17 to 20 (DIO0 to DIO3). As presented in the Fig. 5, the pin number (17) is the one used for receiving the lectures from the LM335 sensor.

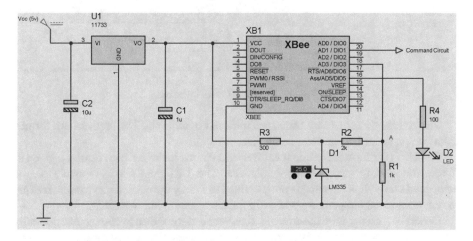

Fig. 5. Circuit of the sensor board released under "Proteus".

3   Software Development

The raspberry pi is programmed as the home embedded server. Different programming languages can be used in the case of the raspberry pi board such as C, C++, Python or java. In the proposed architecture, java language is used to power the embedded server. The JES (Java Embedded Suite) is a lightweight API performed for the embedded environment, which permits creating a java data base (DB) and providing web services for remote control purposes based on the REST architecture. The Glassfish server is the platform adopted for the proposed architecture.

The REST architecture provides web services for users based on the client/server architecture. Controlling of equipment, managing the network and monitoring of temperature are considered as services that are provided by the smart home system.

The Fig. 6 shows a dynamic UI which is coded in HTML with JavaScript. This UI permits to users to monitor and control the household equipments in easy and flexible way.

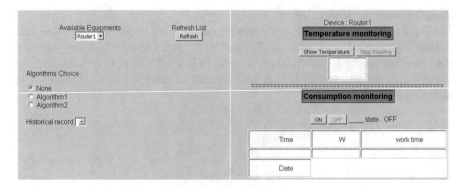

Fig. 6. The HTML user interface.

The actions performed by users, are traduced in requests. The requests are formed from two parts, an URI and a specific data.

The URI (Uniform Resource Identifier) is a unique identifier for a resource given as follow: http://<ip>:<port>/path_to_resouce. The URI points to a resource on the embedded server, this resource is programmed to satisfy the wanted service (ex. reading the environmental temperature where the selected smart plug is located).

Generally, resources on the embedded server need some data to be executed correctly (ex. the MAC address of the smart plug of the equipment that the user wants to turn ON or OFF). REST is representation-oriented architecture, hence, the data that is exchanged with the server have the format of JSON (JavaScript Object Notation) and XML (eXtensible Markup Language) or in another formats types.

All requests goes through the embedded server based on the client/server architecture and this is done locally via the Wi-Fi or LAN network or remotely via internet. Received requests by the embedded server will be transferred to the XBee coordinator via the serial RS232 protocol. The coordinator will interrogate or order to the smart plug to execute the intended commands.

4   System Evaluation

The XBee devices are used in order to form the ZigBee mesh network, and are configured using the X-CTU software. The XBee coordinator is configured as the trust center where commands goes through this node. The 'XBee Java Library', available in the web page of Digi International, is exploited to ensure the communication between the coordinator and the home embedded server.

The proposed architecture relies between the UI (user interface), the embedded server and the ZigBee home network. The executed commands in the real home environment are initiated by the user through the UI, received by the embedded server and then transferred as commands to the XBee coordinator. The coordinator sent commands to a specifics smart plug within the home network. The commands are executed by smart plugs and then the information are sent back to the server. This proposed architecture shows a reliable control system for smart home.

Some proposed control systems are based on GSM or Microcontrollers that have the disadvantage of the lack of flexibility, where the user have to remember commands or codes for controlling equipments within the home environment. In this paper, a user interface is coded in HTML and JavaScript which permit dynamic web page. This UI can be consulted via PCs, mobiles or other smart devices, which increases in flexibility of the control system.

The test conducted shows that the proposed system is a flexible and reliable smart home control system. Smart devices as mobile or GSM modem can be connected to the Raspberry pi, which gives the possibility to an extension of the system. Thus, the Raspberry pi represents a good choice for extensible smart home control system.

5   Conclusion

With the emerging of technologies of communication and information, the development of control systems for household appliances become a necessity. Several control systems have been proposed, which presents some advantages and disadvantages. In this paper, a new control system of household appliances is presented. The proposed system is based on an embedded server, which has the role of a gateway that permits to users to control their equipments via Wi-Fi, LAN network or internet. The Raspberry pi B+ is used and programmed as the embedded server, which is powered by java. The REST architecture is used to provide web services which facilitate controlling and monitoring operations. The communication within the home network, under the supervision of the central unit, is based on the ZigBee standard which is adopted for controlling and monitoring applications. The proposed system shows the benefit of using the newest technologies to increase the flexibility and reliability in the smart home system. Such proposed control system can be incorporated in the future smart home that will be connected to the smart grid.

References

1. Toschi, G.M., Campos, L.B., Cugnasca, C.E.: Home automation networks: a survey. Comput. Stand. Interfaces **50**, 42–54 (2017). https://doi.org/10.1016/j.csi.2016.08.008
2. Robles, R.J., Kim, T.: Review: context aware tools for smart home development. Int. J. Smart Home **4**(1), 1–12 (2010)
3. Nichols, J., Myers, B.A.: Controlling home and office appliances with smart phones. IEEE Pervasive Comput. **5**(3), 60–67 (2006). https://doi.org/10.1109/MPRV.2006.48
4. Sikandar, M., Khiyal, H., Khan, A., Shehzadi, E.: SMS based wireless home appliance control system (HACS) for automating appliances and security. Issues Sci. Inf. **6**, 887–894 (2009). https://doi.org/10.28945/3304
5. Kaur, I.: Microcontroller based home automation system with security. IJACSA Int. J. Adv. Comput. Sci. Appl. **1**(6), 60–65 (2010). www.ijacsa.thesai.org. Accessed
6. Sriskanthan, N., Tan, F., Karande, A.: Bluetooth based home automation system. Microprocess. Microsyst. **26**(6), 281–289 (2002). https://doi.org/10.1016/S0141-9331(02)00039-X
7. Wang, L., Peng, D., Zhang, T.: Design of smart home system based on WiFi smart plug. Int. J. Smart Home **9**(6), 173–182 (2015). https://doi.org/10.14257/ijsh.2015.9.6.19

8. Zhang, S., Xiao, P., Zhu, J., Wang, C., Li, X.: Design of smart home control system based on Cortex-A8 and ZigBee. In: Proceedings of the IEEE International Conference on Software Engineering and Service Sciences, ICSESS, pp. 675–678 (2014). https://doi.org/10.1109/ICSESS.2014.6933658

9. Vujovic, V., Maksimovic, M.: Raspberry Pi as a sensor web node for home automation. Comput. Electr. Eng. **44**, 153–171 (2015). https://doi.org/10.1016/j.compeleceng.2015.01.019

An Indoor Positioning System Based on Visible Light Communication Using a Solar Cell as Receiver

Ameur Chaabna[1], Abdesselam Babouri[2(✉)], and Xun Zhang[3]

[1] LABCAV, Department of Electrical Engineering,
Faculty Science and Technology, University of 8 May 1945 Guelma,
BP 401, 24000 Guelma, Algeria
chaabna.ameur@gmail.com
[2] LGEG, Department of Electrical Engineering,
Faculty Science and Technology, University of 8 May 1945 Guelma,
BP 401, 24000 Guelma, Algeria
abdesselam.babouri@gmail.com
[3] ISEP, Paris, France
xun.zhang@isep.fr

Abstract. This paper studies an Indoor Positioning System (IPS) based on Visible Light Communication (VLC) using the solar cell as an optical receiver unlike conventional receivers. Due to the advantage of the solar cell such as: low cost, flexibility, and high light sensitivity, the proposed system is capable of simultaneous communication and energy gathering. The studied system is considered like an environmentally friendly and a promising technology in the next years.

Keywords: Visible light communication · IPS · Solar cell · Energy gathering
Trilateration

1 Introduction

The operation of positioning people and objects has always been important and will be more important in the next years. The recent technology to detect targets is the visible light communication (VLC), where it's very remarkable [1] as a new type of wireless communication technology with less energy consumption. In addition, the no effected to Electromagnetic Interference (EMI) allows VLC to be applied in many sectors like hospitals, airplane, smart cities, smart homes, offices, etc., where the radio frequency (RF) communication is in interference with equipment's signals. Hence, VLC has attracted many interests recently [2–5]. Classical receivers used in VLC systems are (PIN) photo-diode (PD) or avalanche photodiode (APD). Energy harvesting and signal detecting system is a new conception which was proposed in [6] as a solar-panel VLC receiver system and in [7] the authors used a solar cell as a simultaneous receiver of solar power and visible light communication (VLC) signals. Besides, the modulated VLC optical signal can converted into electrical data signal without having to supply external power, by the solar cell arrays or solar panels Rx. This electrical signal can be

© Springer International Publishing AG 2018
M. Hatti (ed.), *Artificial Intelligence in Renewable Energetic Systems*, Lecture Notes in Networks and Systems 35, https://doi.org/10.1007/978-3-319-73192-6_5

used to fill the battery of the receiver. It should be noted that the silicon-based solar cells can receive VLC data and recover energy at the same time. In our work, to locate the receiver we uses Trilateration technique and to estimate the receiver's distance from transmitters on the ceiling, the information from the received signal is used. Unlike conventional VLC positioning system, a solar cell is used as a positioning receiver. The Field Of View (FOV), light sensitivity and detection area are significantly enhanced compared to the performance of PIN Photodiode (PD). In addition to needless of external power supply, it can provide energy efficiency to the receiver side [8]. This work can be exploited for asset and people tracking in several indoor sectors as tracking patients in hospitals or security guards in malls. A study of I-V and P-V curves of solar cell under visible light is done, a similar model to the real lighting conditions are solved with MATLAB simulations. The results indicate that we can receive data and collect energy by the same solar cells.

2 Proposed Model of Visible Light Positioning System

2.1 Description Model

The proposed model of the visible light positioning system is shown in Fig. 1, where a four LEDs function as transmitters placed on ceiling. The LEDs transmit a unique ID signal to the receiver using the On-Off-Keying (OOK) modulation. The location of the receiver can be estimated after demodulating the ID signal from four LEDs. When the target moved, a new ID signal will be detected, hence a new position can be estimated due to the varying of the received power with the distance between receiver to LEDs. The assumed distance between the 2 × 8 solar cell array and the 15 W LED is 2 m.

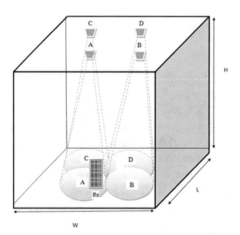

Fig. 1. Proposed model.

2.2 Model Analysis

In order to analyze relationship of the LED's transmitter model and the solar cell detector an approximation is formulated.

2.3 LED Light Model

According to [9], the channel DC gain is given by

$$H(0)_{LOS}= \begin{cases} \frac{m+1}{2\pi d^2} A_s \cos^m(\phi)\cos(\psi)T_s(\psi)g(\psi), & 0\leq\psi\leq\psi_c \\ 0, & \psi>\psi_c \end{cases} \tag{1}$$

Where A_s is the physical area of the photodetector, ψ is the angle of incidence with respect to the receiver axis, ψ_c is the field of view (FOV) of detector, ϕ is the angle of irradiance with respect to the transmitter perpendicular axis and d is the distance between transmitter and receiver. The Lambertian order m is given by: $m = -\dfrac{\ln 2}{\ln\left(\cos\phi_{\frac{1}{2}}\right)}$

and $\phi_{\frac{1}{2}}$ is the half power angle of the LED bulb. $T_s(\psi)$ is the gain of optical filter, $g(\psi)$ is the optical concentrator gain.

The total received optical power of i LEDs is given by

$$P_{rx,LOS} = \sum_{i=1}^{LEDs} P_{tx}H_{LOS}^i(0) \tag{2}$$

2.4 Solar Cell Model

The model of the solar cell shown in Fig. 2 can be expressed as [10]

$$I = N_pI_p - I_0\left\{\exp\left[\frac{q(V+IR_s)}{N_sAKT}\right] - 1\right\} - C\frac{dV}{dt} - \frac{V+IR_s}{R_{th}} \tag{3}$$

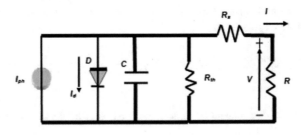

Fig. 2. Typical solar cell model.

Where N_p is the number of solar cells in parallel, N_s is the series number, I_p is the light current, I_0 is the diode saturation current, V is the output voltage of solar cell, I is the output current, and A is a constant which is typically in the range 1 to 3, assuming that

$$K_0 = \frac{AKT}{q} \tag{4}$$

As $R_{th} \gg R_s$, (3) can be written by

$$I = N_1 I_p - I_0 \left\{ \exp\left[\frac{(V + IR_s)}{N_2 K_0} \right] - 1 \right\} - C\frac{dV}{dt} \tag{5}$$

I_p is positively proportional with received illuminance power for solar cell, hence,

$$I_p = \frac{S}{1000} I_{sc} \tag{6}$$

Where S is the illuminance power of solar cell. I_{sc} is the short circuit current. All the parameters are considered in Standard Test Conditions (STC). The solar cell worked in the open state if $I = 0$, (5) becomes

$$N_p I_p = I_0 \left[\exp\left(\frac{U_{oc}}{K_0 N_s} \right) - 1 \right] \tag{7}$$

Hence,

$$I_0 = \frac{N_p I_p}{\exp\left(\frac{U_{oc}}{K_0 N_s} \right) - 1} \tag{8}$$

2.5   Overall System Model

In the case where $P_r = S$, the two precedent models can be connected. Hence, by combining (5), (6) and (7) the final relationship between U of solar cell side and LED power side can be expressed as

$$U = \begin{cases} P_t \frac{m+1}{2\pi d^2} \cos^m(\phi) \cos(\psi) T_s(\psi) g(\psi) \frac{1}{K_1} \left[1 - \frac{1}{K_2 R_h} \exp\left(\frac{U}{K_3} \right) \right], & 0 \leq \psi \leq \psi_c \\ 0, & \psi > \psi_c \end{cases} \tag{9}$$

Where R_h is the load resistance of solar cell, K_1, K_2 and K_3 are constants depended of parameters N_p, N_s, I_{sc} and U_{oc}.

3 Principle of Positioning

Using Trilateration technique in order to locate the target by calculating the accurate position and to achieve the distance a received signal power is exploited. Applying the linear least square estimation for 2-D positioning, the receiver location is obtained [11].

$$(x - x_A)^2 + (y - y_A)^2 = d_A^2 \tag{10}$$

$$(x - x_B)^2 + (y - y_B)^2 = d_B^2 \tag{11}$$

$$(x - x_C)^2 + (y - y_C)^2 = d_C^2 \tag{12}$$

$$(x - x_D)^2 + (y - y_D)^2 = d_D^2 \tag{13}$$

Where $[x_A, x_B, x_C, x_D]$ and $[y_A, y_B, y_C, y_D]$ are the coordinates of LEDs, $[d_A, d_B, d_C, d_D]$ are the horizontal distances from the receiver to LEDs, (x, y) is the receiver's position to be estimated.

4 Results and Discussion

The solar cell AM-5308 parameters are chosen for this work, according to Eqs. (5), (6) and (7) we represent our curves. Figure 3(a) represents I-V curves for different solar cell arrays, noting that the Open Circuit Voltage U_{oc} for 4 × 4 array is half of values for 4 × 8 U_{oc} and the short circuit current I_{sc} for 2 × 8 array also is half of values for 4 × 8 of I_{sc}. The values of U_{oc} of 2 × 8 and 4 × 8 are between 3 to 3.5 V, which it's enough to charge Lithium battery. Figure 3(b) shows I-V curves of 2 × 8 array under different illumination 300, 500 and 1000 Lx, which is according to the International Organization for Standardization (ISO) from 300 to 1500 Lx, where it is sufficient for office work [12].

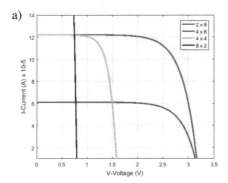
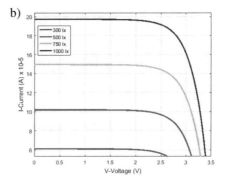

Fig. 3. (a) I-V curves for different solar cell arrays, (b) I-V curves with different illumination of 2 × 8 array (T = 298 K).

In Fig. 4(a) a P-R curves for different solar cell arrays are illustrated. The resistance corresponding for output power 1.4×10^{-4} W under 300 Lx is 50 $k\Omega$. Power properties of 2×8 array under different illumination values are simulated in Fig. 4(b). The amplitudes recorded from the four frequencies transmitted by LEDs are used to get the distance between Rx and each LED. In order to obtain the results of positioning it is necessary to solve the equations system (10), (11), (12) and (13). The estimated position of the target at the center is illustrated in Fig. 5. Due we need a value at 500 Lx of the illuminance of the typical room, an enhance of illuminance of LED lamps must done.

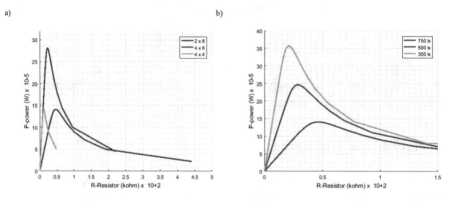

Fig. 4. (a) P-R curves for different solar cell arrays, (b) P-R curves with different illumination of 2×8 array (T = 298 K).

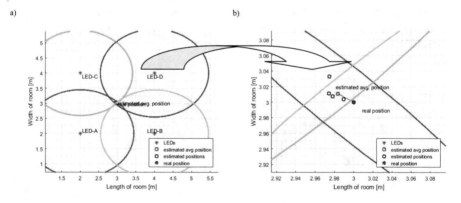

Fig. 5. Results using Trilateration positioning method at the center: (a) Global view, (b) Eye diagram

5 Conclusion

In our work we studied an indoor positioning system using solar cell as an optical receiver unlike conventional VLC positioning system. As we know the solar cell is a passive component, hence it does not require an external power supply and this is the difference compared with the PIN photodiode PD. Energy gathering and VLC signal detection simultaneously can be realized. The Trilateration technique and the ID signal information are used in this paper to achieve a suitable accuracy. The studied receiver can be integrated with wearable device to obtain an indoor positioning device with low cost and eco-friendly. Many factors were ignored in this model like response of solar cell to frequency and the channel impact, in the next works we take them into consideration and an optimization of the model will be done.

References

1. Jovicic, A., Li, J., Richardson, T.: Visible light communication: opportunities, challenges and the path to market. IEEE Commun. Mag. **51**(12), 26–32 (2013)
2. Komine, T., Nakagawa, M.: Fundamental analysis for visible-light communication system using LED lights. IEEE Trans. Consum. Electron. **50**, 100–107 (2004)
3. Afgani, M., Haas, H., Elgala, H., Knipp, D.: Visible light communication using OFDM. In: Proceedings of 2nd International Conference on Testbeds and Research Infrastructures, Development of Networks and Communities, pp. 129–134 (2006)
4. Vucic, J., Kottke, C., Nerreter, S., Langer, K.D., Walewski, J.W.: 513 Mbit/s visible light communications link based on DMT modulation of a white LED. J. Lightwave Technol. **28**, 3512–3518 (2010)
5. Chow, C.W., Yeh, C.H., Liu, Y., Liu, Y.F.: Digital signal processing for light emitting diode based visible light communication. IEEE Photon. Soc. Newslett. **26**, 9–13 (2012)
6. Wang, Z., Tsonev, D., Videv, S., Haas, H.: Towards self-powered solar panel receiver for optical wireless communication. In: Proceedings of IEEE International Conference on Communication, ICC 2014, pp. 3348–3353 (2014)
7. Kim, S.-M., Won, J.-S., Nahm, S.-H.: Simultaneous reception of solar power and visible light communication using a solar cell. Opt. Eng. **53**(4), 046103 (2014)
8. Liu, Y., Chen, H.Y., Liang, K., Hsu, C.W., Chow, C.W., Yeh, C.H.: Visible light communication using receivers of camera image sensor and solar cell. IEEE Photon. J. **8**(1), 1–7 (2016). Article No. 7800107
9. Kahn, J.M., Barry, J.R.: Wireless infrared communications. Proc. IEEE **85**(2), 265–298 (1997)
10. Koutroulis, E., Kalaitzakis, K., Voulgaris, N.C.: Development of a microcontroller-based, photovoltaic maximum power point tracking control system. IEEE Trans. Power Electron. **16**(1), 46–54 (2001)
11. Zhang, W., Chowdhury, M.I.S., Kavehard, M.: Asynchronous indoor positioning system based on visible light communications. Opt. Eng. **53**(4), 045105 (2014)
12. Wang, L., Wang, C.H., Chi, X., Zhao, L., Dong, X.: Optimizing SNR for indoor visible light communication via selecting communicating LEDs. Opt. Commun. **387**, 174–181 (2017)

Energy-Saving Through Smart Home Concept

Rabea Cheggou[✉], El-Hadi Khoumeri, and Kamila Ferhah

LTI Laboratory, Ecole Nationale Supérieure de Technologie, ENST, Algiers, Algeria
{rabea.cheggou,elhadi.khoumeri,kamila.ferhah}@enst.dz

Abstract. Smart Home has attracted significant attention from the research community in the recent years. Considerable progress has been made with the miniaturization of electronic devices for home automation and with the development of the technology of embedded system. The present work designs a smart and secure system based on various sensors and Raspberry Pi via the Internet of Things (IoT). This paper focuses on two aspects of smart home i.e. home security and home automation. This system is also designed to assist and support persons at home. It is intended to control appliances and the effective interactions to the persons. The user can visualise and control devices using internet, a webpage interface has been developed to help users to save the electrical energy by regular monitoring of home devices or the proper switching ON/OFF scheduling of them for more energy efficiency.

Keywords: Smart Home · Raspberry Pi · Sensors · IoT · Energy efficiency

1 Introduction

Smart Home has the capacity to increase the comfort of the inhabitant, for example, natural interfaces to control light, temperature or different electronic devices. The management of energy resources is another issue for smart homes. It is possible to put on standby the heating devices that are not in use when the users are absent, or automatically adapt the use of electrical resources according to the needs of the residents in order to reduce the wastes of energy resources. In addition, another essential goal of the application of information technologies to homes is the protection of persons. This is made possible by systems able to anticipate potentially dangerous situations or respond to events that endanger the integrity of individuals. Various intelligent home systems have been developed where control is via Bluetooth [1], internet [2] and Android applications [3] and short message services (SMS) [4]. In [5] they presented a Home Automation System based on ZigBee. This system consists of a home network unit and a gateway. [6] Proposed a home energy management focused home gateway, which connects the home network with the Internet. [7] Implements the initial provisioning function for home gateway based on open service gateway initiative platform. [8] Implemented the Home Gateway and GUI for control the home Appliance. Our project involves the realization of an intelligent system to control the home by using the data received from sensors. The controller (Raspberry Pi) is the main component of the home automation system; it has the essential role of implementing mechanisms necessary to

M. Hatti (ed.), *Artificial Intelligence in Renewable Energetic Systems*, Lecture Notes in Networks and Systems 35, https://doi.org/10.1007/978-3-319-73192-6_6

act in response to the needs of the user. The objectives of our project are Smart Home, energy saving, security, and monitoring of typical data using a web service to collect and manage of harvested data.

2 Proposed System

2.1 Internet of Things

Internet of Things (IoT) is a concept that encompasses various objects and methods of communication to exchange information. Today IoT is more a descriptive term for a vision that everything must be connected to the internet. IoT will be fundamental for the future, as the concept opens up possibilities for new services and innovations. All objects will be connected and able to communicate with each other. IoT is based primarily on sensors to collect information. Sensors are already used in everyday life. Different types of smartphones contain sensors, such as accelerometers, cameras, and GPS receivers [8].

2.2 Objectives of the Smart Home

The functions that are covered by home automation are classified into three categories:

- Support for basic housing functions: These functions are directly related to the comfort and optimization of the consumption of energy resources. This is to facilitate or automate the control of basic functions such as heating, ventilation, light, or the use of electronic devices.
- Security: The security of people and goods is one of the stakes of the products using the technologies of the information. Security functions can be related to physical access to the environment (anti-burglary systems), or health (prevention and surveillance).
- Support for autonomy: In the context of an aging population, the objective is to develop tools to improve the autonomy and independence of frail, disabled and elderly people. These tools must provide, additional to surveillance, means to compensate for possible disorders (e.g. auditory or visual impairment) and improve communication with the outside world [10].

2.3 Functional Expressions of Need

The needs will have to be expressed as functions in order to improve the system, i.e. why it is designed.

Control and Monitoring of Lights: Home automation allows intelligent lighting management based on parts and needs. It saves energy because its system can turn off a light, let it light up or illuminate when passing through a corridor. In a safe logic, it can also simulate presence during travel.

Security of Property and Persons: There are several more or less widespread devices that prevent the intrusion of malicious individuals into a habitat.

Presence Detection: Motion (or presence) detection is often in the heart of safety devices, because the sensor plays the role of safety in the field of safety and is often the cause of the triggering of others devices.

Soil Contact Detection: Floor contact sensors are often installed in the bottom of garage doors. This device detects its opening and triggers an alarm. The opening contact consists of a magnet creating the magnetic field and a contact whose state is modified by the presence of this field.

Energy Management: Natural inputs can be managed according to the thermal envelope of the home, air conditioning, ventilation, lighting, opening and closing of Shutters (depending on the sunshine or time of day).

   Our system is based on two architectures, the hardware architecture that represents the different constituent modules and the software architecture that defines the communication mode.

2.4   Hardware Architecture

The proposed Smart Home System consists of several modules as shown in Fig. 1, each module performing one or more functions of its own. The technical process is the source and destination of the technical data. The central module reads the state of the process by means of the sensors and influences this state by means of the actuators.

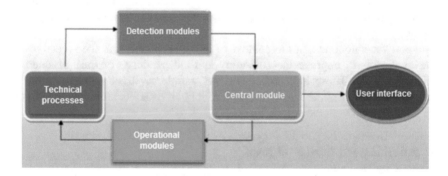

Fig. 1.   Structure of the intelligent home system.

2.4.1   The Central Module

It supports the command, receives the instructions from other modules and manages the operation of the system. In our case, the Raspberry Pi card is used. The Raspberry Pi is a small and low cost computer (the size of a credit card). It was created in order to encourage learning computer programming. Basic, the Raspberry Pi is supplied without case, power supply, keyboard, mouse or monitor. This is done in order to minimize costs and to be able to recycle other materials. Motherboard-compatible enclosures (some of which are original) are available at most sites selling Raspberry Pi.

By using the USB port, you can connect many devices: external hard drives, Blue-ray player, Wi-Fi module, webcam, and printer. The serial input ports can allow interfacing with other devices, other electronic circuits (robots, etc.). Figure 2 shows Raspberry Pi 2 B architecture with GPIO connectors. The CPU is an ARM processor, 900 MHz clock It has a variety of interfacing peripherals, including HDMI port, USB port, 1 GB RAM, SD card storage and 40 pin GPIO port for expansion.

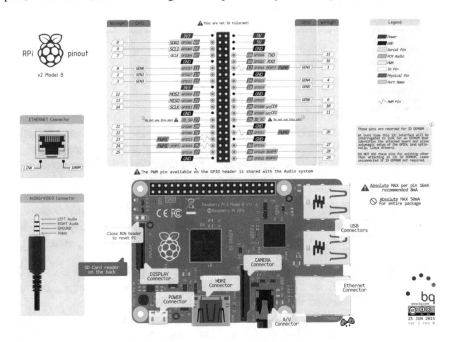

Fig. 2. Raspberry Pi 2 B [11].

Monitor, the keyboard and the mouse can be connected to the Raspberry Pi 2 through HDMI and USB connectors and it can be used like a desktop computer. It supports many of operating systems like a Debian based Linux distro, Raspbian which is used in this design. Raspberry Pi 2 can be connected to a local area network by using an Ethernet cable or USB Wi-Fi adapter, and then it can be accessed through remote login. Functional building blocks of the base station, including gateway application, database, and web application [12].

2.4.2 Detection Modules

They give information about the state of different devices and send information like opening doors, rising temperature or motion. In this project we use:

Photoresistor: Its resistance varies as a function of the luminous intensity applying to its surface. The module in Fig. 3 is designed for this purpose.

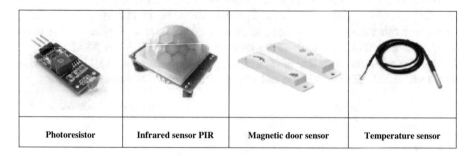

| Photoresistor | Infrared sensor PIR | Magnetic door sensor | Temperature sensor |

Fig. 3. Detection modules.

Infrared Sensor PIR: Infrared sensors PIR (Passive Infrared) provide very simple solutions for motion detection. All objects at home has low level infrared radiation, however anybody exposed to a heat source emits more radiation. PIR sensors are capable of detecting any change in radiation level within their detection area. For example, when a person enters a room, the sensor detects the change in radiation level.

Magnetic Door Sensor: The magnetic contact is also called the opening contact. It lands on doors and windows. Once a door or window is opened, this magnetic field is broken and the information is transmitted to the central module of the system.

Temperature Sensor: This waterproof temperature sensor based on a DS18B20 is anti-rust due to the high quality stainless steel tube. The connection is via a one wire bus and it can be connected to any microcontroller.

We used other components operative module like a fan which is intended to create an air conditioner. A buzzer that is used to trigger an alarm in case of intrusion. LEDs: Used in the lighting system. A servo motor is used to move the window blind. The last component is the TV.

2.5   Software Architecture

The communication mode of the system is the client/server environment through a network between several programs or software. The server waits for requests sent by client and responds to them. In our case, the computer on which the server is running is our Raspberry Pi 2. The exchange of information between client and server is summarized in Fig. 4.

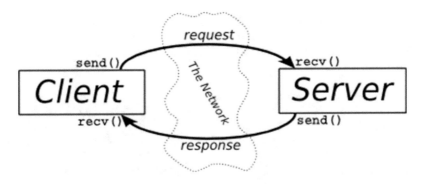

Fig. 4. Client/server system [13].

We use the Raspberry Pi card as a web server rather resorting to web hosting providers, from an economic point of view, it is important to know that web hosting services are not free, unlike the Raspberry Pi card that just requires an internet connection. The Apache web server is installed, this software allows the machine to analyse a user's queries in http form, and return the corresponding file to the query (or an error If the file is not found, or if the query is incorrectly formulated), PHP is installed which mainly make webpage dynamic, i.e. the user sends information to the server that in return send it the results modified according to this information.

The web interface makes the control of devices easy and gives more efficiency and mobility. We have installed the Apache/PHP server on the Raspberry Pi. The server can receive web pages. PHP and JavaScript manage the interactions with the user and the animation of the HTML page. The server is available only locally, i.e. we cannot access it from any other network. An access to our server via a URL is recommended. For this purpose we have the DynDNS [14] is a service that allows a user to have a dynamic IP so that he is able to use a URL. We will retrieve a URL from noip.com.

3   Implementation Results and Discussion

The system is designed to be secure; the user needs to use login and password to access the application as seen in Fig. 5. This step highlights the security aspect. The following

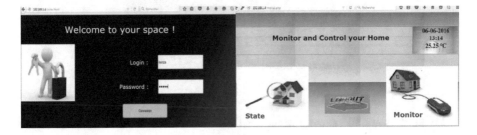

Fig. 5. Authentication page & home page

figure shows the main interface that allows us the access to the control interface or to the status interface, or to disconnect from the application.

Control interface: This interface allows enabling or disabling the various systems of the home (Fig. 6 - Right)

Fig. 6. Control & status interface

Status interface: This interface shows the state of the systems, either ON or OFF (Fig. 6 - Left).

The light is up if the sensor detects that it is dark and the opposite if it is light. The fan will be active once the room temperature is high. The awning of the window is controlled directly from the interface. The TV can be controlled and switched ON/OFF. It is clear that all these reduce the electricity consumption. The proposed system has been successfully implemented and tested for the desired functionalities. The final prototype of the system is illustrated in Fig. 7.

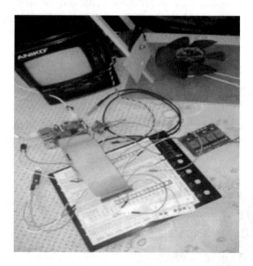

Fig. 7. Implemented proposed system

Figure 8 shows an example when opening the door, the alarm is triggered (alarm status ON) and we have all other sensor OFF.

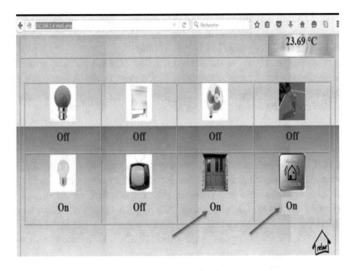

Fig. 8. Testing of Proposed System

4 Conclusion

In this paper, we have introduced the design of a low-cost and flexible solution of Smart Home. The system is very useful for the control of the home. Energy saving and safety are the most important point of our application. We have presented a functional and promising prototype of a remote home monitoring and control system. Indeed, every user has the ability to remotely control his home in a reliable and efficient way. The prospects are to feed the sensors in solar system for better energy consumption.

References

1. Yan, M., Shi, H.: Smart living using Bluetooth-based Android smartphone. Int. J. Wirel. Mob. Netw. **5**(1), 65 (2013)
2. Kovatsch, M., Weiss, M., Guinard, D.: Embedding internet technology for home automation. In: 2010 IEEE Conference on Emerging Technologies and Factory Automation (ETFA) (2010)
3. Gurek, A., et al.: An android based home automation system. In: 2013 10th International Conference on High Capacity Optical Networks and Enabling Technologies (HONET-CNS) (2013)
4. Elkamchouchi, H., ElShafee, A.: Design and prototype implementation of SMS based home automation system. In: 2012 IEEE International Conference on Electronics Design, Systems and Applications (ICEDSA) (2012)
5. Gill, K., Yang, S.-H., Yao, F., Lu, X.: A ZigBee-based home automation system. IEEE Trans. Consum. Electron. **55**(2), 422–430 (2009)
6. Kushiro, N., Suzuki, S., Nakata, M., Takahara, H., Inoue, M.: Integrated home gateway controller for home energy management system. In: IEEE International Conference on Consumer Electronics, pp. 386–387 (2003)

7. Ok, S., Park, H.: Implementation of initial provisioning function for home gateway based on open service gateway initiative platform. In: The 8th International Conference on Advanced Communication Technology, pp. 1517–1520 (2006)

8. Yoon, D., Bae, D., Ko, H., Kim, H.: Implementation of home gateway and GUI for control the home appliance. In: The 9th International Conference on Advanced Communication Technology, pp. 1583–1586 (2007)

9. https://ido2016.sciencesconf.org/

10. Friedewald, M., Da Costa, O., Punie, Y., Alahuhta, P., Heinonen, S.: Perspectives of ambient intelligence in the home environment. Telematics Inform. **22**, 221–238 (2005)

11. https://www.megaleecher.net/Raspberry_Pi_2_Schematic_And_Pinout_Diagram

12. Ferdoush, S., Li, X.: Wireless sensor network system design using Raspberry Pi and Arduino for environmental monitoring applications. In: The 9th International Conference on Future Networks and Communications (FNC-2014), Department of Electrical Engineering, University of North Texas, Denton, Texas, 76203, USA (2014). Procedia Comput. Sci. **34**, 103–110 (2014)

13. https://raspberrypi.stackexchange.com/questions/12105/how-to-communicate-between-raspberry-pis-using-wifi

14. https://account.dyn.com/entrance/

Uses and Practices of E-health Environments: An Interactive Architecture for Effective Real-Time Monitoring of Patients

Soundouss Ismahane Talantikite[✉] and Salah Chaouche[✉]

Department of Architecture, Faculty of Architecture and Urbanism, University of Constantine 3, Constantine, Algeria
talantikite.s.i@outlook.com, Salahchaouche@yahoo.fr

Abstract. The impact of new information and communication technologies on our societies is universal. They have the advantage of being transversal and affecting all sectors, notably health.

E-health, a multidisciplinary field of research, is growing significantly and is nowadays necessary for systems of healthcare. The determinants of this break-through in healthcare remain the age demographic of the population coupled with the inadequacy of the health workforce, without ignoring the rising costs of care. This is reflected in the continuing rise in the number of people with addictions, as chronic diseases become more acute, creating a huge need for appropriate technologies. The circumstances are therefore conducive to reflection leading to the modeling of more innovative solutions for effective patient care.

The NICT's incursion in the medical field is most often seen as the introduction of tools designed to guarantee or even improve the quality of healthcare services, so our research combines crosses and puts into dialogue the relationship between space and the psychiatric patient. To do this, we approach space as a determining factor in helping people with psychiatric illnesses, and as an interactive, real-time approach to the needs of the sick.

Keywords: E-health · NICT · Ambient intelligence · Interactive space
Psychiatric patients · Hospital

1 Introduction

To live is to move in, to develop, to move, to appropriate, to desert, to walk, but above all to live the space [1]. We ourselves are elements of space. By our daily actions, our simple presence, acts and transforms the space spontaneously, generating imperatively a new atmosphere in space. In addition, "The notion of atmosphere engages a sensible relationship with the world" [2].

At the beginning of the 20th century, Ernst Mach examined the relationship between the physical and the psychic state of individuals through the perception of sensations. He then highlighted the role of movement in any perception [3].

© Springer International Publishing AG 2018
M. Hatti (ed.), *Artificial Intelligence in Renewable Energetic Systems*, Lecture Notes in Networks and Systems 35, https://doi.org/10.1007/978-3-319-73192-6_7

Then Erwin Straus, in his critical work on the Pavlovian reigning approach, shows that every perception engages an action, that there is no feeling without a movement [4]. "It is not the physiological functions of the sensory organs that make a being a sentient being, but rather the capacity to approach and the latter does not belong to sensation or movement alone." [5].

Later, Merleau-Ponty emphasizes that it is not the object that would be felt and approached, but the sensible quality of the very place of perception, as the place of encounter with the world. Feeling would be the approach and perception would then be movement: "Therefore, if perception as an approach is open to the world itself, each experience is a presentation of this world and not the apprehension of an object" [6].

Gibson, for his part, shows that space is not qualitatively neutral, but that it presents the potentials of actions at all times. Space gives infinite possibilities of diverse and ephemeral actions, or "affordances" (offerings).

Many architects claim an attitude that takes into account the senses and compose it with the social context as well as with the physical environment. The space in which we live is multi-sensory, dynamic, and relational, and contemporary architecture feels explicitly concerned with these aspects.

On this subject, architects possess knowledge and expertise, often intuitive, which can produce architectures of high quality, both from a sensible point of view and from a point of view of usage.

The perception of environmental factors is a subject that is beginning to be relatively documented in social psychology or in the psychology of space, including in terms of health. The present study was based on several references that precisely introduce the impact of environmental factors on the health of patients or people (patients). Osmond's (1957) study on the impact of noise, Ulrich's (1981) research, which deals precisely with the impact of perception of the atmosphere (for example induced by plants) Post-operative hospitalization. In the same vein, research by Tennessen and Cimprich (1995), Kaplan (2001) and Raanaas, Gustave Nicolas Fischer and Virginie Dodeler (2009), Patil and Hartig (2010) report the impact of the space environment on patient health. These authors also assume that the NICT could become tools to aid in the therapy of patients by offering "a solution for the control of atmospheres".

2   The Effect of Architecture on the Clinical Health of Patients

Gustave Nicolas Fischer and Virginie Dodeler, lecturers in social psychology at French universities, in their book Health Psychology and the Environment (2009), discuss the influence of the architectural characteristics of a health building on clinical outcomes. They distinguish five measures related to health and stress in hospital buildings: the level of stimulation, coherence, affordance, control and the reconstituting qualities of space.

Indeed, to give patients the possibility to control their environment through techno-logical devices with the use of home automation but also with furniture adapted to their handicap (removable walls when patients are in multiple rooms etc.) allows one to bring a feeling of well-being and security. The latter can be seen as a helping tool in the healing process of the sick.

3   The Introduction of NICT's in Health Spaces: E-Health

The medical sector has undergone profound changes and has a new face. The rise of digital technology, through the incursion of the New information and communication technologies (N.I.C.T.), has revolutionized the approach to health. Innovation has largely contributed to the upheaval of the medical world and health in general. The spread of technologies has made it possible to provide a greater number of healthcare services that are more accessible to all, while improving medical performance and the quality of healthcare services (integration of virtual reality and "intelligent scalpels", telemedicine and tele-surgery, transplants, surgical robots, endoscopy...) according to Christensen in Blagg (2009). E-health, the concept to which we refer, encompasses multiple components. On the basis of World Health Organization (WHO) resolution WHA58.25 (2005), supplemented by WHO Resolution 148/17 (2011), e-health can be defined as "the use of information and communication technologies in support of health action and related areas, including health care services, health surveillance, health literature and education, knowledge, and research".

3.1   Ambient Intelligence

Computer science is a universe in constant evolution on several levels: hardware, software, and architecture. Today, computing opens up to a new era: that of ambient intelligence - help to the human being - which, through home automation, is gradually changing our habits, improving our daily life but is also upsetting our societies. Certain societal progress is made against a backdrop of questions about the man-machine relationship. Artificial intelligence wants to develop a chain of events from perception, analysis to reflection and ultimately action. Some architectures take into account intentional attitudes such as desires and beliefs and integrate "practical rationality" as a "medium-fine" reasoning.

Multi-agent systems are interacting sets of intelligent agents. This discipline is a sub-branch of artificial intelligence also called distributed artificial intelligence. Intelligent agents can be virtual entities (for example, software) or real beings. In this case, the objective of the SMA (multi-agent system) is the simulation of a set of autonomous and in interaction with the environment. For example, multi-agent systems are an interesting way of modeling societies, and as such have broad applications, extending to the human sciences.

Ambient intelligence is therefore at the heart of current technological advances. In parallel, it allows social and societal studies but whose invasiveness must be regularly re-evaluated. Technological evolution is therefore no longer the object of attention of engineers and scientists alone.

Ambient intelligence is therefore a project of the future, which must take into account the software, the human-machine interface.

4   Methodology

Our research is a first attempt to understand the relationship between space and the psychiatric patient in Algeria. To do this, we approach space as a determining factor in helping people with psychiatric illnesses, as an interactive, real-time approach to the needs of the sick.

In order to do this, we carried out a practical study at the psychiatric hospital (240 beds) Er Rasi of Annaba and the Neurology Department Ibn Rochd Annaba, Eastern Town of Algeria.

A questionnaire was adopted for a sample of 25% of inpatients, 50% of whom were female. In addition, a long interview with the heads of services and the nursing staff of the services, was conducted with a guided tour of the places. The rooms of the patients have been carefully selected according to their locations to the different sources of light, noise and the opposite.

Our cross-data collection methodology includes a direct approach of the field of ethnographic type, observations and interviews.

We aim at the collection of "inside" data that is difficult to access by external quantitative observations. They affect the subjectivity of the users and the feelings of the people. The desire to have access to all the spaces of the hospital and to the different uses made us opt for an immersion of two weeks. We thus circulated in the different places, clothed in a white coat provided by the doctor (Fig. 1).

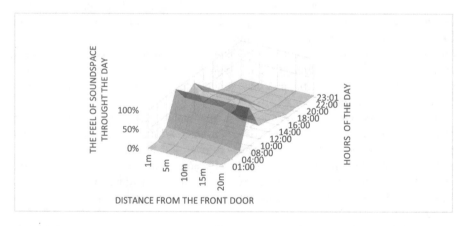

Fig. 1. *The feel of the soundscapes throughout the day, and the distance from the front door by the patients. Source: Author 2017*

This point is important because it allowed us to move freely but, above all, it affected the type of relationship with the various actors: assimilated to the nursing staff, we established with the latter an easy contact but it was also reassuring On the meaning of our presence spontaneously associated with an evaluation by management; On the patients' side, the white coat gave us a form of authority and legitimacy but also led them to monitor their remarks (Fig. 2).

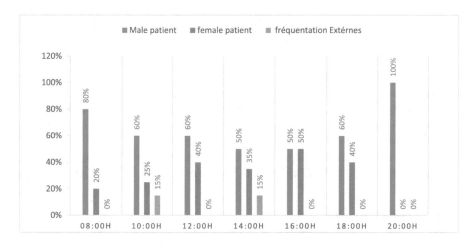

Fig. 2. The *attendance of the patients in the garden. Source: Author 2017*

In the first days, we were perceived as members detached by the hierarchy, to limit this inconvenience, we proscribed the notes taken in the presence of the users. Over the course of the day, in fact very quickly, we have gained the confidence of the staff of which we have become a sort of confidant on all the problems related to the architecture of the building or even beyond (Fig. 3).

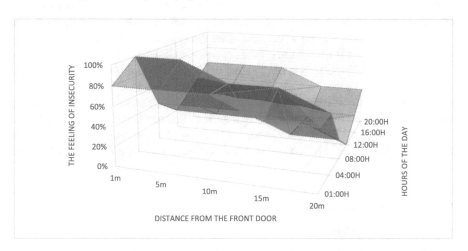

Fig. 3. The feeling of insecurity in the case of females patients. Source: Author 2017

4.1 Spatial Observations

Our observations focused on the different ways of using space for patients and caregivers. It was observed every day in order to be able to carry out a map of the displacements. Some of these observations were still going on at the same time, except for appointments, and I reported at lunchtime and in the evening, on various maps, the observations on the movements and the arrangements.

4.1.1   Regarding Patients

– The different furniture fittings of the rooms were made by the patients. This shows how patients can personalize their rooms, what they allow themselves, and what their room for maneuver is. Through visits with doctors, nurses or nursing aides. We have thus been able to draw up three major planning scenarios.
– Places of socialization and their use. What are these places, how to appropriate them, which were created by the architect and which are the fruit of the patients?

4.1.2   Concerning Staff

– For the staff, three themes emerged. The manner in which the various staff members use the rooms (doctors during visits, nurses, nursing aides), the way with which they appropriate the corridors (according to the schedules it is the territory of the aides/ nurses, and at the time of the visits it is that of the doctors), and, finally, their places of relaxation.

4.1.3   Results

Women raise the problem of the interdependence of the spaces of circulation that is confirmed by the disposition of the rooms. The issue of insecurity was mentioned by 1/3 of the women because of the people from outside and the mixity created by the design of the building. The tunneling effect caused by the narrow corridor serving the rooms and its elongation gives half of the patients (women and men) continuous anxiety from 2 PM until 6 PM. The presence of cladding on the windows of the rooms led all the patients to interpret space as a place of penance. The noise generated by the morning visit of the medical staff is perceived as a source of nuisance for 4/5 of the patients. The hospital has a garden frequented by more than ¾ of the men from 11 AM to 1 PM and from 3 PM to 5 PM. On the other hand, 1/3 of the women use only one corner of the garden from 2 PM to 4 PM. The layout of this garden remains inappropriate for its use for both sexes.

4.2   The Report on NICT

4.2.1   Regarding Patients

• During the interviews, 80% of the patients revealed that they had a smartphone, 50% of them with an average age of 35 years were constantly connected to the Internet network through the 3G or 4G services.
• 90% of patients reported having an internet connection at their homes.

4.2.2   Concerning Staff

– During the days spent in the hospital, we noted that 100% of the medical staff automatically connected their smartphones to the Internet once they were on a break in their places of relaxation, using the services of 3G or 4G.
– The registration of patients and the processing of files as well as their classification is always done manually in registers, the telephone line cut; doctors, nurses and nursing aides denounce using their personal lines at their expense.
– Lack of connection with other services, and to the internet.
– Doctors and nurses seem to have a general idea of technological advances in the field of health.

4.2.3   Results

By addressing the concept of e-health, its protocol, instructions for use, and the diversity of offers and services it offers, 100% of the patients questioned as well as 100% of the personnel who compose the medical body say they are favorable for its experimentation. Indeed, they shout their confusion as regards the logistics, which they considered outdated and dilapidated (Fig. 4).

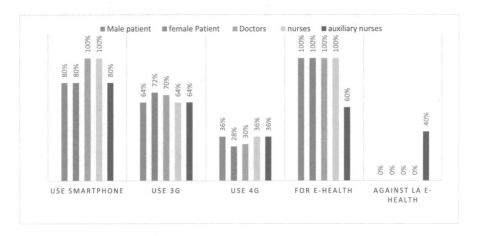

Fig. 4.   The report of the medical staff with NICT. Source: Author 2017

5   Conclusion

The users' acceptance of space and the notion of quality of life cannot be foreseen. The gap between the space conceived and the space experienced experiences a discrepancy between the sensitive created by architecture and the sensory needs of individuals, more specifically in our case, that of psychiatric patients.

The ambient intelligence made possible thanks to the NICT is one of the keys that could improve the offer of healthcare in Algeria. Indeed, e-health has revolutionized the approach to health. The spread of technologies has made it possible to provide a greater number of healthcare services that are more accessible to all, while improving medical performance and the quality of healthcare services (integration of virtual reality and "intelligent scalpels", telemedicine and tele-surgery, transplants, surgical robots, endoscopy, etc.). The latter will then be optimal by integrating its notions in an intelligent building that responds to a flexible and interactive architecture based on intelligent networks, ambient intelligence and cognitive radio technologies.

The introduction of NICT's in the health sector in Algeria is still very precarious. In order to create a connected hospital that meets international standards, it is imperative to boost the use of NICT's in the various services to realize the different spots.

References

1. Georges, P.: Espèces d'espaces, pp. 49–50. Éd. Galilée, Paris (1974)
2. Amphoux, P., et al.: La notion d'ambiance: Une mutation de la pensée urbaine et de la pratique architecturale. Éditions Plan Urbanisme Construction Architecture, Paris (1998)
3. Ernst, M.: L'analyse des sensations - Le rapport du physique au psychique. Éd. Jacqueline Chambon, Nîmes (1996). 1ère édition 1922
4. Erwin, S.: Du sens des sens. Éd. Jérôme Millon, Grenoble (1989). 1ère édition 1935
5. Erwin, S.: p. 378 (1989)
6. Erwin, S.: Le visible et l'invisible, p. 71. Éd. Gallimard, Paris (1989). Explication à propos de Merleau-Ponty Maurice (1964)
7. Jean-François, A.: La vue est-elle souveraine dans l'esthétique paysagère ? In: Le Débat, Paris, no. 65, pp. 51–59 (1991)

Using a Hybrid Approach to Optimize Consumption Energy of Building and Increase Occupants' Comfort Level in Smart City

Brahim Lejdel[1]([⊠]) and Okba Kazar[2]

[1] University of El-Oued, El-Oued, Algeria
Lejdel82@yahoo.fr
[2] University of Biskra, Biskra, Algeria
kazarokba@yahoo.fr

Abstract. Energy consumption in city is increasing because the number of population is increasing. Also, this energy consumption differs according to the weather data, inhabitants of buildings and type of building; commercial, residential or administrative. Whereas, the citizen needs to have a compromise between the energy consumption, the economic cost, the comfort and the environmental impact of the building. In this paper, we will propose a smart model which permits to manage, control and regulate the consumption of energy according to some criteria. Thus, this model allows figuring, regulating, optimizing energy consumption and satisfying the occupants' comfort in real time. Thus the citizen does not need to read electricity metrics or wait the billing period to know its energy consumption. Also, this approach allows saving the energy resources and increasing the system productivity even in peak demand hour.

Keywords: Energy consumption · Multi-agent system · Genetic algorithm
Peak demand hour · HVAC-L system

1   Introduction

The objective of Smart Cities (SC) is enhancing the life quality of citizen. It has been gaining increasing importance in the agendas of policy makers [12]. Thus, the City is an urban area which is composed of many factors that are in interactions as the population, many networks as energy or water network, buildings…etc. actually, the smart city is designed to optimize resources as energy, water or Internet connection. To optimize the energy consumption, we can set sensors and streetlights which can gather and send information. Thus, we need to connect streets in smart city. These smart connected streetlights are the core of smart cities.

In a city, there are buildings which have peak-off hours. For example, if taking the residential buildings as student dormitories, during the day, they are in their classes or studying, therefore the demand of energy is lower. But, this is the contrast in the evening. In the commercial or administrative buildings, they demand a great energy during the day but in the evening, the consumption will be lower [11]. And, in the same building,

© Springer International Publishing AG 2018
M. Hatti (ed.), *Artificial Intelligence in Renewable Energetic Systems*, Lecture Notes in Networks and Systems 35, https://doi.org/10.1007/978-3-319-73192-6_8

we can find a peak-off hour and low consumption. Thus, we must create a smart model which can manage, regulate and optimize the energy consumption of all these types of building. Thus, smart buildings require indoor environment control system to control and manage the energy consumption while maintaining the comfort level of their occupants [7].

In this paper, we propose to use a multi-agent system which allows to distribute the different tasks between the agents when each agent can perform Genetic Algorithms to optimize energy consumption in real-time, thus adapting rapidly to building consumption. Thus, we will use some agents as meter agent which represents the meter in the building. Also, we attribute to each energy resource which provide the energy to streets; an agent called resource agent that can interact with the other actors or meter agents to regulate and optimize the energy consumption. Thus, all agents can cooperate and negotiate to find the best solution which can regulate the energy consumption for all types of buildings. Then, we develop a GIS system which allows to know the position of buildings and all data associated with it as building-id, energy consumption, the billing of this building etc.

This paper is organized as the following. Firstly, we will present a state of the art review for optimization of energy consumption. Then, we describe our proposed approach which is based on two approaches, the Multi Agent System and Genetic Agent (MAS-GA). Finally, we add a conclusion and some future works.

2   Related Works

The two last decades see a great development in intelligent technologies which merged in the city to improve the citizen livings. These intelligent technologies permit to create a smart building that can facilitate the life of their occupants'. Generally speaking, smart buildings are expected to address both intelligence and sustainability issues by using advanced computing systems and intelligent technologies to achieve the optimal combinations of overall citizen comfort and energy consumption [10]. In this context, we can cite that a heating, ventilation and air-conditioning (HVAC) systems and lighting systems are the main energy consumers in residential, administrative or commercial buildings; so this demand a driven control measures such as turning off/dimming smart lighting systems, controlling ventilation, as well as heating and cooling supplied to buildings using actual building occupancy information contributes towards improving the energy performance in buildings [5]. These systems consume up to 60% of the energy for buildings. The rest of energy is consumed in different kinds of equipment depending on the functionality of the building [6].

The multi-agent systems are an ideal approach to model the complex system because they can address distributed and adaptive situations. In this paper, we will use this approach to manage, regulate and optimize the energy consumption in residential building to control environmental parameters and solve possible conflicts which can occur in energy consumption and customers comfort. Generally, an agent is defined as a software (or hardware) entity that is located in a certain environment and is able to autonomously react to changes in that environment.

In this section, we present a review of state of art that using multi-agent system to find a compromise between energy consumption and building occupants' satisfaction.

Davidsson and Boman describe a decentralized system consisting of a collection of software agents that monitor and control an office building. It uses the existing power lines for communication between the agents and the electrical devices of the building, such as sensors and actuators for lights and heating. The objectives are both energy saving and increasing customer satisfaction through value added services [2].

Hagras *et al.* propose to use a system to learn the thermal responses of a building to external climatic factors as well as to internal occupancy loads to reduce building energy in condition to maintain occupants' comfort [3].

Then, Liao and Barooah develop a Multi-Agent System to simulate the behaviors of all the occupants of a building, and extract reduced-order graphical models from simulations of the agent-based model [5].

Also, Joumaa *et al.* developed a Multi-Agent System to manage anticipatory and reactive control of HVAC (heating, ventilation and air conditioning) and lighting systems for buildings [4]. They use the agent based approach because building energy consumption can be modeled by distributed systems. These systems rely on integration with facility systems and appliances through HVAC-L system and sensors. We can cite also, the work of [1] which develops a new Agent based approach that allows simulating the consumption energy of in commercial building. This simulation considers occupants with different energy usage characteristics as well as potential changes to occupant using due to their interactions with the building environment and with each other.

Finally, Al-Daraiseh et al. present an intelligent agent based system to optimize energy consumption of HVAC system in Higher Educational Institution (HEI) buildings. The developed system employs artificial intelligence techniques to predict the demand of the system and optimize energy consumption of the HVAC system [8].

All these above-mentioned multi-agent systems try to find the potentially promising opportunity to reduce building energy through direct cooperation and coordination with building occupants in addition to improving control of building systems and energy resources. But there are not approaches developed to optimize the energy consumption and the satisfaction of the occupants' demand comfort. In this paper, we will propose a hybrid approach of multi agent system and genetic agent (MAS-GA) which permits to each agent to find the optimal plan that can increase the occupants' comfort level and optimize the energy consumption.

3   Building Agent's Optimizer

As previously said, the building Agent has an optimizer and simulator that are used together to discover the values of HVAC-L system that they can optimize the energy consumption in the prevailing conditions and also satisfy the comfort level of occupants. The use of genetic algorithm has a major advantage over systems that rely on predefined values, as each building agent enables a genetic algorithm to discover the values of HVAC-L system that they may not resemble any predefined values, but they may be optimal values for the current indoor conditions. The optimizers should achieve a satisfactory balance between discovery time of solutions and consumption of energy. Thus, each building agent executes a genetic algorithm to find the optimal values of HVAC-L systems that can be attributed to each system to perform optimal energy consumption and increasing the occupants' comfort level.

3.1   Chromosome Structure

To apply the genetic algorithm, we should define the genes and the chromosome structure. The gene of can be characterized by its identifier, and a set of values of HVAC-L systems that can be applied to perform the optimal energy consumption and satisfy the occupants' comfort level. We use multiple forms to coding the genes. Firstly, we use the strings to encode the identifiers, and then we use real number to encode the values of temperatures, the ventilation, the air conditioning and lighting system. Figure 1 presents the structure of the gene.

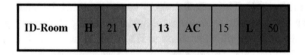

ID-Room : room identify in the building,

H: heating system and its value,

V : ventilation system and its value,

AC: Air Conditioning system and its value,

L: light System ands its value.

Fig. 1.   Gene structure's of room.

3.2   Genetic Algorithm Steps

(a) Initialization

The initialization operator determines how each chromosome is initialized for participating in the population of the genetic algorithm. Here, the chromosome is filled with the genetic material from which all new solutions will evolve. In this work, we will use the Steady State to initial the generation process and select the population of genetic algorithm for next generation [9]. First, Steady State creates a population of individuals by cloning the initial chromosomes. Then, at each generation during evolution, it creates a temporary population of individuals, adds these to the previous population and then removes the worst individuals in order that the current population is returned to its original size. This strategy means that the newly generated offspring may or may not remain within the new population, dependant upon how they measure up against the existing members of the population.

(b) **Crossover**

The crossover operator defines the procedure for generating a child from two parent chromosomes. The crossover operator produces new individuals as offspring, which share some features taken from each parent. The probability of crossover determines how often crossover will occur at each generation. In this approach, we will use the single point crossover strategy was adopted for all experiments. In this paper, the results for all experiments presented were generated using a crossover percentage of 50%, which is to say that at each generation 50% of the new population were generated by splicing two parts of each chromosome's parents together to make another chromosomes. Figure 2 shows the crossover operator.

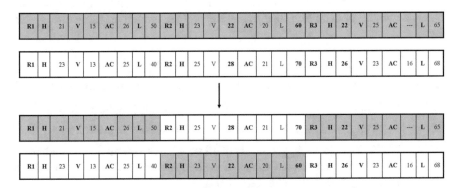

Fig. 2. Crossover operator.

(c) **Mutation**

The mutation operator is very important. It defines the procedure for mutating the chromosome. Mutation, when applied to a child, randomly alters a gene with a small probability. It provides a small amount of random search that facilitates convergence at the global optimum. The probability of mutation determines how much of an each genome's genetic material is altered, or mutated. If mutation is performed, part of chromosome is changed. The mutation should not occur too often as this would be detrimental to the search exercise. In this work, the results presented here were generated using a 1% mutation probability, which was determined experimentally, utilizing a single case of vector HVAC-L system. We present the random mutation in Fig. 3.

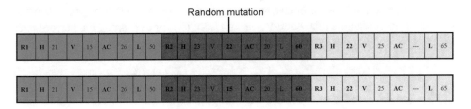

Fig. 3. Mutation operator.

(d) **Evaluation of solutions**

We can say that the success of any discrete optimization problem rests upon its objective function, the purpose of which is to provide a measure for any given solution that represents its relative quality. In our resolution method of consumption energy problem in buildings, the objective function used here works by calculating and summing the penalties associated with the temperature, the illumination, the indoor air quality and ventilation within our state representation. Thus, we will use the objective functions to evaluate solutions of the energy consumption problem and examines the weighted relationship between the actual measured values of the temperature, the ventilation, the indoor air quality, illumination level and values of occupants' comfort level according to these four parameters. The objective functions used to evaluate solutions requires a number of definitions that model the problem underlying structure, specifically:

- $R = \{R_1, R_2, R_3, \ldots \ldots .. R_n\}$ is the set of all rooms in the building,
- $H = \{H_1, H_2, H_3, \ldots \ldots .. H_n\}$ is the set of all heating systems in the building,
- $L = \{L_1, L_2, L_3, \ldots \ldots .. L_n\}$ is the set of all illumination systems in the building,
- $A = \{A_1, A_2, A_3, \ldots \ldots .. A_n\}$ is the set of all air conditions in the building,
- $V = \{V_1, V_2, V_3, \ldots \ldots .. V_n\}$ is the set of all ventilation system in the building,
- H_m, L_m, A_m, V_m are the measured values of the temperature, the illumination, and the indoor air quality and ventilation respectively.
- H_c, L_c, A_c, V_c are the comfort values of the temperature, the illumination, and the indoor air quality, respectively.
- *N1, N2, N3, N4* is the all number of the temperature, the illumination, the indoor air quality and ventilation system respectively.
- $[T_{min}, T_{max}]$ represent the interval time where the values of three parameters were measured.
- $[C_{min}, C_{max}]$ represent the comfort range. This range can be defined by customers.
- $[E_{min}, E_{max}]$ represent the consumption energy range.

Two important parameters are in our MAS-GA, the assigned energy to the HVAC system E_H and the assigned energy to the lighting system E_L.

In this context, we have mainly two important functions $f(C)$ and $f(E)$ which permits to evaluate the performance and efficiency of the proposed approach. These two functions are calculated by the Building Agent.

The objectives of this optimization mechanism is to maximize occupants' comfort $f(C)$ and to minimize the total energy consumption $f(E)$ for evaluating the performance and the efficiency of our system. Firstly, we have

$$f(C) = C_1 \times H_c/H_M + C_2 \times L_c/L_m + C_3 \times A_c/A_m \qquad (1)$$

C_1, C_2 and C_3 are the user-defined weighting factors, which indicate the importance of three comfort factors and resolve the possible equipment conflicts. These factors take values in the range of [0, 1]. Occupants can set their own preferred values in different situations according the season or the occupancy period. As previously said, since occupancy period has a profound influence on energy savings, it should be taken into account in the control strategy design. Generally speaking, in the occupied hours, the building agent activates the optimizer to tune the set point in order to obtain the acceptable indoor visual comfort with minimized energy. Otherwise, the agent building turns off all the resource lights and keep the blind position to save energy if there are no occupants in the building. The objective function is defined in Eq. (1), and the optimization goal is to maximize these objective functions. Since the ratio between the measured value and comfort value determined by occupants plays via graphic interface, it has an important role in achieving the control goal. Thus, it allows increasing the occupants comfort level and optimizing the energy consumption.

The second objective function permits to control the energy consumption in indoor environment of building. The objective of this function consists to minimize the total energy consumption of HVAC-L system. Thus, we can define this objective function as the following.

$$f(E) = E_{HVAC} + E_L \qquad (2)$$

E_{HVAC} and E_L represent the consumption energy of HVAC system and the lighting system, respectively.

4   Experimentation and Results

In this section, we present a case study that illustrates how to design the different agents of our system and show collaboration between them. We use Jade (http://jade.tilab.com/) to implement the different agents, room agent, profile agent and building agent. Also, we use Java (https://www.java.com/fr/) to implement the different steps of genetic algorithm as crossover operator, mutation operator and the evaluation function. Thus, the smart building used is a residential building; it aims at providing a comfortable environment for all occupants inside the building. Firstly, the room agents use the sensor to learn the HVAC-L data which can be used as input in the genetic algorithms. The occupants can introduce their preferences in the profile agent via a graphic interface. The building agent runs a genetic algorithm that can find the optimal values of HVAC-L system that allow optimizing the energy consumption and increasing the occupants' comfort level. The results are regarded as a seasonal. In Table 1, we introduce the different intervals of occupants' satisfaction and the energy consumption.

Table 1. Intervals of occupants' satisfaction and the energy consumption

Evaluation parameters	Unacceptable	Less satisfaction	Highly satisfaction
Occupants' satisfaction	[1, 4.5]	[5, 8]	[8.5, 10]
Energy consumption (Kwh)	[250, 300]	[200, 240]	[150, 195]

To control the different systems, room agent uses some data of the HVAC-L system. As we know, to maintain a higher occupants' comfort level, we must increase the energy consumption. Whereas, the smart building agent tries to find a compromise between the energy consumption and the higher occupants' comfort level. Thus, it should find the optimized values to determine energy consumption dispatched to both the HVAC system and the lighting system. Remember that the objectives of this optimization mechanism are to maximize occupants' comfort level and to minimize the total energy consumption of the smart building. In the Fig. 4, we state that there is difference in occupants' comfort level in the two approaches, with MAS-GA and without MAS-GA. With MAS-GA, the system achieves a higher occupants' comfort level compared to second approach, without MAS-GA. Thus, the occupants' comfort level in the MAS-GA has been improved rapidly compared to the second approach.

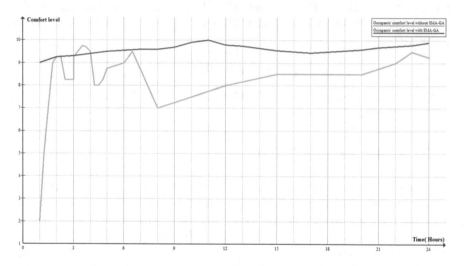

Fig. 4. Occupants' comfort level with and without MAS-GA.

Figure 5 shows that when we use our proposed approach, the energy consumption has been improved compared to classic approach which can be used to decrease the consumption energy. Thus, when we use MAS-GA approach, we can higher minimize the energy consumption thus, the MAS-GA approach permits to optimize the energy consumption compared to the classic approach, without MAS-GA.

The MAS–GA is designed to enable the interactions between the occupants and the environment by learning the occupants' behaviors. According to the case studies and simulation results, the proposed MAS-GA is capable of managing, regulating and controlling the building effectively to satisfy occupant's comfort and to optimize energy consumption.

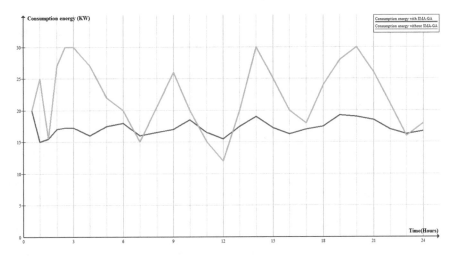

Fig. 5. Consumption energy with MAS-GA and without MAS-GA.

5   Conclusion and Future Works

In this work, a MAS-GA is developed to manage, control and regulate the building indoor area via emerging different technologies as genetic algorithm and multi-agent system. The multi-agent system has been embedded with a genetic algorithm for optimizing consumption energy of the buildings and increase occupants' comfort level. The simulator of agent building can execute a simulation which allows to find the optimal scheme that can saving the energy resources in building and increase theirs performances. Also, the occupants of building can introduce their preferences and a substantial degree of intelligence enhances the operability of the control system via the graphic interface. Our proposed approach provides a strong and open architecture in which agents can be easily configurated, and new agents can be added without changing the whole architecture system. Thus, the proposed approach has attained the compromise between the energy consumption and occupants' comfort level. The main problem of the genetic process is consumption time. In a future work, it would be of great importance to analyze the execution time of different tasks in the genetic process. Also, we can propose to use the same approach in the water consumption because it presents a great problem for the citizen in the smart cities.

References

1. Azar, E., Menassa, C.: An agent-based approach to model the effect of occupants' energy use characteristics in commercial buildings, pp. 536–543. American Society of Civil Engineers (ASCE) (2011)
2. Davidsson, P., Boman, M.: Distributed monitoring and control of office buildings by embedded agents. Inf. Sci. **171**, 293–307 (2005)

3. Hagras, H., Packharn, I., Vanderstockt, Y., McNulty, N., Vadher, A., Doctor, F.: An intelligent agent based approach for energy management in commercial buildings. In: IEEE International Conference on Fuzzy Systems, FUZZ-IEEE 2008, IEEE World Congress on Computational Intelligence, pp. 156–162 (2008)

4. Joumaa, H., Ploix, S., Abras, S., De Oliveira, G.: A MAS integrated into home automation system, for the resolution of power management problem in smart homes. Energ. Procedia **6**, 786–794 (2011)

5. Liao, C., Barooah, P.: An integrated approach to occupancy modeling and estimation in commercial buildings. In: American Control Conference (ACC), pp. 3130–3135 (2010)

6. Spataru, C., Gauthier, S.: How to monitor people 'smartly' to help reducing energy consumption in buildings? Archit. Eng. Des. Manag. **10**(1–2), 60–78 (2014)

7. Labeodann, T., Aduda, K., Boxem, G., Zeiler, W.: On the application of multi-agent systems in buildings for improved building operations, performance and smart grid interaction – a survey. Renew. Sustain. Energ. Rev. **50**, 1405–1414 (2015)

8. Al-Daraiseha, A., El-Qawasmeha, A., Shah, N.: Multi-agent system for energy consumption optimisation in higher education institutions. J. Comput. Syst. Sci. **81**, 958–965 (2015)

9. Syswerda, G.: Uniform crossover in genetic algorithms. In: Shaeffer, J.D. (ed.) Proceeding of 3rd International Conference on Genetic Algorithms and Their Applications, pp. 2–9. Morgan Kaufmann Publishers Inc., San Francisco (1989)

10. Wang, Z., Wang, L., Dounis, A.I., Yang, R.: Multi-agent control system with information fusion based comfort model for smart buildings. Appl. Energ. **99**, 247–254 (2012)

11. Yang, R., Wang, L.: Development of multi-agent system for building energy and comfort management based on occupant behaviors. Energ. Build. **56**, 1–7 (2013)

12. Neirotti, P., De Marco, A., Cagliano, A.C., Mangano, G., Scorrano, F.: Current trends in SmartCity initiatives: some stylised facts. Cities **38**, 25–36 (2014)

Smart Power Management Hybrid System PV-Fuel Cell

Amar Ben Makhloufi[1(✉)], Mustapha Hatti[2], and Taleb Rachid[1]

[1] LGEER Laboratory, Hassiba Benbouali University, Chlef, Algeria
amar.ben.makhloufi@gmail.com, rac.taleb@gmail.com
[2] Unité de Développement des Equipements Solaires,
Bou Ismail, Tipaza, Algeria
musthatti@ieee.org

Abstract. In this paper, we are developed a novel strategy for a hybrid energy management system consisting of a photovoltaic (PV) array, a polymer electrolyte membrane fuel cell (PEM-FC) as energy sources, the purpose of this system is to balance the power supply with load demand fluctuation and to ensure its long-term sustainability. Therefore, we used multiple forms of power supply such as a Fuel cell energy storage system, PV generator, and an intelligent energy management system to control puissance's of system. DC-DC converters are used to interface PV-FC combination. Energy management between the two sources is done by operating them in three different modes based on power requirement of load. In the case of additional energy directed to the electrolyzer, to be converted to hydrogen and stored for the time of need.

Keywords: Intelligent management · Energy storage · Fuel cell
Hydrogen · Photovoltaics · Artificial neural network

1 Introduction

Energy is one of the keys to the development of nations and society. Civilization is dependent on a constant, consistent supply of energy; globally, the demand for energy has been increasing consistently in parallel with growth in population and economic consumption [1]. Stringent energy-related problems have led to an increased interest in renewable energy sources. Due to the intermittent nature of those sources, the energy production varies significantly according, for example, to the hour of the day and the period of the year. Therefore, it is necessary to store the produced energy allowing its use after production. The criteria for the selection of solutions for the storage of renewable energy are still under debate, both for stationary and mobile applications [2]. A subject of numerous researches]. Apart from conventional ways to store the energy like lead-acid batteries and pumped hydro-storage [3], other methods including super-capacitors [4], flow batteries, hydrogen [5] and renewable power methane production are considered.

Hydrogen could provide additional utility-scale energy storage options and unique opportunities to integrate the transportation and power sectors. Although hydrogen is currently a high-cost option, it offers some advantages over competing technologies,

© Springer International Publishing AG 2018
M. Hatti (ed.), *Artificial Intelligence in Renewable Energetic Systems*, Lecture Notes
in Networks and Systems 35, https://doi.org/10.1007/978-3-319-73192-6_9

including that it has a high storage energy density and a potential for co-firing in a combustion turbine with natural gas to provide additional flexibility for the storage system [4].

In addition, the analysis shows that the cost of producing larger volumes of hydrogen to support both grid energy storage and hydrogen fuel cell vehicles could be competitive with the cost of producing hydrogen for fuel cell vehicles in a dedicated hydrogen production facility. R&D efforts focused on lowering the cost of electrolyzer and fuel cell technologies and increasing roundtrip efficiency of storage have potential to improve value and environmental performance. Additional analysis is required to understand the potential contribution of hydrogen storage to power system operating reserves [6].

Fuel cells that are currently being developed can be used as possible substitutes for the internal combustion engine in vehicles as well as in stationary applications for power generation. A fuel cell is an electro-chemical device which produces electricity without any intermediate power conversion stage [7]. The most significant advantages of fuel cells are low emission of greenhouse gases and high-power density. The energy density of a typical fuel cell is 200 Wh/l, which is nearly ten times that of a battery. The efficiency of a fuel cell is also high, in the range of 40% to 60%. If the waste heat generated by the fuel cell is used for cogeneration, the overall efficiency of such a system could be as high as 80% [8].

2 Structure of the Hybrid Power System

The hybrid power system comprises a PV panel, a PEM fuel cell, which are connected to the same DC voltage bus through appropriate dc-dc power converters and controls. Figure 1 illustrates the structure of the proposed hybrid power system. There are two main sources of energy: PV panel and fuel cell. Although the fuel cell is an energy storage device by hydrogen, it is also a source of energy when the load demands excess energy. The PV panel provides as much power as possible to the load. The function of the fuel cell is to supply to the load the rest of the average power that the PV panel cannot meet. The fuel cell supplies transient power to peak load demands or absorbs transient power from the main sources [10, 11].

2.1 Circuit Topology of the Proposed PV Power System

The proposed PV power system consists of a dc/dc converter with MPPT, and a dc/dc converter with voltage regulation and controller, as shown in Fig. 2.

A dc/dc converter adopt cuk converter coupled PV array, Normally DC-DC converter used for regulating the output voltage for change in input voltage for various load condition. and the MPPT control algorithm is used to extract the maximum power from the PV arrays [13].

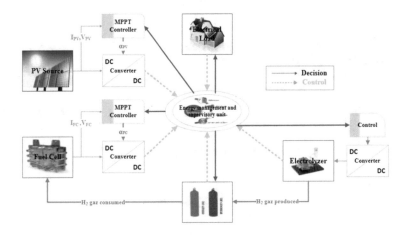

Fig. 1. Structure of the hybrid power system.

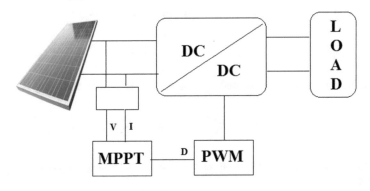

Fig. 2. PV system.

2.2 Fuel Cells

Fuel cells (FC) have received a major boost in recent years, as a result of the growing demand by a number of sectors. For this reason, this technology is rapidly expanding and there are many different research lines associated with the various sectors. A particular case which has excited much interest is the integration of hydrogen-based fuel storage systems with renewable energies in many applications [7].

Fuel cells are electrochemical devices that convert the chemical energy of a gaseous fuel, usually hydrogen, directly into electricity (Fig. 3) without any mechanical work. The only by-product from their operation is heat and water [9].

The partial pressure and molar flow of any gas is related by equation: [9]

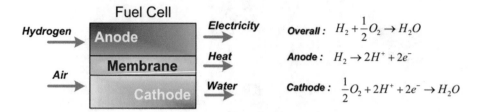

Fig. 3. Fuel cell and energy conversion.

$$\frac{q_{H_2}}{P_{H_2}} = \frac{K_{am}}{\sqrt{M_{H_2}}} = qK_{H_2} \tag{1}$$

Where K_{H2} is the hydrogen valve constant and M_{H2} is molar mass of hydrogen (kg/kmol).

The relationship between flow rate of hydrogen that has reacted and the fuel cell system current (I_{fc}) is given by the electrochemical equation:

$$q_{H_2}^r = \frac{N_o \times I_{fc}}{2.F} = 2K_r I_{fc} \tag{2}$$

Also, the partial pressure of hydrogen gas is given by the equation: [9]

$$p_{H_2} = \frac{K_{H_2}}{1 + T_{H_2^s}} \times \left(q_{H_2}^{in} - 2K_r I_{fc} \right) \tag{3}$$

Where activation over voltage (η_{act}) and ohmic over voltage (η_{ohmic}), time constant of hydrogen:

$$T_{H_2} = \frac{V_{an}}{K_{H_2} RT} \tag{4}$$

R = universal gas constant (kmol/s.atm), T = Absolute temperature (K). Similarly, the partial pressure of water and oxygen can be written. The output voltage of fuel cell system is expressed as the sum of Nernst voltage (E): [9]

$$V_{cell} = E + \eta_{act} + \eta_{ohmic} \tag{5}$$

$$\eta_{act} = -B \times \ln(C \times I_{fc}) \tag{6}$$

$$\eta_{ohmic} = -R_{int} \times I_{fc} \tag{7}$$

$$E_o = N_o \times \left[E_o + \frac{RT}{NF} \times log\left(\frac{p_{H_2} \times \sqrt{p_{O_2}}}{p_{H_2O}} \right) \right] \qquad (8)$$

Where B and C are constants, Eo = No load voltage (V), No = Number of series cells and F = Faraday's constant (C/kmol) [9].

2.3   Electrolyzer System

An electrolyzer is an electrochemical device, which decompose water into hydrogen and oxygen. The production rate of H2 in an electrolyzer cell is directly proportional to the transfer rate of electrons at the electrodes, which in turn equivalent to the electrical current in the circuit, as expressed in Eq. (9), and the ratio between the actual and the theoretical maximum amount of H2 produced in electrolyzer is known as Faraday efficiency, which is also expressed in Eq. (9) as, [11].

$$m_{H_2} = \frac{n_F N_{i_e}}{2F} \quad and \quad n_F = 96.5 \left(e^{\frac{0.09}{i_e} - \frac{75.5}{i_e^2}} \right) \qquad (9)$$

2.4   Hydrogen Storage Tank

The dynamics of H2 storage tank system can be expressed in Eq. (8) as,

$$P_b - P_{bi} = \frac{ZN_{H_2}RT_b}{m_{H_2}V_b} \qquad (10)$$

2.5   Power Management

The proposed PV power management system includes four dc/dc converters connected in parallel to supply power to the load. Its operational modes can be divided into two modes. Note that PPV is the output power of the PV arrays, PFC is that of the fuel cell and PL is the load power. Moreover, "1" represents the power which is generated by PV arrays or is dissipated by load, while "0" is the power which is not generated or is not dissipated. According to the power management of the proposed PV power system shown in Table 1, all operational modes are described.

E.1. Operational Mode I
In the operational mode, the dc/dc converter with P_V is adapted to supply power to the load. when the load power $P_V > P_L$, the proposed P_{FC} power system is shut down. When the load power $P_L < P_V$.

E.2. Operational Mode II
In operational mode II, when $P_V < P_L$, the power source used to supply power to the load, is fuel cell with PV arrays. If $P_L < P_V$, the proposed P_{FC} power system is shut down.

3 Simulation Results

To investigate the performances of the proposed hybrid photovoltaic – fuel cell power system, simulation studies were conducted in programme MATLAB, which is a powerful simulation tool for complex advanced power systems. Figure 4 shows the schematic view of a 1 kW hybrid power system under study. The power system was assumed was operated from 8:00AM to 00:00PM. The sky was assumed to be clear and the ambient temperature was assumed to be variable during the operation (Fig. 5).

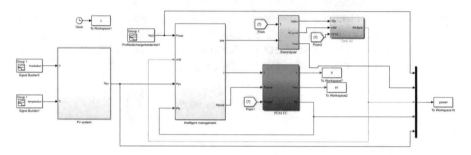

Fig. 4. Simulation of hybrid system.

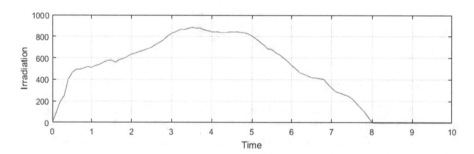

Fig. 5. Irradiation.

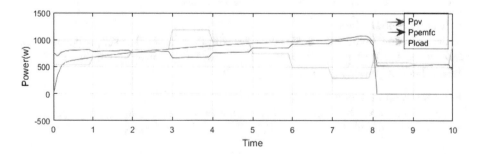

Fig. 6. Power system.

The simulation results are shown in next Figure. shows the power from the PV panel, the fuel cell stack, and the load. Note that the power of fuel cell is change with power load.

The simulation results are shown in Fig. 6. shows the power from the PV panel, the fuel cell stack, and the Load. Note that the power Fuel cell is with power load. as the Power of the fuel cell increases if the power load is less than the power of the PV. In the opposite case, the fuel cell helps to achieve the power required by the load.

4 Conclusion

This paper presents an effective power management system for a hybrid photovoltaic - fuel cell power system so that the combination can be used as a reliable power source. The hybrid power system comprises a PV panel, a PEM fuel cell stack, which are connected to the same DC voltage bus through appropriate dc-dc power converters and controls. The PV panel powers the load and charges the fuel cell through a buck converter which acts as a maximum power point tracker. A boost converter is selected to adapt the low DC voltage output from the fuel cell to the regulated bus voltage. The Fuel cell is directly connected to the voltage bus. The power may flow through the Fuel cell in both directions. The fuel cell boost converter is controlled to maintain the battery to a given state of charge (or voltage). To investigate the performance of the proposed hybrid power system, simulation studies were conducted in MATLAB. Simulation studies demonstrate that the power management system controls the power of each source of energy properly and the hybrid power system works reliably. Simulation results show that the developed control strategies are effective for the hybrid power systems.

References

1. McLellan, B., Zhang, Q., Farzaneh, H., Utama, N.A., Ishihara, K.N.: Resilience, sustainability and risk management: a focus on energy. Challenges **3**, 153–182 (2012)
2. Eslami, M., Bahrami, M.A.: Sensible and latent thermal energy storage with constructal fins. Int. J. Hydrogen Energy **42**(28), 17681–17691 (2017)
3. Ma, T., Yang, H., Lu, L.: Feasibility study and economic analysis of pumped hydro storage and battery storage for a renewable energy powered island. Energy Convers. Manag. **79**, 387–397 (2014)
4. San Martín, I., Ursúa, A., Sanchis, P.: Integration of fuel cells and supercapacitors in electrical microgrids: analysis, modelling and experimental validation. Int. J. Hydrogen Energy **38**, 11655–11671 (2013)
5. Valdés, R., Rodríguez, L.R., Lucio, J.H.: Procedure for optimal design of hydrogen production plants with reserve storage and a stand-alone photovoltaic power system. Int. J. Hydrogen Energy **37**, 4018–4025 (2012)
6. Hou, P., Hu, W., Chen, Z.: Operational optimization of wind energy based hydrogen storage system considering electricity Marlet's influence. In: 2016 IEEE PES Asia-Pacifc Power and Energy Conference, Xi'an, China (2016)

7. Hatti, M., Tioursi, M., Nouibat, W.: Neural network approach for semi-empirical modelling of PEM fuel-cell. In: 2006 IEEE International Symposium on Industrial Electronics, vol. 3, pp. 1858–1863 (2006)

8. Motapon, S.N., Dessaint, L.A., Al-Haddad, K.: A comparative study of energy management schemes for a fuel-cell hybrid emergency power system of more-electric aircraft. IEEE Trans. Ind. Electr. **61**(3), 1320–1334 (2014)

9. Jayalakshmi, N.S., Nempu, P.B., Shivarudraswamy, R.: Control and power management of stand-alone PV-FC-UC hybrid system. J. Electr. Eng. **25**, 45–60 (2012)

10. Tamalouzt, S., et al.: Performances analysis of WT-DFIG with PV and fuel cell hybrid power sources system associated with hydrogen storage hybrid energy system. Int. J. Hydrogen Energy **41**(45), 21006–21021 (2016)

11. Tiang, T.L., Ishak, D.: Modeling and simulation of deadbeat-based PI controller in a single-phase H-bridge inverter for stand-alone applications. Turk. J. Electr. Eng. Comput. Sci. **22**, 43–56 (2014)

12. Khalilnejad, A., Sundararajan, A., Sarwat, A.I.: Performance evaluation of optimal photovoltaic-electrolyzer system with the purpose of maximum Hydrogen storage. In: 2016 IEEE/IAS 52nd Industrial and Commercial Power Systems Technical Conference (I&CPS) (2016)

13. Ben Makhloufi, A., Hatti, M., Rachid, T.: Comparative study of photovoltaic system for hydrogen electrolyzer system. In: 2017 6th International Conference on Systems and Control (ICSC). IEEE Conference Publications (2017)

Renewable Resources

Design of Array CSRRs Band-Stop Filter

Kada Becharef[✉], Keltoum Nouri, Boubakar Seddik Bouazza,
Mahdi Damou, and Tayeb Habib Chawki Bouazza

LTC Laboratory, Department of Electronic, Faculty of Technology,
University Dr. Moulay Tahar de Saida, Saida, Algeria
becharef_kada@yahoo.fr, keltoum_nouri@yahoo.fr

Abstract. Metamaterials are artificial pseudo-homogeneous structures with electromagnetic properties not found in nature. This paper presents a microwave band-stop filter making use of complementary split-ring resonator (CSRR). This filter combines a conventional bandstop filter characteristics and negative permittivity metamaterial to establish a metamaterial filter. This structure is designed in the X-band [8.2–12.4] GHz, using relatively dielectric constant substrate material (RO4003 $\varepsilon r = 3.38$ and tangential losses (tg(δ) = 0.0027)), Numerical calculations using the Finite Element Method MEF based the High Frequency Structure Simulator (HFSS) software was used to design this filter.

Keywords: Metamaterials · Complementary split ring resonators (CSRRs)
Band-stop filter design

1 Introduction

Left-handed metamaterials with simultaneously negative permittivity and permeability have received considerable attention in both scientific and engineering communities [1–3]; it is due to the possibility for the design of novel microwave components with more compact size and better performance [10]. One component of a communication system which needs often an improvement is a microwave filter. Because of the increasing demand on size miniaturization and low cost, many microstrip filters designs have been proposed for size miniaturization [2]. The aim of this paper is to apply a bandstop filter synthesis proposed in the literature based on CSRRs, to design a very compact microwave bandstop filter [6]. Recently, complementary split ring resonators proposed by pendry and al [2], attracted much attention as a canonical metamaterial structure that gives rise to an effective magnetic response without the need for magnetic materials.

These resonators can be derived from the SRR structure in a straight forward way by using the concepts of duality and complementary SRR have been successfully applied to the fabrication of left-handed metamaterials [1–4], the rings are made of nonmagnetic metal like copper and have a small gap between them as shown Fig. 1.

© Springer International Publishing AG 2018
M. Hatti (ed.), *Artificial Intelligence in Renewable Energetic Systems*, Lecture Notes in Networks and Systems 35, https://doi.org/10.1007/978-3-319-73192-6_10

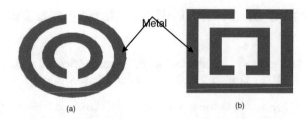

Fig. 1. Split Ring Resonators (a) circular and (b) square

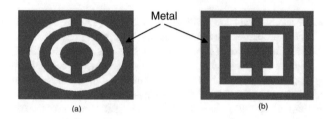

Fig. 2. Two different kinds of CSRR (a) Circular CSRR and (b) Square complementary split ring resonator.

The complementary split-ring resonator structure is achieved by SRR in the ground plane, structures complementary to double split rings [5], in this way structures with apertures in metal surface are obtained, as shown Fig. 2.

In this paper, we will do a detailed investigation of CSRR based stopband filters: starting with a single CSRR etching in the ground plane, finding its stopband characteristics, after that the effect of number of CSRRs and periodicity on the stopband filter performance is investigated.

2   Extraction of Effective Parameters

The extraction of the effective parameters from the coefficients of reflection and transmission, is known as the method name of Nicolson-Ross-Weir (NRW) [15], To obtain the effective electromagnetic parameters of the structure, a theory of homogenization is used. The main purpose of this theory is to describe in a simple and macroscopic way the microscopic complexity of the response of objects to an incident electromagnetic radiation. Indeed, the idea was to model the metamaterial as an isotropic homogeneous slab, and to calculate the effective parameters ε, and μ of the homogenous slab from the transmission and reflection coefficients obtained by simulations under Matlab.

For an isotropic homogeneous slab in a vacuum space, the transmission t, and reflection r have the following relations with the refractive index n, and the impedance z, of the slab [14]:

$$t^{-1}\left[\cos(nkd) - \frac{i}{2}\left(z + \frac{1}{z}\right)\sin(nkd)\right] \tag{1}$$

$$\frac{r}{t} = \left[-\frac{i}{2}\left(z - \frac{1}{z}\right)\sin(nkd)\right] \tag{2}$$

Where k and d are the wave vector and the thickness of the slab respectively.

Equations (1) and (2) can be inverted to calculate n and z from t, and r. By completing this inversion, we obtain:

$$z = \pm\sqrt{\frac{(1+r)^2 - t^2}{(1-r)^2 - t^2}} \tag{3}$$

$$\cos(nkd) = \frac{1}{2t}\left(1 + t^2 - r^2\right) \tag{4}$$

By using the S_{ij} parameters, the effective material
Parameters can be extracted [13]:

$$z = \pm\sqrt{\frac{(1+S_{11})^2 - S_{21}^2}{(1-S_{11})^2 - S_{21}^2}} \tag{5}$$

$$re(n) = \pm Re\left[\frac{\cos^{-1}\left(\frac{1}{2S_{21}}\left(1 - S_{11}^2 + S_{21}^2\right)\right)}{kd}\right] \tag{6}$$

$$im(n) = \pm im\left[\frac{\cos^{-1}\left(\frac{1}{2S_{21}}\left(1 - S_{11}^2 + S_{21}^2\right)\right)}{kd}\right] \tag{7}$$

The ambiguity on the signs of the Eqs. (5), (6) and (7) is prevented if account is held owing to the fact that the real part of the impedance is positive if it is about a passive medium, and the imaginary part of the refractive index is positive to ensure that the incident wave amplitude decreases inside the structure.

Then, the effective permittivity and permeability can be computed from the equations:

$$\varepsilon = \frac{n}{z} \tag{8}$$

$$\mu = nz \tag{9}$$

3   The Design of Complementary Split Ring Resonators

CSRR is dual counterparts of SRR. Therefore a dual electromagnetic behaviour for them is expected according to the duality theorem. The incident electric field needs to be polarized in the axial direction of the resonator. In this way, CSRRs are etched on center line of the microstrip technology [7]. This arrangement makes sure that the CSRRs are properly exited by the electric field applied parallel to the ring axis. Since CSRRs are excited by the electric field, they produce negative effective permittivity Re $(\varepsilon_{eff}) < 0$. The CSRR topology and equivalent circuit model are illustrated in Fig. 3. The CSRR unit cell was designed to operate around 9.5 GHz. The geometry of the cell is as follows: c = d = 0.33 mm, g = 0.33 mm, m = 3 mm and the global size is 3.63 mm × 3.63 mm. The substrate used is a RO4003C having the following characteristics (relative permittivity ε_r = 3.38, loss tangent tg(δ) = 0.0027 and thickness h = 0.81 mm). The resonator is simulated by using a commercially available 3D full-wave solver (Ansoft HFSS).

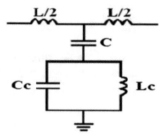

Fig. 3. (a) CSRR particle with relevant dimensions; (b) unit-cell equivalent circuit of a CSRR-loaded microstrip line

The CSRR itself can be described as an LC resonant; the resonant frequency is described by the following expression [7]:

$$f = \frac{1}{2\pi\sqrt{L_c C_c}} \qquad (10)$$

Figure 4 shows the S_{ij} parameters simulated results. It shows a rejected frequency band around the designed frequency F = 10.22 GHz of the CSRR resonator explained by a transmission of about −24 dB.

As shown in Fig. 5, the real part of the permittivity shows Lorentz response behavior, it is negative in the frequency range between 1 GHz and 20 GHz

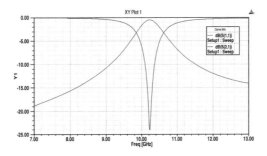

Fig. 4. Simulated S_{ij} parameters square complementary split ring resonator

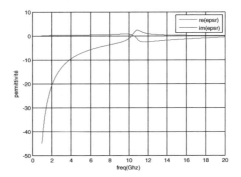

Fig. 5. Real and imaginary parts of permittivity

4 CSRRs Band Stop Filter Design

4.1 Low Pass to Band Stop Transformation

The low-pass filter elements are thus converted to series resonant circuits (having low impedance at resonance) in the series arms and to parallel resonant circuits (having high impedance at resonance) in the shunt arms. Notice that both series and parallel resonator elements have a resonant frequency of ω_0 [12]

$$\omega \rightarrow -\Delta \left(\frac{\omega'}{\omega_0} - \frac{\omega_0}{\omega'} \right)^{-1}$$

Then series inductors of the low-pass prototype are converted to parallel LC circuits having element values given by [9, 11]

$$C'_k = \frac{1}{\Delta L_k \omega_0} = \frac{1}{\Delta g_k \omega_0}$$

$$L'_k = \frac{\Delta L_k}{\omega_0} = \frac{\Delta g_k}{\omega_0}$$

$$\Delta = \frac{\omega_2 - \omega_1}{\omega_0}$$

To determine the parameters g_k, it must first seek the order n we need from the data specifications, before using the following equations:

$$g_1 = \frac{2}{\eta} \sin\left(\frac{\pi}{2n}\right)$$

$$\eta = \sinh\left[\frac{1}{n} \arg\sinh\left(\frac{1}{\xi}\right)\right]$$

$$g_k g_{k+1} = \frac{4 \sin\left[\frac{2k-1}{2n}\pi\right] \sin\left[\frac{2k+1}{2n}\pi\right]}{\eta^2 + \sin^2\left(\frac{k\pi}{n}\right)}$$

The shunt capacitor of the low-pass prototype is converted to series LC circuits having element values given by

$$C'_k = \frac{\Delta C_k}{\omega_0} = \frac{\Delta g_k}{\omega_0}$$

$$L'_k = \frac{1}{\Delta C_k \omega_0} = \frac{1}{\Delta g_k \omega_0}$$

The transformation of the elements shown in Fig. 6:

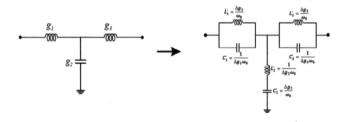

Fig. 6. Transformation elements low-pass band stop

4.2   Simulations

Therefore, if an array of CSRRs is etched on the ground plane aligned with the strip, a strong electric coupling with the desired polarization is expected. A conventional bandstop filter based on a 50 Ω line and the CSRR etched in the conductor strip [2] is designed in the microstrip technology.

First, we present a bandstop structure obtained by a network of 1 × 3 CSRRs. The substrate employed is the RO 4003CR with a copper layer thickness of 35 microns on each side the geometry shown in Fig. 7, The transmission line has a width of 1.858 mm and a length of 16.52 mm. The periodicity of CSRRs network is 3.63 mm and thus.

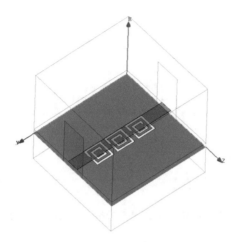

Fig. 7. Bandstop filter topology using microstrip line loaded with 3 CSRRs

The results show a bandstop behavior over a wider band than that obtained with the configuration of Fig. 8 despite using fewer CSRRs. This is because a better electrical coupling is created when the strip is centered over the row of CSRRs. This coupling extends the most intense band to lower frequencies.

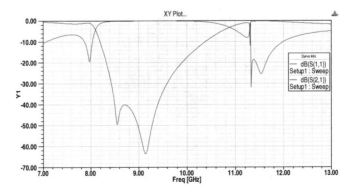

Fig. 8. Response frequency Microstrip filter loaded with 3CSRRs

The negative band for the refractive index approximately lies between 1.05 GHz and 4.63 GHz, 7.46 GHz and 8.22 GHz, 8.55 GHz and 10.31 GHz, 12.51 GHz and 15 GHz; it is the frequency range where the permittivity and the permeability are simultaneously negative as it appears on Fig. 9.

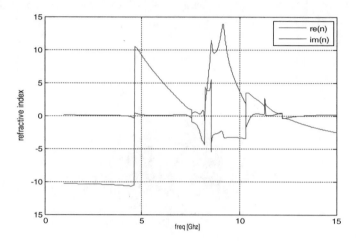

Fig. 9. Results obtained for effective refractive index.

If we insisted to pass face which includes the fourth CSRRs, as shown in Fig. 10.

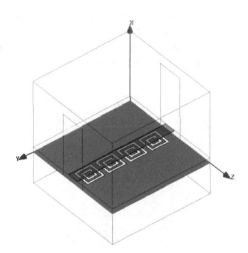

Fig. 10. Bandstop filter using the microstrip line loaded 4 CSRRs

Figure 11 gives the results of simulation parameters S_{ij}. These results show a behavior band stop in the frequency range [8.25–10.82]GHz. At frequencies around 9.5 GHz, the transmission $_{21}$ is approximately −47 Db.

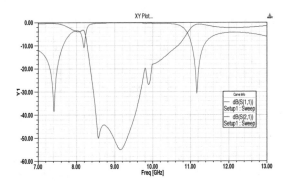

Fig. 11. Response frequency of Microstrip bandstop filter loaded with 4 CSRRs

The CSRRs structure in the ground plane is further increased to twelve as shown in Fig. 12 and simulated in the same frequency range of 7 GHz to 13 GHz.

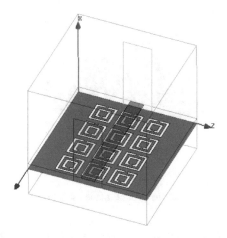

Fig. 12. Bandstop filter using the microstrip line loaded with 12 CSRRs.

Figure 13 shows the simulation results of the stopband filter structure shown in Fig. 12. The results are plotted for the scattering parameters (S_{11} and S_{21}).

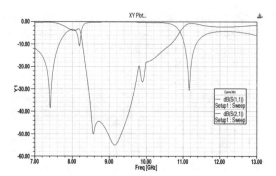

Fig. 13. Scattering parameters S_{11} and S_{21} of Microstrip line loaded with 12 CSRRs

Figure 14 shows a conventional bandstop filter design; it consists of 16 rings CSRRs. Further increase of rings would have just an increase of reject level.

The network area for $13,89 \times 13,89 \text{ mm}^2$. Therefore, a length of 1,315 mm is left on each side of the network on the tape.

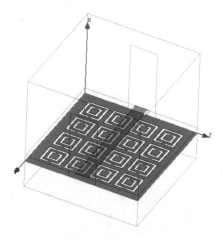

Fig. 14. CSRRs etched into the ground plane of the microstrip band stop filter.

The structure is simulated for the frequency band [7–13] GHz, and the results are displayed in Fig. 15. A bandstop behavior is obtained in the band [8.76–10.42] GHz around the resonant frequency of CSRRs. The transmission response shows steep slopes on each side cuts and insertion losses of −2 dB outside this band.

When increasing the number of cell CSRRs we noticed the difference between the level of rejection (S_{21}) and rejected Strip.

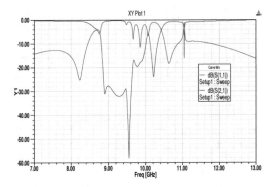

Fig. 15. Response frequency Microstrip filter loaded with 16CSRRs

5 Conclusions

In this paper, a compact stop band microstrip filter based on CSRRs has been proposed, successfully designed and simulated. The resulting device is very compact, produces very high rejection with sharp cut-offs in the forbidden band, and exhibits a flat and lossless pass band. This behaviour has been interpreted as corresponding to a frequency band with negative valued permittivity and permeability provided by the CSRRs. The size of the structure could be further reduced by tailoring SRR and CSRR dimensions, using a properly modified version of the CSRR. CSRRs are usually etched in the ground plane of the substrate. So compared with the SRR, the CSRR does not occupy extra space and for this reason it is highly suitable for designing of size miniaturized microwave devices.

References

1. Vesalago, V.G.: The electrodynamics of substances with simultaneously negative values of permittivity and permeability. Sov. Phys. Usp. **10**, 509–514 (1968)
2. Pendry, J.B., Holden, A.J., Robbins, D.J., Stewart, W.J.: Magnetism from conductors and enhanced non linear phenomena. IEEE Trans. Microw. Theory Tech. **47**(11), 2075–2084 (1999)
3. Pendry, J.B.: Negative refraction makes a perfect lens. Phys. Rev. Lett. **85**, 3966–3969 (2000)
4. Smith, R., Padilla, W.J., Vier, D.C., Nemat-Nasser, S.C., Schultz, S.: Composite medium with simultaneously negative permeability and permittivity. Phy. Rev. Lett. **84**, 4184–4187 (2000)
5. Hayt, W.H., Buck, J.A.: Engineering Electromagnetics, 6th edn. McGraw-Hill, New York (2001)
6. Palandöken, M., Henke, H.: Compact LHD-based Band-Stop. IEEE (2010)
7. Falcone, F., Lopetegi, T., Baena, J.D., Marqués, R., Martín, F., Sorolla, M.: Effective negative epsilon stop-band microstrip lines based on complementary split ring resonators. IEEE Microw. Wirel. Compon. Lett. **14**, 280–282 (2004)
8. Whitaker, J.: The Electronics Handbook, 2nd edn. CRC Press, Boca Raton (2005)

9. Pozar, M.: Microwave Engineering, 3rd edn. Wiley, New York (2005)
10. Engheta, N., Ziolkowski, R.W.: Electromagnetic Metamaterials: Physics and Engineering Explorations. Wiley-IEEE Press, Hoboken (2006)
11. Salleh, M.K.M.: Contribution à la synthèse de résonateurs pseudo elliptiques en anneau. Application au filtrage planaire millimétrique. Thèse de Doctorat en Électronique de l'Université de Toulouse, Octobre 2008
12. Pozar, D.: Microwave Engneering, 4th edn. Wiley, New York (2011)
13. Sabah, C.: Tunable metamaterial design composed of triangular split ring resonator and wire strip for S and C microwave bands. Prog. Electromagn. Res. B **22**, 341–357 (2010)
14. Smith, D.R., Schultz, S., Markos, P., Soukoulis, C.M.: Determination of effective permittivity and permeability of metamaterials from reflexion and transmission coefficients. Phys. Rev. **65**, 195104 (2002)
15. Nicolson, M., Ross, G.F.: Measurement of the intrinsic properties of materials by time-domain techniques. IEEE Trans. Instrum. Meas. **19**(4), 377–382 (1970)

Improvement of DTC with 24 Sectors of Induction Motor by Using a Three-Level Inverter and Intelligent Hysteresis Controllers

Habib Benbouhenni[1], Rachid Taleb[2(⌧)], and Fayçal Chabni[2]

[1] Electrical Engineering Department, ENP d'Oran, Oran, Algeria
habib0264@gmail.com
[2] Laboratoire: LGEER, Electrical Engineering Department,
Hassiba Benbouali University, Chlef, Algeria
rac.taleb@yahoo.fr, chabni.fay@gmail.com

Abstract. The main objective of this paper is to control the flux and torque of an induction motor by using three-level DTC with 24 sectors. Direct torque control has been widely used due to its advantages of less parameter dependences and faster torque response. However, in conventional DTC, there are obvious torque and flux ripples. In this present paper, we propose to reduce the electromagnetic torque ripple, flux ripple, and THD of stator current of the induction machine controlled by three-level DTC with 24 sectors, by using artificial intelligence techniques, fuzzy logic and neural networks. In the other hand, we propose to replace conventional hysteresis controller of torque by fuzzy logic, and hysteresis controller of flux by neural networks. Simulation results are presented and show the effectiveness of the proposed hysteresis.

Keywords: Induction motor · DTC · Neural hysteresis · Fuzzy hysteresis
Three-level inverter · THD

1 Introduction

Multilevel voltage source inverters are applicable for high voltage and medium power standard drive applications. Multilevel inverters are available for medium voltage industrial applications. In order to limit the motor winding insulation stresses and to reduce the harmonics, the inverter output has more number of levels [1].

Since the advantages of multilevel inverters and induction machine complement each other. In the other hand, multilevel inverter fed electric machine systems are considered as a promising approach to achieving high power/high voltage ratings [2]. Moreover, multilevel inverters have the advantages of overcoming voltage limit capability of semiconductor switches and improving 2 harmonic profiles of output waveforms. The output voltage waveform approaches a sine wave, thus having practically no common-mode voltage and no voltage surge to the motor windings. Furthermore, the reduction in dv/dt can prevent motor windings and bearings from failure [2].

Direct torque control has been actively investigated during the last decade in the area of AC drives for induction motors. This control strategy was first introduced by

© Springer International Publishing AG 2018
M. Hatti (ed.), *Artificial Intelligence in Renewable Energetic Systems*, Lecture Notes
in Networks and Systems 35, https://doi.org/10.1007/978-3-319-73192-6_11

Takahashi in 1986 [3]. In this method, stator voltage vectors are selected according to the differences between the reference and actual torque and stator flux linkage [4].

In the other hand, the multilevel direct torque control (DTC) of electrical drives has become an attracting topic in research and academic community over the past decade [2]. DTC provides a very quick response with the simple control structure and hence, this technique is gaining popularity in industries. Though, DTC has high dynamic performance, it has few drawbacks such as the high ripple in torque, flux, current and variation in switching frequency of the inverter [5]. The effects of flux and torque hysteresis band amplitudes in the induction motor drive performance have been analyzed in [6].

The application of artificial neural networks and fuzzy logic control attracts the attention of many scientists formal over the world. The reason for this trend is the many advantages which the architectures of ANN have over traditional algorithmic methods [7].

In this paper, the techniques of fuzzy logic and neural networks to select the state of the controller's hysteresis of torque and flux. Used to minimize torque and flux fluctuations manifested in the case of a classical three-level DTC. The performance intelligent three-level DTC associated with fuzzy and neural hysteresis is examined by extensive modeling and simulation studies.

2   Modeling of the Induction Motor

The dynamic behavior of an induction machine is described by the following equations written in terms of space vectors in a stator reference frame [8]:

$$\overline{Vs} = Rs.\overline{Is} + \frac{d\overline{\Phi s}}{dt} \tag{1}$$

$$0 = Rr.\overline{Ir} + \frac{d\overline{\Phi r}}{dt} - jWm.\overline{\Phi r} \tag{2}$$

$$\overline{\Phi s} = Ls.\overline{Is} + M.\overline{Ir} \tag{3}$$

$$\overline{\Phi r} = Lr.\overline{Ir} + M.\overline{Is} \tag{4}$$

Where Rs and Rr represent the stator and rotor resistances, Ls, Lr and M represent the self and mutual inductances, and w_m represent the rotor angular speed expressed in electrical radians. The electromagnetic torque is expressed in terms of stator and rotor fluxes as [9]:

$$C_e = p \frac{M}{\sigma L_r L_s} \Phi_s \Phi_r \sin(\overset{\wedge}{\overline{\Phi_s}\ \overline{\Phi_r}}) \tag{5}$$

Where p is the pole pair number. The elimination of $\overline{Is}$ and $\overline{Ir}$ from (1) to (4) leads to the state variable form of the induction machine equations with stator and rotor fluxes as state variables:

$$\begin{bmatrix} \frac{d\overline{\Phi s}}{dt} \\ \frac{d\overline{\Phi r}}{dt} \end{bmatrix} = \begin{bmatrix} \frac{-1}{\sigma.Ts} & \frac{M}{\sigma.Ts.Lr} \\ \frac{M}{\sigma.Ts.Tr} & j.Wm - \frac{-1}{\sigma.Ts} \end{bmatrix} \begin{bmatrix} \overline{\Phi s} \\ \overline{\Phi r} \end{bmatrix} + \begin{bmatrix} 1 \\ 0 \end{bmatrix} Vs \tag{6}$$

With $Ts = \frac{Ls}{Rs}$, $Tr = \frac{Lr}{Rr}$ and $\sigma = 1 - \frac{M^2}{Ls.Lr}$

3   Modeling of the Three-Level Inverter

Neutral point clamped (NPC) topology where six diodes and four insulated gate bipolar transistors (IGBTs) are used to construct a half bridge (Fig. 1). The breakdown voltage class selection for the elements depends on the voltage applied to the capacitors.

When the output voltage is positive, depending on the phase current directions, the commutation will be either between outer switches or clamp diodes or between outer diodes and inner switches. As a result, the turned off switches will always be clamped to the neutral point voltage which is half of the dc-link voltage allowing the use of lower [10].

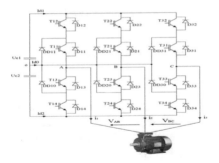

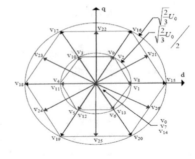

Fig. 1. Schematic diagram of a three-level inverter

Fig. 2. Space vector diagram of three-level inverter

The representation of the space voltage vectors of a three-level inverter for all switching states is given by Fig. 2.

4   Direct Torque Control Principle

DTC is actually direct torque and flux control, with two parameters involved in the control strategy, so it is also named as direct torque and flux control (DTFC) in some publications. DTC is a control method that directly selects inverter states based on the torque and flux errors. Hysteresis (relay) controllers are employed, and no current controllers are present. The voltage source inverter shown in Fig. 1 in the general block diagram of direct torque and flux control system will be used in all systems considered in this thesis. Figure 3 depicts a detailed DTC system diagram [11].

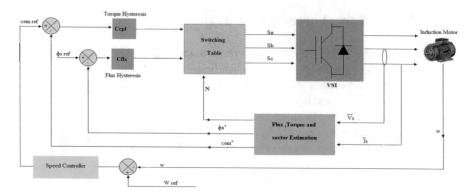

Fig. 3. Direct torque and flux control system with a speed control loop

5 Direct Torque Control Based on Fuzzy Logic and Neural Networks Strategy's

The principle of fuzzy logic and neural networks direct torque control is similar to traditional DTC. However, the hysteresis controllers are replaced by fuzzy logic and neural networks controllers. The structure of three-level DTC based on fuzzy and neural controllers of the induction motor is shown in Fig. 4.

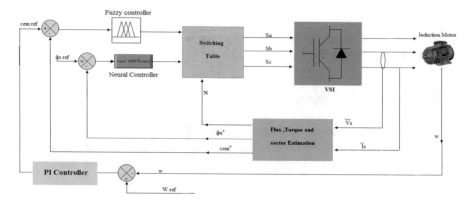

Fig. 4. Direct torque control based on fuzzy and neural controllers

The structure of the neural hysteresis to perform the three-level DTC applied to induction motor satisfactorily was a neural network with 1 linear input node, 3 neurons in the hidden layer, and 1 neuron in the output layer, as shown in Fig. 5. The general structure of the fuzzy hysteresis of the three-level DTC for induction motor is represented by Fig. 6. The architecture of fuzzy hysteresis of three-level DTC is shown in Fig. 7.

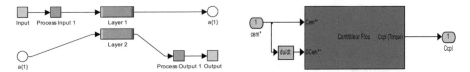

Fig. 5. Neural hysteresis structure for three-level DTC with 24 sectors

Fig. 6. Fuzzy hysteresis structure for three-level DTC

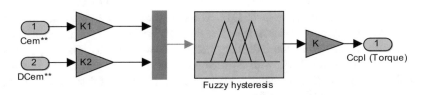

Fig. 7. Architecture of fuzzy hysteresis controller

In Table 1 and Figs. 8 and 9 shows the membership functions of input and output variables respectively. The rules were formulated using analysis data obtained from the simulation of the system using different values of torque hysteresis band [5, 12].

Table 1. Fuzzy rules of torque hysteresis controller

Cem^{**} / ΔCem^{**}	NL	NM	NP	EZ	PS	PM	PL
NL	NL	NL	NL	NL	NM	NP	EZ
NM	NL	NL	NL	NM	NP	EZ	PS
NP	NL	NL	NM	NP	EZ	PS	PM
EZ	NL	NM	NP	EZ	PS	PM	PL
PS	NM	NP	EZ	PS	PM	PL	PL
PM	NP	EZ	PS	PM	PL	PL	PL
PL	EZ	PS	PM	PL	PL	PL	PL

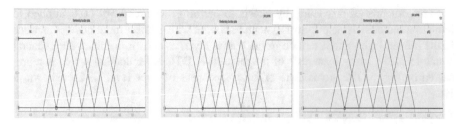

Fig. 8. Input variable membership functions   **Fig. 9.** Output variable membership function

6 Simulation Results

The simulations of the DTC induction motor drive were carried out using the Matlab/Simulink simulation package. A 3-phase, 3 poles, induction motor with parameters of $Fs = 60$ Hz, $Rs = 0.228$ Ω, $Rr = 0.332$ Ω, $Ls = 0.0084$ H, $Lr = 0.0082$H, $M = 0.0078$ H, $J = 20$ kg.m2, and $F = 0.008$ are considered.

The performance of three-level DTC is used as classical hysteresis, fuzzy and neural hysteresis controllers separately. The dynamic performance of three-level DTC control with 24 sectors of the induction motor is shown Fig. 10. The dynamic performance of three-level DTC with 24 sectors based on fuzzy and neural hysteresis controllers for the induction motor is shown Fig. 11. Torque and stator flux linkage comparing curves are shown in Fig. 12.

The induction motor is accelerating from standstill to reference speed 1000 r.p.m, and reference torque 0 N.m. Afterwards, a step variation in the load torque ($Tr = 6500$ N.m) is applied at time t = 0.8 s. And then a sudden reversion in the reference torque from 6500 N.m to −6500 N.m was introduced at 1.8 s.

Torque response comparing curves are shown in Fig. 12 (a and b). See figure the torque ripple is significantly reduced when fuzzy and neural hysteresis is in use.

Figure 12 (a and b) show the stator flux responses of both the conventional and fuzzy-neural DTC schemes it is found that the proposed variable band torque and flux hysteresis controllers-based DTC scheme exhibits smooth response and lesser ripple in flux as compared to the classical DTC with 24 sectors.

The obtained results shown that the three-level DTC with based on fuzzy and neural hysteresis for induction motor ensures good decoupling between stator flux linkage and torque. Also, it can decrease the torque ripple and flux ripple in comparison to the classical three-level DTC of the induction motor.

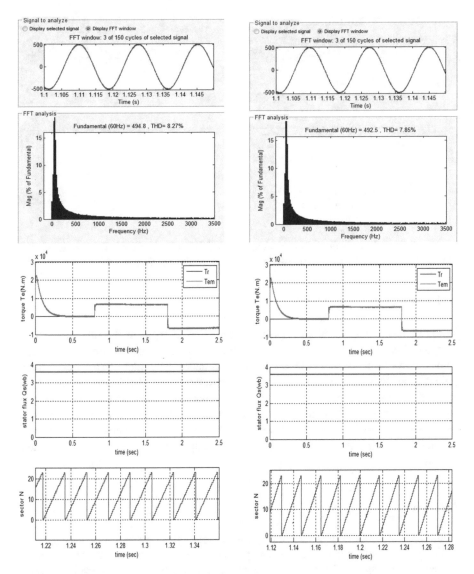

Fig. 10. Dynamic responses of classical three-level DTC with 24 sectors

Fig. 11. Dynamic responses of classical three-level DTC with controllers hysteresis based on fuzzy and neural controllers

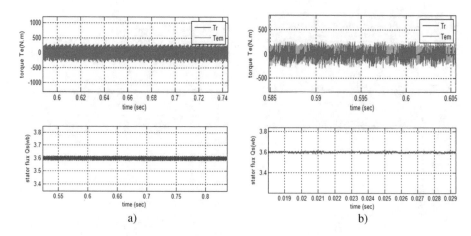

Fig. 12. Comparaison results: (a) classical three-level DTC with 24 sectors, (b) three-level DTC with 24 sectors for controllers hysteresis based on fuzzy and neural networks

7   Conclusion

The direct torque control is an important alternative method for the induction motor drive, with its high performances and simplicity. In this paper, a three-level direct torque control with 24 sectors based on artificial neural network and fuzzy controllers applied to an induction motor. This controller determinates the desired amplitude of flux and torque hysteresis band. It is shown that the proposed scheme results in improved stator flux and torque responses under steady state condition the advantage is the improvement of THD of current, torque and flux ripple characteristics at low-speed region, this provides an opportunity for motor operation under minimum switching loss and noise.

References

1. Pradeep Sagar, G., Roopa, K., Rajasheker Reddy, M.: Simulation and comparison of symmetrical and asymmetrical 3-phase H-Bridge multilevel inverter for DTC induction motor drives. Int. J. Adv. Res. Electr. Electron. Instrum. Eng. **2**(10), 5012–5023 (2013)
2. Benyoussef, E., Meroufel, A., Barkat, S.: Three-level DTC based on fuzzy logic and neural network of sensorless DSSM using extende Kalman filter. Int. J. Power Electron. Drive Syst. (IJPEDS) **5**(4), 453–463 (2015). ISSN: 2088-8694
3. Kadri, F., Drid, S., Djellal, F.: Direct torque control of induction motor by three-level NPC inverter using fuzzy logic. In: International Conference on Systems and Processing Information, Guelma, Algeria, May 15–17 2011
4. Toufouti, R., Meziane, S., Benalla, H.: Direct torque control for induction motor using intelligent techniques. J. Theor. Appl. Inf. Technol. (JATIT) **3**, 35–44 (2007)
5. Idir, I., Kidouche, M.: Direct torque control of three phase induction motor drive using fuzzy logic controllers for low torque ripple. In: International Conference on Control, Engineering & Information Technology (CEIT 2013), Tunisie (2013)

6. Lascu, C.: A modified direct torque control (DTC) for induction motor sensorless drive. In: Proceedings of Conference on Rec, pp. 415–422. IEEE (1998)
7. Hassan Adel, A., Abo-Zaid, S., Refky, A.: Improvement of direct torque control of induction motor drives using neuro-fuzzy controller. J. Multidiscip. Eng. Sci. Technol. (JMEST) 2(10), 2913–2918 (2015)
8. Fethi Aimer, A., Bendiabdellah, A., Miloudi, A., Mokhtar, C.: Application of fuzzy logic for a ripple reduction strategy in DTC scheme of a PWM inverter fed induction motor drives. J. Electr. Syst. Spec. Issue 1, 13–17 (2009)
9. Riad, T.: Contribution à la commande directe du couple d'une machine asynchrone triphasée. Thése de Doctorate, université Mentouri, Constantine (2008)
10. Obdan, H., Cemozkilic, M.: Performances comparison of 2-level and 3-level converters in a wind energy conversion system. Rev. Roum. Sci. Techn. Electrotechn. Et Énerg. 61(4), 388–393 (2016). Boucarest
11. Zolfaghari, M., Moosavi, K., Abedi, M., Rahimi Khoei, H.: Twelve sector based direct torque and flux control of induction motor. In: 4th International Conference on Electrical, Computer, Mechanical and Mechatronics Engineering (ICE 2016), Dubai, Emirates, 4–5 February 2016
12. Abdelhafidh, M.: Stratégie de commande DTC-SVM et DPC appliquées à une MADA utilisée pour la production d'énergie élioienne. Thése de Doctorat, Ecole Nationale Polytechnique, Alger (2014)

Energy Control Strategy Analysis of Hybrid Power Generation System for Rural Saharan Community in Algeria

Fadhila Fodhil[1]([✉]), Abderrahmane Hamidat[2], and Omar Nadjemi[1]

[1] LabSET, Département d'Electronique, Faculté de Technologie,
Université Saad Dahlab Blida, Route de Soumaa, 09000 Blida, Algeria
fadfodl@yahoo.fr, omarnadjemipro@gmail.com
[2] Centre de Développement des Energies Renouvelables,
Route de l'Observatoire, 16340 Bouzaréah, Algiers, Algeria

Abstract. The paper deals with design and sizing of hybrid PV/diesel/battery energy system using two energy management strategies: load following (LF) and cycle charging strategy (CC). The hybrid system is designed to electrify 25 households, primary school, and small dispensary in the Saharan and rural village of Moulay Lahcen, in the province of Tamanrassat, Algeria. A comparative analysis has been elaborated using HOMER in terms of cost of energy (COE), net present cost (NPC), renewable fraction (RF), fuel consumption, battery autonomy and GHG emissions (mainly carbon emissions) between the two control strategies. The results showed that cycle charging strategy is more cost effective compared to load following in both households and school, in the case of dispensary the load following is more cost effective than the cycle charging strategy. The LF strategy has less GHG emissions and more PV penetration than CC strategy in all cases.

Keywords: Energy management · Photovoltaic · Diesel · Hybrid
Load profile

1 Introduction

Promoting national and regional rural electrification plans in the rural development programs will stimulate the socio- economic development of the rural communities. The electricity generation option in both village electrification schemes, microgrids and small standalone home power supply systems, can be either conventional using fossil fuels like diesel, a renewable energy source, or some hybrid combination of the two. Renewable energy based electrification can be the adequate solution to sustain rural electrification programmes in developing countries. In Algeria, the rural population was last reported at 29.27% of total population according to a World Bank report published in 2015 [1]. Access to electricity in rural and remote areas like Sahara regions in Algeria has been achieved mostly by decentralized installations based mainly on diesel gensets within few rural locations. The installed capacity of diesel units which usually serve these areas is about 175 MW. It is estimated that 183 units with individual rating varying between 0.35 MW and 8 MW have been installed in the South and feed separate networks [2].

© Springer International Publishing AG 2018
M. Hatti (ed.), *Artificial Intelligence in Renewable Energetic Systems*, Lecture Notes in Networks and Systems 35, https://doi.org/10.1007/978-3-319-73192-6_12

The progress made by Algeria in the field of rural electrification is based on a set of national and regional programs funded by the state. Rural electrification policy consists of national and regional programs developed by the Ministry of Energy with the participation of Wilayas (provinces) and Sonelgaz (National Society for Electricity and Gas, the state-owned utility in charge of electricity and natural gas distribution in Algeria). With the growth in electric demand the Algerian government realized that the renewable energy projects could contribute significantly in the rural electrification strategy and sustainable development of rural communities. Almost all the renewable installations in rural and Saharan region are PV standalone systems without the use of another energy source. It has been demonstrated that hybrid energy like PV/Diesel/battery systems can significantly reduce the total lifecycle cost [3] of standalone power supplies in many situations, while at the same time providing a more reliable supply of electricity through the combination of energy sources [4–6]. The subject of this paper is to analyze two energy management strategies: load following (LF), and cycle charging (CC) in the design and sizing of hybrid PV/diesel/battery energy system, the hybrid system is used to supply electricity to 25 un-electrified households, a school and dispensary in the rural Saharan village of Moulay Lahcen, province of Tamanrasset. HOMER software is used to simulate and to find the most optimized configuration of components based on the cost of energy (COE) and net present cost (NPC) for each dispatch strategy. Furthermore, a technical, economic and environmental comparison between the two control strategies has been achieved.

2 Modeling and Methods

The basic schematic of the hybrid PV/diesel/battery energy system is illustrated in Fig. 1, the main components of hybrid system are: PV array, bidirectional inverter, battery bank and diesel generator. In this section the mathematical models of the components used in the simulation and control strategies are summarized.

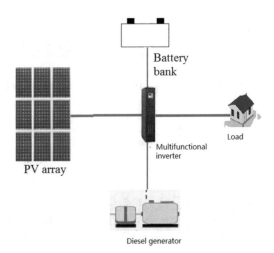

Fig. 1. The basic schematic of the hybrid PV/diesel/battery energy system.

2.1 PV Model

HOMER models the PV array as a device that produces DC electricity in direct proportion to the global solar radiation incident upon it, independent of its temperature and the voltage to which it is exposed. The power output of the PV array is calculated using the following equation [7]:

$$P_{PV} = f_{PV} \cdot Y_{PV} \cdot \left(\frac{G_T}{G_{T,STC}}\right) \cdot \left[1 + \alpha_p \left(T_c - T_{c,STC}\right)\right] \tag{1}$$

Where, f_{PV} is the PV derating factor, Y_{PV} the rated capacity of the PV array (kW), G_T the global solar radiation (beam plus diffuse) incident on the surface of the PV array (kW/m^2), and $G_{T,STC}$ (W/m^2) is irradiation at standard test conditions (i.e. 1000 W/m^2), α_p (%/°C) is the temperature coefficient, Tc (°C) is the PV cell temperature, and $T_{c,STC}$ (°C) is the PV cell temperature under standard conditions. The cell temperature can be calculated using the energy balance for the PV array as follows:

$$T_c = T_a + G_T \left(\frac{\tau \cdot \alpha}{U_L}\right) \left(1 - \frac{\eta_C}{\tau \cdot \alpha}\right) \tag{2}$$

Where T_a (°C) is the ambient temperature, τ is the transmittance of the cover over PV array, α is the solar absorptance of the PV array, U_L (W/m^2 K) is the coefficient of heat transfer to the surrounding, and η_c is the electrical conversion efficiency of the PV array. The cell temperature under the condition of $G_T = 800$ W/m^2, $T_a = 20$ °C, and $\eta_c = 0$ (no load operation) is called nominal operating cell temperature (NOCT) and is reported by manufacturers.

$$\frac{\tau\alpha}{U_L} = \frac{T_{c,NOCT} - T_{a,NOCT}}{G_{T,NOCT}} \tag{3}$$

And

$$T_c = T_a + G_T \left(\frac{T_{c,NOCT} - T_{a,NOCT}}{G_{T,NOCT}}\right) \left(1 - \frac{\eta_C}{\tau\alpha}\right) \tag{4}$$

The inverter converts DC electricity of PV panels to AC electricity with an efficiency of η_{inv} as follows:

$$P_{inv,Out} = \eta_{inv} \cdot P_{PV} \tag{5}$$

2.2 Battery Model

HOMER uses the kinetic battery model, which treats the battery as a two tank system. Modelling the battery as a two-tank system rather than a single-tank system has two effects. First, it means the battery cannot be fully charged or discharged all at once. it means that the battery's ability to charge and discharge depends not only on its current state of charge, but also on its recent charge and discharge history. HOMER calculates the life of the battery bank in years as [8]:

$$C_{bw} = \frac{C_{rep,bat}}{N_{bat} \cdot Q_{lifetime} \cdot \sqrt{\eta \cdot r \cdot t}}$$ (6)

Where $C_{rep,bat}$ is the replacement cost of the battery bank (dollars), N_{bat} is the number of batteries in the battery bank, $Q_{lifetime}$ is the lifetime throughput of a single battery.

2.3 Diesel Generator

The principal physical properties of the generator are its maximum and minimum electrical power output, its expected lifetime in operating hours, the type of fuel it consumes, and its fuel curve, which relates the quantity of fuel consumed to the electrical power produced. HOMER modeled the fuel curve by a linear curve characterized by a slope and intercept at no load and uses the following equation for the generator's fuel consumption [8]:

$$F = F_0 \cdot Y_{gen} + F_1 \cdot P_{gen}$$ (7)

Where F_0 is the fuel curve intercept coefficient, F_1 is the fuel curve slope, Y_{gen} the rated capacity of the generator (kW), and P_{gen} the electrical output of the generator (kW). The units of F depend on the measurement units of the fuel. If the fuel is denominated in liters, the units of F are L/h. If the fuel is denominated in m^3 or kg, the units of F are m^3/h or kg/h, respectively.

2.4 Dispatch Strategies and System Control

HOMER program control strategies are based on the strategies described by Barley in 1995 [9].

• Load Following Dispatch Strategy

In the load following strategy, when the generator operates, it produces only enough power to meet the primary load. Charging the battery bank or serving the deferrable load are left to the renewable power sources.

There is a "Frugal" option that can be applied in all the strategies. The Critical Discharge Load (L_d) is the net load above which the marginal cost of generating energy with the Diesel generator is less than the cost of drawing energy out of the batteries. If the Frugal option is applied, then the Diesel generator meets the net load whenever the net load is above the critical discharge load, regardless of whether or not the battery bank is capable of meeting the net load. The cost of generating energy with the diesel generator and the cost of drawing energy out of the batteries are equal when the net load is [10]:

$$B \cdot P_{Ngen} \cdot Pr_{fuel} + C_{O\&Mgen} + C_{rep\_gen\_h} + A \cdot Pr_{fuel} \cdot L_d = \frac{C_{cycling\_bat} \cdot L_d}{\eta_{inv}} \quad (8)$$

Than L_d can be calculated as follows:

$$L_d = \frac{\eta_{inv} \left(B \cdot P_{Ngen} \cdot Pr_{fuel} + C_{O\&Mgen} + C_{rep\_gen\_h} \right)}{C_{cycling\_bat} - \eta_{inv} \cdot A \cdot Pr_{fuel}} \quad (9)$$

Where $C_{O\&Mgen}$ is the diesel generator's hourly operation and maintenance cost ($/h), Pr_{fuel} is the fuel price ($/l), $A = 0,246$ l/kWh and $B = 0,08415$ l/kWh are the fuel curve coefficients [11]. The fuel cost of 1 h DG running, C_{fuel} is:

$$C_{fuel} = Pr_{fuel} \cdot \left(B \cdot P_{Ngen} + A \cdot P_{gen} \right) \quad (10)$$

Where, P_{gen} is the diesel generator output power in this hour (kW). $C_{rep\_gen\_h}$ ($/h) is the DG hourly replacement cost:

$$C_{rep\_gen\_h} = \frac{C_{gen}}{life_{gen}} \quad (11)$$

Where, C_{gen} is the diesel generator acquisition cost plus O&M cost throughout diesel generator lifetime (€) and $Life_{gen}$ is the diesel generator lifetime (h), $C_{cycling\_bat}$ ($/kWh) is the cost of cycling energy through the batteries [10]:

$$C_{cycling\_bat} = \frac{C_{bat}}{C_N \cdot N_{bat\_P} \cdot U_{DC} \cdot N_{cycles\_eq} / 1000} \quad (12)$$

Where, C_{bat} is the batteries bank acquisition cost plus O&M cost throughout batteries lifetime (€), C_N is the nominal capacity of one battery (Ah), N_{bat_p} is the number of batteries in parallel, and N_{cycles_eq} is the number of full cycles of battery life. We have assumed that the batteries can cycle a certain amount of energy, which divided by its nominal capacity, gives the equivalent cycles (full cycles). It is true that the energy that a battery can cycle depends on the depth of discharge, but is almost constant if the discharge is never allowed to fall below SOC_{min}, this being greater than 20% [10].

- Cycle Charging Dispatch Strategy

In cycle charging strategy, whenever the generators operate, they produce more power than required (or at a rate not exceeding the maximum energy that batteries are capable of absorbing) to serve the load with surplus electricity going to charge the battery bank [8]. If a SOC set point is applied, the diesel generator will continue running until the batteries reach this SOC set point. The Frugal option also can be applied in this strategy [10].

2.5   Economic Models

To calculate the total net present cost the following equation has been used:

$$C_{NPC} = \frac{C_{ann,tot}}{CRF\left(i, R_{proj}\right)} \tag{13}$$

Where, $C_{ann,tot}$ is the total annualized cost, i the annual real interest rate(the discount rate), R_{proj} the project lifetime, and CRF is the capital recovery factor, given by the following equation:

$$CRF(i, N) = \frac{i(1+i)^N}{(1+i)^N - 1} \tag{14}$$

Where, i is the annual real interest rate and N is the number of years.

Levelized cost of energy is calculated as follows:

$$COE = \frac{C_{ann,tot}}{E_{prim} + E_{def} + E_{grid,sales}} \tag{15}$$

Where, $C_{ann,tot}$ is the total annualized cost, E_{prim} and E_{def} are the total amounts of primary and deferrable load, respectively that the system serves per year, and E_{grid}, sales is the amount of energy sold to the grid per year (when the system is connected to the grid).

3   Data and Resources

This section presents the case study, the proposed load profiles, the meteorological data and the different technical and economic input data.

3.1   Case Study

For our study, a rural remote village. is chosen, namely Moulay Lahcen, which is Saharan village with 25 un-electrified households, a school and dispensary. The village is located a long distance from the grid in district of In-Amguel, province of Tamanrasset at latitude 24°42′ N 4°39′ E at an altitude of 981 m. Moulay Lahcen has a hot desert climate, with very hot summers and mild winters, and very little precipitation throughout the year. The Fig. 2 shows the geographical location of the village.

3.2   Meteorological Data

To analyze and evaluate the performance of hybrid PV/diesel/battery system, an appropriate meteorological data of the study area is required. The solar radiation of the location was sourced from s the NASA Surface Meteorological dataset, the other set of data containing the air temperature was collected directly from NASA Surface

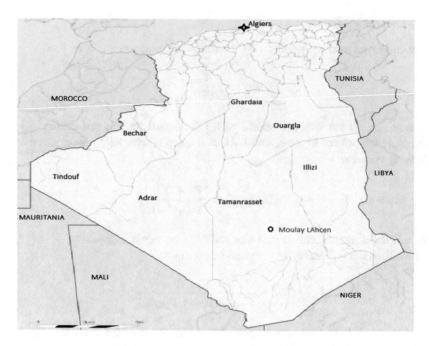

Fig. 2. Geographical location of the village.

Meteorological dataset [12] and implemented in HOMER database. The Fig. 3 shows the meteorological characteristics of site (the global horizontal radiation and temperature). Moulay Lahcen is Saharan village with desert climate therefore, there is considerable variation in temperature between the day and the night, and the temperature is high particularly in the summer. It is observed that the village of Moulay Lahcen has high solar radiation with an annual average of 6.17 kWh/m^2/d.

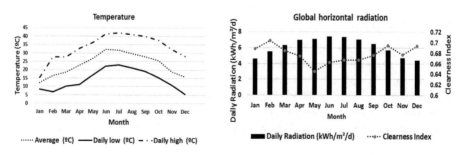

Fig. 3. The meteorological characteristics of the site.

3.3 Load Profile

The proposed load profiles for our study are categorized in three types: a household load, school load and dispensary load. The load profiles were proposed based on an investigation nearby the inhabitants of the rural villages in the southern of Algeria. the following assumptions were considered: The rural household consists of 2 bedrooms, a sitting room, a kitchen, a hallway, a bathroom and a toilet. The school consists of 6 classrooms, 2 offices, a cafeteria and 6 toilets. The dispensary is a small health center with an examination room, a waiting area, nurse room, 3 toilets. The Figs. 4 and 5 illustrate the daily and the monthly load profile for a household, a dispensary and school respectively.

- *Household Load Profile*

   The load profile of household covers the basic power needs (lighting, refrigeration, TV, radio, ventilation). We considered the following hypotheses:

- The dwelling is occupied continuously throughout the year.
- Utilization of low energy electrical appliances.
- Two patterns are considered for winter and summer.

   As shown in Fig. 4, the two daily patterns present a prominent peak in the evening corresponding to lighting use, a midday peak and base peak in the morning, and during the night the power consumption is very low.

- *School Load Profile*

The school requires electricity for (lighting, refrigeration, ventilation, and computers). During the holidays and weekends, we assumed that the consumption of school is constant (equal to the consumption of the fridge).

- *Dispensary Load Profile*

The load profile of our typical dispensary covers (lighting, medical tools, refrigeration, ventilation). During weekends, we assumed that the consumption of dispensary is constant (equal to the consumption of the fridge).

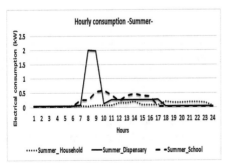 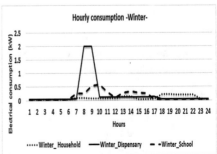

Fig. 4. Daily load profile for a household, a dispensary and school.

The load profiles for school and dispensary have the largest demand during working hours (daytime). The school and the dispensary have their peaks power in the morning between 8 AM and 10 AM (see Fig. 4). The monthly electrical consumptions of school and dispensary are considered small compared to the monthly electrical consumption of household (see Fig. 5).

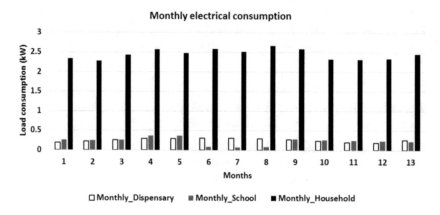

Fig. 5. Monthly load profile for a household, a dispensary and school.

3.4   Hybrid System Components

- PV Generator

The capital cost of PV panels is 2300 $/kW, the operating and maintenance cost is 23 $ (1% of the capital cost), the replacement cost is specified as zero, lifetime is 20 years, no tracking system, and the effect of temperature is considered.

- Battery Bank

The capital cost of the battery is 0.4 $/Wh, the replacement cost is 0.4 $/Wh, the operating and maintenance cost is 10 $/year, the bus voltage is 48 V, minimum battery life is considered 6 years. Hoppecke batteries were chosen in this study.

- Bidirectional Converter

The capital cost of the converter is 0.711 $/W, the replacement cost is 0.711 $/W, the operating and maintenance cost is 7 $/year, lifetime is 15 years, and the efficiency is 95% as a rectifier and 98% as an inverter.

- Diesel Generator

The capital cost of the diesel generator is considered 300 $/kW, the replacement cost is 300 $/kW, the operating and maintenance cost is 0.05 $/hr, lifetime (operating hours) is 15000 h, and the fuel price is specified as 0.17 $/L [13].

The project life time is 20 years and the annual real interest is 6% also the capacity shortage penalty, the system fixed capital cost and the system fixed O&M cost are not considered.

4 Results and Discussion

HOMER simulates all system configurations in search of feasible systems that satisfy the technical constraints at the lowest life-cycle cost [8]. The hybrid system is simulated under both load following and cycle charging strategies in school, dispensary and household. A sample of hourly simulation results for a household in both dispatch strategies LF and CC are illustrated in Figs. 6 and 7.

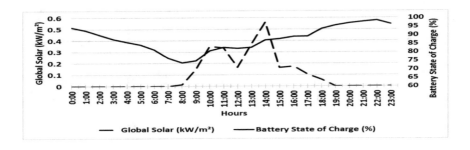

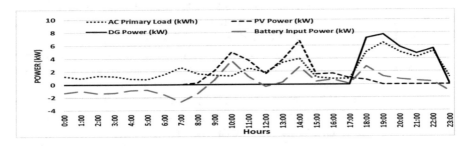

Fig. 6. Sample of hourly simulation results for a household in cycle charging strategy (CC).

In the cycle charging strategy (CC), when the PV generator and the battery bank are unable to cover the load consumption, the diesel generator supplies enough energy to meet the load and to charge the battery until the battery bank become able to cover the load consumption, this behavior clearly appears between 17 h and 21 h.

In the load following strategy (LF), when the PV generator and the battery bank are unable to cover the load consumption, the diesel generator only covers the load without charging the battery bank. We can notice this behavior between 17 h and 22 h.

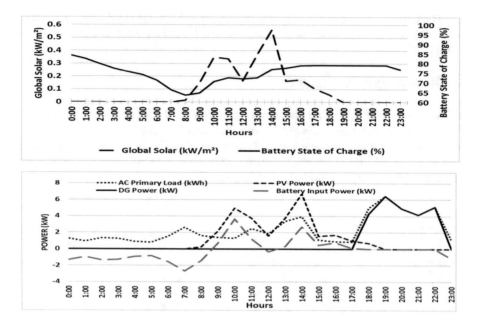

Fig. 7. Sample of hourly simulation results for a household in load following strategy (LF).

The hybrid system is simulated for both load following and charge control dispatch strategies. The technical, economic and environmental results of the optimal solutions for the households, the school and dispensary are summarized in Table 1.

For the households, the battery autonomy and the sizes of PV generator, diesel generator, battery bank and converter are the same in both strategies, the PV penetration in the LF case is more than the CC (69% in LF and 64% in CC), which leads to: less fuel consumption (3036 L/yr in LF and 3434 L/yr in CC), less excess of electricity (4.2% in LF and 11.3% in CC), and more battery life (18.8 yr in LF and 18.1 yr in CC). From the economic results, it can be noticed that the CC strategy is the more cost effective than the LF strategy (with COE = 0.354 $/kWh in CC and COE = 0.357 $/kWh in LF, and NPC = 87350 $ in CC and NPC = 88098 $ in LF). Reducing fuel consumption in LF strategy means less GHG emissions from the diesel generator (CO_2 emissions = 7990 kg/yr in LF and CO_2 emissions = 9042 kg/yr in CC).

For the school, the size of components of the optimal solution in CC and LF strategies are different, the PV generator is bigger in LF case, leading to: more PV penetration (80% in LF and 68% in CC), more battery life (14.2 yr in LF and 11.9 yr in CC), more excess of electricity (20.8% in LF and 18.6% in CC), and less fuel consumption (252 l/yr in LF and 321 l/yr in CC). The CC strategy is more cost effective than LF strategy (COE = 0.420 $/kWh in CC and COE = 0.429 $/kWh in LF, and NPC = 9740 $ in CC and NPC = 9927 $ in LF). The LF strategy is more environmental effective than CC strategy with less emissions of CO_2, PM and NO_X.

For the dispensary, the battery autonomy (34.9 h) and the size of the components remains the same for both strategies, but the PV penetration is slightly more in LF

Table 1. Results of the optimal solution for household, school and dispensary for both CC and LF strategy.

Control strategy	Household		School		Dispensary	
	CC	LF	CC	LF	CC	LF
Technical						
PV Array (kW)	9	9	1	1.2	1.5	1.5
DG (kW)	10	10	2	2	2	2
Battery (kW)	72	72	4.8	4.8	9.6	9.6
Converter (kW)	10	10	1.2	1.2	3	3
Electrical production PV (%)	64	69	68	80	96	97
Electrical production DG (%)	36	31	32	20	4	3
Fuel consumption (L/yr)	3434	3036	321	252	36.2	23.5
Excess of electricity (%)	11.3	4.2	18.6	20.8	32.9	32.2
Battery Autonomy (Hours)	20.5	20.5	14.6	14.6	34.9	34.9
Battery life (Years)	18.1	18.8	11.9	14.2	16.5	16.6
Economic						
NPC ($)	87350	88098	9740	9927	15021	14653
COE ($/kWh)	0.354	0.357	0.42	0.429	0.777	0.758
Capital cost ($)	61410	61410	5673	6133	10023	10023
Replacement cost ($)	14200	15608	1314	1193	2354	2353
O&M cost ($)	13884	14567	2517	2644	4135	3802
Fuel cost ($)	7877	6965	625	492	70	46
Salvage ($)	−10021	−10452	−389	−535	−1562	−1571
GHG emissions (kg/year)						
Carbon dioxide	9042	7996	844	664	95.2	61.8
Carbon monoxide	22.3	19.7	2.08	1.64	0.235	0.153
Unburned hydrocarbons	2.47	2.19	0.231	0.182	0.026	0.017
Particulate matter	1.68	1.49	0.157	0.124	0.018	0.012
Sulfur dioxide	18.2	16.1	1.7	1.33	0.191	0.124
Nitrogen oxides	199	176	18.6	14.6	2.1	1.36

strategy (97% in LF and 96% in CC), leading to less fuel consumption (23.5 l/yr in LF and 36.2 l/yr in CC), less excess of electricity (32.2% in LF and 32.9% in CC), and more battery life (16.6 yr in LF and 16.5 yr in CC). In this case, the LF strategy is more cost effective than the CC strategy (with COE = 0.777 $/kWh in CC and COE = 0.758 $/kWh in LF, and NPC = 15021 $ in CC and NPC = 14653 $ in LF). Also the LF strategy has less GHG emission than CC strategy.

5   Conclusion

This paper analyzed the design and sizing of the hybrid PV-Diesel-battery energy system to supply electricity to rural Saharan village of Moulay Lahcen in the province of Tamanrasset. The hybrid system is designed under two control strategies, the load following and cycle charging strategy. The results showed that the CC strategy is more cost effective in both households and school, furthermore the LF is more cost effective in the case of dispensary. The LF strategy has the lower GHG emissions and fuel consumption, also LF strategy has more PV penetration and battery life than CC strategy in all cases.

Acknowledgements. This work is performed in collaboration with the Electricity and Gas Distribution Company SDC (La Société de Distribution de l'Electricité et du Gaz du Centre, Sonelgez, Blida). The authors would like to thank Mme Ramdani, Mme Thabet and Mr Kobbi, (SDC) for her helpful assistance and providing data and related information about the site.

References

1. Bank, W.: Rural population in Algeria (2015). http://data.worldbank.org/indicator/SP.RUR. TOTL.ZS?end=2015&locations=DZ&start=2015&view=bar. Accessed 2017
2. Tsikalakis, A., Tomtsi, T., Hatziargyriou, N., Poullikkas, A., Malamatenios, C., Giakoumelos, E., Jaouad, O.C., Chenak, A., Fayek, A., Matar, T.: Review of best practices of solar electricity resources applications in selected Middle East and North Africa (MENA) countries. Renew. Sustain. Energ. Rev. **15**(6), 2838–2849 (2011)
3. Karki, R., Billinton, R.: Reliability/cost implications of PV and wind energy utilization in small isolated power systems. IEEE Trans. Energ. Convers. **16**(4), 368–373 (2001)
4. Rehman, S., Al-Hadhrami, L.M.: Study of a solar PV–diesel–battery hybrid power system for a remotely located population near Rafha, Saudi Arabia. Energy **35**(12), 4986–4995 (2010)
5. Yamegueu, D., Azoumah, Y., Py, X., Zongo, N.: Experimental study of electricity generation by Solar PV/diesel hybrid systems without battery storage for off-grid areas. Renew. Energ. **36**(6), 1780–1787 (2011)
6. Rezzouk, H., Mellit, A.: Feasibility study and sensitivity analysis of a stand-alone photovoltaic–diesel–battery hybrid energy system in the north of Algeria. Renew. Sustain. Energ. Rev. **43**, 1134–1150 (2015)
7. HOMER, HOMER help manual, August 2016. http://www.homerenergy.com/pdf/ HOMERHelpManual.pdf. Accessed 2017
8. Lambert, T., Gilman, P., Lilienthal, P.: Micropower system modeling with HOMER. Integr. Altern. Sources Energ. **1**(15), 379–418 (2006)
9. Barley, C.D., Winn, C.B., Flowers, L., Green, H.J.: Optimal control of remote hybrid power systems. Part 1: Simplified model. National Renewable Energy Lab., Golden, CO, United States (1995)
10. Dufo-López, R., Bernal-Agustín, J.L.: Design and control strategies of PV-Diesel systems using genetic algorithms. Solar Energ. **79**(1), 33–46 (2005)
11. Skarstein, Ø., Uhlen, K.: Design considerations with respect to long-term diesel saving in wind/diesel plants. Wind Eng. **13**(2), 72–87 (1989)
12. SMSE, NASA surface meteorology and solar energy (2017). http://eosweb.larc.nasa.gov/sse/
13. Algeria Diesel prices (2016). http://www.globalpetrolprices.com/Algeria/diesel_prices/

Improving the Electrical Stability by Wind Turbine and UPFC

Djamel Eddine Tourqui[1(✉)], Meryem Benakcha[2], and Tayeb Allaoui[1]

[1] L2GEGI, Laboratory of Energy Engineering and Computer Engineering,
University Ibn Khaldoun of Tiaret, Tiaret, Algeria
tourqui.djamel@gmail.com, allaoui_tb@yahoo.fr
[2] 2LGE, Laboratory of Electrical Engineering,
University Mohamed Boudiaf of M'Sila, M'Sila, Algeria
benakchameryem@yahoo.fr

Abstract. The aim of this work is the evaluation of wind power presented by DFIG and FACTS device which is the UPFC on the critical clearing fault time in multi-machine power system. The performance of wind power integrated into power system is investigated by searching the optimal location can be the DFIG generators installed for enhancing the critical fault-clearing time (TCid). This performance can be also improvement using one of powerful FACTS device on transient stability improvement. The simulation of our system models has been prepared by using MATLAB/SIMULINK software where the IEEE 3 machines 9 buses (Western System Coordinating Council) is taken as a test system and the obtained results showed the performance of wind power and UPFC to increase the fault critical clearing time of power system.

Keywords: Wind power · FACTS · UPFC · Fault-clearing time

1 Introduction

The global policy to combat increasing greenhouse gas emissions has been debated since the early 1990s. The 195 countries participating in the Paris Climate Conference (COP21) in December 2015, Limiting global warming between 1.5 °C and 2 °C by 2100 (international agreement). Moreover, the beginning of the importance of this agreement to increase the progressive elimination of fossil fuels such as oil and coal is responsible for high levels of climate because it favors the use of renewable energy sources in various fields. Among the most common uses of this clean energy are integrated into power grids. The integration of distributed energy resources (DER) as wind turbines and photovoltaic farm to the electricity networks are transforming the traditional model in which electricity is produced in power plants, transported on transmission lines to the electric power consumption centers.

Wind turbine is one of the renewable energy power plant developing fast [1, 2] such that the integration of wind farms in the system of electrical grids is increasing day by day largely. As one of the most important types of wind parks, Doubly Fed Induction Generators (DFIGs) are widely used for their flexibility in active/reactive power control and low cost [3, 4]. However, this energy integration can affect the

© Springer International Publishing AG 2018
M. Hatti (ed.), *Artificial Intelligence in Renewable Energetic Systems*, Lecture Notes in Networks and Systems 35, https://doi.org/10.1007/978-3-319-73192-6_13

stability of the system because of non-optimal location of the wind turbine in the power grid may essentially cause the increase in transmission loss, power flow and the level of voltage lower than the legal limit.

Conventional means of network control (tap transformer adjustable load, phase-shifting transformers, countervailing serial or parallel switched by circuit breakers, changing production orders, network topology change and action on the excitement generators) could in the future be too slow and insufficient to effectively respond to system disturbances, especially given the new constraints such as the integration of renewable sources. It will therefore supplement them by implementing power electronic devices with short response time, known under the name FACTS (Flexible Alternative Current Transmission Systems) for control of networks [5]. Among them, the Unified Power Flow Controller (UPFC) [5–7] is an important FACTS device, which is used extensively for improving the utilization of the existing transmission system.

The UPFC power flow controller is considered one of the most powerful and most advantageous FACTS, insofar as it has the ability to control the power flow in the transmission line in order to improve the transient stability, reduce oscillations of the system and support the supply voltage [8, 9]. This reduces the negative impacts of wind energy on the stability of power systems on one hand, and on the other hand, it improves the transit of powers active and reactive, through the power lines.

The rest of the paper is organized as follows: Sect. 2 presents an overview on the modeling of a wind turbine of type doubly fed induction generators (DFIGs). Section 3 overviews the description and operation of Unified Power Flow Controller (UPFC). Section 4 describes the studied system used for the analysis of transient stability. Finally, the conclusion and future action of the paper is summarized in Sect. 5.

2 Wind Energy

The wind turbine is a device that transforms the kinetic energy of the wind into mechanical energy. The kinetic energy resulting from the displacement of an air mass with a velocity is:

$$E_C = \frac{1}{2} \cdot m \cdot V_V^2 \tag{1}$$

With
m: Mass of air volume (in kg).
V_V: Instantaneous wind speed (in m/s).

A power is the variation of energy in a time t:

$$P_V = \frac{\Delta E_C}{\Delta t} \tag{2}$$

The mass of air 'm', which moves for a time t through a section S, is expressed as follows:

$$\frac{\Delta m}{\Delta t} = M = \rho \cdot S \cdot V_V \tag{3}$$

With

ρ: Air density of about 1.225 kg.m^3, which decreases with altitude and varies in changes in temperature or humidity [10].

Therefore, the power of the wind is expressed [11, 12]:

$$P_V = \frac{1}{2} \frac{\Delta m}{\Delta t} V_V^2 = \frac{1}{2} M \cdot V_V^2 \tag{4}$$

$$P_V = \frac{1}{2} \rho \cdot S \cdot V_V \cdot V_V^2 \tag{5}$$

$$P_V = \frac{1}{2} \rho \cdot S \cdot V_V^3 \tag{6}$$

The German Albert Betz demonstrated in 1919 that the maximum recoverable power (Pm) that can be generated by a wind turbine could in no case exceed 59.3% of the kinetic power of the air mass circulating in the second. In this way, the maximum theoretical power coefficient (Cp_{max}) is defined as [13]:

$$Cp_{max} = \frac{P_m}{P_V} = 0.593 \tag{7}$$

The limit of the power coefficient Cp is calculated as 16/27 (59.3%) as a function of the Betz law [10, 11]. Therefore, the mechanical power is:

$$P_m = \frac{1}{2} \rho \cdot S \cdot V_V^3 \cdot Cp \tag{8}$$

Cp: Coefficient of power (aerodynamic efficiency).

The Cp_{max} does not take into account the energy losses caused by the conversion of mechanical wind energy into electrical energy. This is why the maximum efficiency of wind turbines is only 60 to 70% of the Betz limit.

The equation of the power coefficient is a polynomial dependent on two terms λ and β [14, 15]:

$$Cp(\lambda, \beta) = C_1 \left(\frac{C_2}{\lambda_i} - C_3\beta - C_4 \right) e^{\frac{-C_5}{\lambda_i}} + C_6\lambda_i \tag{9}$$

Where

λ: Tip speed ratio.

β: Angle of inclination of the blades

Using the following representative values for the coefficients: $C_1 = 0.5176, C_2 = 116$, $C_3 = 0.4, C_4 = 5, C_5 = 21$ and $C_6 = 0.0068$ is expressed as [16, 17].

The peak speed ratio is defined by Ref. [18, 19]:

$$\lambda = \frac{R \cdot \Omega_t}{V_V} \qquad (10)$$

Where

Ω_t: Angular speed of the turbine.

R: Radius of the section formed by the blades

The terms λ and β, are generally modeled by the following equation:

$$\frac{1}{\lambda_i} = \frac{1}{\lambda + 0.08\beta} - \frac{0.035}{\beta^3 + 1} \qquad (11)$$

The aerodynamic torque:

$$C_{aero} = \frac{P_m}{\Omega_t} \Rightarrow C_{aero} = \frac{\frac{1}{2}\rho \cdot S \cdot V_V^3 \cdot Cp(\lambda, \beta)}{\Omega_t} \qquad (12)$$

3 Description and Operation of UPFC

The UPFC can control independently or separately all parameters affecting the power flow on a transmission line. It can have various modes of operation when the voltage injected in series with different magnitudes and phase angle as showing in Fig. 1. It consists of two voltage inverters (VSC static voltage converters) [19]. The shunt converter is connected in parallel to the network via a three-phase transformer that is used across the DC link to provide the necessary power to the active serial converter. It also performs the power factor correction function since it can supply (provide) or absorb reactive power, regardless of the active power to the grid.

The second serial converter is placed between the starting point and the ending point from three single-phase transformers. It injects an adjustable voltage V and provides active and reactive power needed to serial converter. Can say that the UPFC serves as a combination of a shunt static synchronous converter (STATCOM) and a synchronous static converter series (SSSC) connected by a continuous connection (DC-link). [20] The voltage of the DC-link is supported by a capacitor. Figure 1 shows the operation modes of a UPFC. It can be operated in four modes: Voltage adjustment, series compensation, adjusting the phase angle, automatic mode.

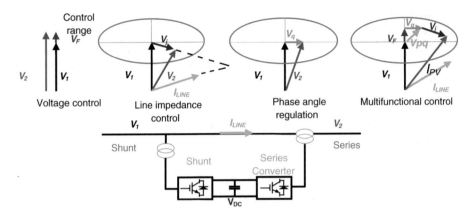

Fig. 1. Schematic representation of UPFC.

4 Studied System

As showing in Fig. 2, the test network is the IEEE 3 m 9 bus power system, where G1, G2, G3 are the generator equipped with classical voltage and speed regulator. A symmetrical three phase to ground fault is applied at bus 7. In this study, we examine

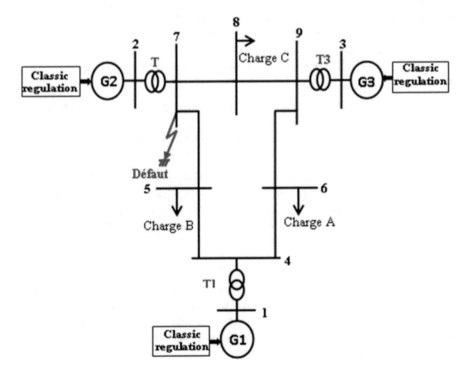

Fig. 2. Schematic of IEEE 3 machine 9-bus power system.

the impact of wind power on transient stability for that a wind generator presented by DFIG is coupled with the power system or the active power produced by the DFIG is 63 MW what means a 20% penetration rates compared to the total power of the network output that is equal to 315 MW. The data of studied system is found at appendix.

The critical time of default isolation (TCid) was taking as index to evaluate the performance of wind power and UPFC for enhancing the transient stability of power system. TCid is the maximum time that our network can support a fault (short-circuit, over-load, over-voltage over-current,) without losing its stability.

Simulations are conducted for following two cases:

(A) Case 1: INFLUENCE OF WIND SOURCE
(B) Case 2: Influence of UPFC & WIND SOURCE.

4.1   Influence of Wind Source

To examine the influence of the wind generator interconnected with the grid on the stability, a comparative study vis-à-vis the values of TCid matching each simulation is carried out. In this part, we'll calculate the critical time of the fault isolation study model with and without wind energy observing transient behavior of this model.

4.1.1   Without the Wind Source

The wind generator need 15 s to achieve the active power desired because of characteristic of **DFIG**, for that we start the fault at 20 s in this case we continues extend the duration of fault time until the hydro generator lose their stability and then calculate the (Tcid) of our system.

The observation of the transient behavior of the system can reveal only two cases to get the TCid are:

- Stable system: if all angular speeds of generators tend towards a fixed value very close to nominal.
- Unstable system: if at least an angular speed of a single generator does not converge to the nominal values.

According to both Figs. 3, we note that the maximum fault time can be supported by the system is 326 ms (Fig. 3a), if we extend the fault time by 1 ms (fault time duration equal to 327 ms). Speed and load angle can no longer lies with nominal values (Fig. 3b) and then we can say that the system loses its stability and is noted TCid = 326 ms.

4.1.2   With the Wind Source

Now, a wind turbine is presented in the test network with an active power generated of 63 MW as we explained above. To well assess the performance of the wind turbine on the behavior of hydro turbine's (speed, load angle), we vary the location of the wind turbine and we note each time the TCid. Table 1 shows the TCid depending on the bus of the installation of the wind turbine.

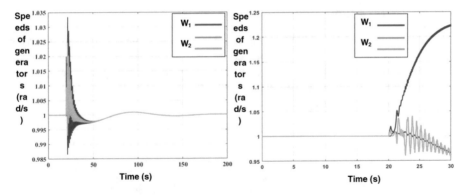

Fig. 3. Evolution of the rotational speed of the generators the without wind source.

It is very clear that the placement of wind turbines in a power grid is more important to examine their influence on the stability of the power system. In this test, we see that the installing of wind farm on bus 2 reduce the TCid from 326 ms to 277 ms, However, installing this latest energy on bus 1 or 4 improves the TCid from 326 ms to 371 ms as showing in Table 1. Therefore, we can deduce that the optimal place for integrating a wind turbine is on bus 1 and 4 where the TCid is 371 ms.

Table 1. TCid depending on the bus of the installation of the wind energy.

BUS	1	2	3	4	5	6	7	8	9
TCid (ms)	**371**	**277**	329	**371**	363	365	326	338	342

Figure 4 shown the speed of generators when the DFIG installed on bus 4. The integration of wind energy in power system enhance its transient stability and increase the max fault time duration can the grid support it.

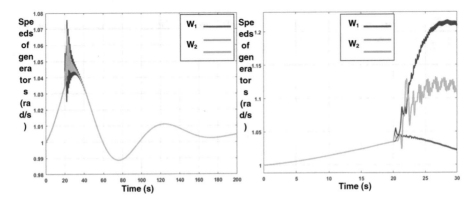

Fig. 4. Evolution of the rotational speed of the generators with the wind source.

4.2 Influence of UPFC and Wind Source

To show the effectiveness of the combination FACTS/renewable energy to maintain the stability of electric networks and improved the critical time to eliminate the fault, we installed an UPFC in our system with a wind turbine placed in the optimal position that is bus 4 as we have shown above. Where the UPFC is installed between bus 4 and bus 5 as shown in the following Fig. 5.

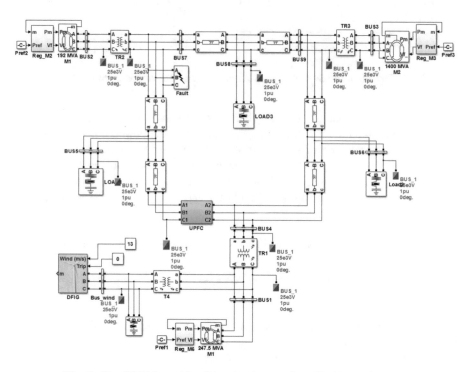

Fig. 5. The IEEE 3 machine 9-bus implemented on Sim-Power-System.

Figure 6 summarize the impact of the UPFC and the DFIG (wind power source) on the stability of test power system.

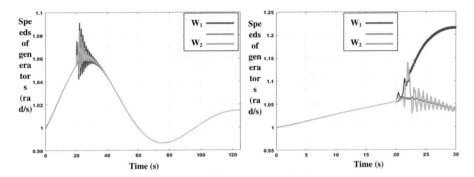

Fig. 6. Evolution of the rotational speed of the generators with wind and UPFC.

We see that the TCid increase from 371 ms (when only wind turbine installed on the optimal place) to 401 ms when the power test system equipped with wind turbine and UPFC. Using FACTS device as UPFC and renewable source such as wind power can enhance the stability of power system and improve the maximal time to eliminate the fault.

5   Conclusion

As a general conclusion of this work, the integration of the renewable energy source (the wind turbine) into the power grids has a beneficial impact on the improvement of the stability when they are integrated into the optimal position. As well as the contribution of FACTS devices to the improvement of the stability of electrical networks, this integration of renewable energies becomes very efficient.

Finally, this work is only a starting point for future contributions and there are still many perspectives to explore:

- A first perspective is the implementation of recent methods of optimization such as the algorithm of the bats (BAT Algorithm) and the optimization by gray wolves (Gray Wolf Optimization).
- As a second perspective, it is possible to optimize the location of FACTS devices and renewable energy equipment to ensure the safe operation and safety of electrical networks.

6   Appendix

6.1   Data of Studied System

f = 60 Hz, Length = 100 km for all line (Tables 2, 3, 4, 5 and 6).

Table 2. Generator parameters.

Generator	1	2	3
Pn (VA)	100 e6	100 e6	100 e6
Vn (Vrms)	16.5 e3	18 e3	13.8 e3
Xl (p.u)	0.05	0.05	0.05
Xd (pu)	0.146	0.8958	1.3125
Xd$^{'}$ (pu)	0.0608	0.1198	0.1813
Xd$^{''}$(pu)	0.005	0.005	0.005
Xq (p.u)	0.0969	0.8645	1.2578
Xq$^{'}$ (p.u)	0.0969	0.1969	0.25
Xq$^{''}$(p.u)	0.005	0.005	0.005
T$^{'}$do (s)	8.96	6.0	5.89
T$^{''}$do (s)	0.01	0.01	0.01
T$^{'}$qo (s)	0.5	0.535	0.6
T$^{''}$qo (s)	0.001	0.001	0.001
H (s)	23.64	6.4	3.01

Table 3. Load parameters.

Load bus	5	6	8
Active power MW	125	90	100
Reactive power MVAR	50	30	35

Table 4. AVR parameters.

GOV	1	2	3
Rs	−0.04	0.05	0.06
Tc(s)	−0.04	0.05	0.06
Ts(s)	−0.04	0.05	0.06

Table 5. DFIG parameters.

Rated voltage	575 V
Stator resistance	0.00706 pu
Stator leakage reactance	0.171 pu
Magnetizing reactance	2.9
Rotor resistance	0.005 pu
Rotor leakage reactance	0.156 pu
Inertia constant	5.04
DC link voltage	1200 V
DC link capacitor	6*10000e−6 μF

Table 6. Line parameters.

Line	r1r0 (Ohms/km)		l1 l0 (H/km)	c1 c0 (F/km)
1	0.0629	0.1573	1.41e−3 3.53e−3	10.47e−9 06.15e−9
2	0.0449	0.1124	1.01e−3 2.02e−3	7.471e−9 04.39e−9
3	0.2063	0.5157	2.38e−3 6.09e−3	17.95e−9 10.55e−9
4	0.1692	0.4232	2.25e−3 5.64e−3	15.34e−9 09.02e−9
5	0.0529	0.1322	1.19e−3 2.38e−3	08.82e−9 05.18e−9
6	0.0899	0.2248	1.29e−3 3.22e−3	7.922e−9 04.7e−9

References

1. American Wind Energy Association website. www.awea.org
2. Heidari, M.: Improving the power quality of wind power plants through modifying the instantaneous active and reactive power theory. Int. J. Renew. Energy Res. **5**(3), 2338–10858 (2015)
3. Guoa, W., Liu, F., Si, J., He, D., Harley, R., Mei, S.: Approximate dynamic programming based supplementary reactive power control for DFIG wind farm to enhance power system stability'. Neurocomputing **170**, 417–427 (2015)
4. Kayıkçı, M., Milanoviç, J.V.: Reactive power control strategies for DFIG-based plants. IEEE Trans. Energy Convers. **22**(2), 389–396 (2007)
5. Tripathy, L.N., Jena, M.K., Samantaray, S.R.: Differential relaying scheme for tapped transmission line connecting UPFC and wind farm. Electr. Power Energy Syst. **60**, 245–257 (2014)
6. Zhou, X., Wang, H., Aggarwal, R.K., Phil, B.: Performance evaluation of a distance relay as applied to a transmission system with UPFC. IEEE Trans. Power Del. **21**(3), 1137–1147 (2006)
7. Amir, G., Babak, M., Mohammad, R.A.: Digital distance protection of transmission lines in the presence of SSSC. Int. J. Electr. Power Energy Syst. **43**(1), 712–719 (2012)
8. Pradhan, P.C., Sahu, R.K., Panda, S.: Firefly algorithm optimized fuzzy PID controller for AGC of multi-area multi-source power systems with UPFC and SMES. Eng. Sci. Technol. Int. J. **19**, 338–354 (2015)
9. Hingorani, N.G., Gyugyi, L.: Understanding FACTS: Concepts and Technology of Flexible AC Transmission System. IEEE Press (2000)
10. SultaN, A.-Y., Charabi, Y., Gastli, A., et al.: Assessment of wind energy potential locations in Oman using data from existing weather stations. Renew. Sustain. Energy Rev. **14**, 1428–1436 (2010)
11. Dai, J., Liu, D., Dai, J.: Research on power coefficient of wind turbines based on SCADA data. Res. Power Renew. Energy **86**, 206–215 (2016)
12. Singer, S., Braunstein, A.: Maximum power transfer from a nonlinear energy source to an arbitrary load. IEEE Proc. **134**(4), 1–7 (1987). Pt G
13. Yang, B., Jiang, L., Wang, L., Yaob, W., Wu, Q.H.: Nonlinear maximum power point tracking control and modal analysis of DFIG based wind turbine. Electr. Power Energy Syst. **74**, 429–436 (2016)
14. Fei, M., Pal, B.: Modal analysis of grid-connected doubly fed induction generators. IEEE Trans. Energy Convers. **22**, 728–736 (2007)

15. Barghi, S., Golkar, M.A., Hajizadeh, A.: Effect of distribution system specifications on voltage stability in presence of wind distributed generation. In: Electrical Power Distribution Networks (EPDC), pp. 1–6 (2011)
16. Qiao, W.: Dynamic modeling and control of doubly fed induction generators driven by wind turbines. In: Power Systems Conference and Exposition, pp. 1–8 (2009)
17. Sajadi, A., Rosłaniecn, L., Kłos, M., Biczelnb, P., Loparo, K.A.: An emulator for fixed pitch wind turbine studies. Renew. Energy **87**, 391–402 (2016)
18. Anaya-Lara, O., Jenkins, N., Ekanayake, J., Cartwright, P., Hughes, M.: Wind Energy Generation, Modelling and Control. Wiley, Chichester (2009)
19. Parvathy, S., Sindhu Thampatty, K.C.: Dynamic modeling and control of UPFC for power flow control. Procedia Technol. **21**, 581–588 (2015)
20. Bhattacharyya, B., Gupta, V.K., Kumar, S.: UPFC with series and shunt FACTS controllers for the economic operation of a power system. Ain Shams Eng. J. **5**, 775–787 (2014)

Photo-Thermal Study of the Optical Properties of Multi-layer Coatings Based on Black Pigmented Coatings/SiO$_2$ Applied for Solar Absorber

Faouzi Haddad[1]($\boxtimes$), Mustapha Hatti[2], R. Zaamoum[1,2],
and Khadidja Rahmoun[1]

[1] Département de Physique, Faculté des Sciences,
Université Aboubakr Belkaid de Tlemcen, Tlemcen, Algeria
faouzi72dz@yahoo.fr, k_rahmoun@yahoo.fr
[2] UDES, Unité de Développement des Equipements Solaires,
UDES/EPST, Bou Ismail, Tipaza, Algeria
musthatti@gmail.com

Abstract. Since solar thermal stations, domestic hot water and heat-process systems all are being to use solar energy by converting the sun's rays into heat. In order for these processes to be effective, they must absorb as much solar radiation as possible while limiting the loss radiative heat from the absorbent surface which has to have selectivity in the wavelengths of the solar spectrum. This selectivity allows it to have a maximum of absorbance with a minimum of thermal radiation emitted. This article deals with the optimization based on artificial neural networks of the solar absorption and infrared emission of the absorbing surface, in terms of absorbance and emissivity, study the possibility of integrating multilayer films deposited on aluminum and cover substrates to describe the properties of a surface. More than 300 different coatings and surface treatments for selective absorption of solar energy have been reported in the scientific and technical literature. Only a few of these have been subjected to detailed theoretical analyses [1]. In our multilayer, we used two materials having, respectively, a high and a low refractive index. We studied two cases: silicone black-pigmented coating/Polyurethane black-pigmented coating and Polyurethane black-pigmented coating/SiO$_2$. The thin films were deposited by Dip-coating technique. Spectroscopic ellipsometry was used to determine the optical constants and the thicknesses of every individual multilayer.

Keywords: Photo-thermal conversion · Solar water heater absorber
Spectral selectivity · Optical properties emission · Light- material interactions

1 Introduction

The first solar water heater with the absorber that consists of an aluminum plate on which is applied a layer of black paint has been developed in 1990 by Development Unit solar equipment (UDES). These coatings generally offer sufficient absorption of the high energy solar radiation but also act as good re-radiators in the thermal-infrared

© Springer International Publishing AG 2018
M. Hatti (ed.), *Artificial Intelligence in Renewable Energetic Systems*, Lecture Notes
in Networks and Systems 35, https://doi.org/10.1007/978-3-319-73192-6_14

spectral region, thus losing a large amount of the initially captured energy (Fig. la) [2]. The aim of our work is to introduce other types of selective black-pigmented coatings paints layers developed with ENAB society, to optimize this layer on the optical, physical and mechanical support perspective on two types of media in this case aluminum and copper. There are also other applications that use electromagnetic radiation interactions with matter: solar cells, smart windows, Infrared detectors, satellite communication ... etc. Despite their diversity, all these applications reply on thin layers on their surfaces whose physical and optical characteristics depend on the wavelengths of the electromagnetic radiation spectrum. solar absorber should possess the maximum possible absorptance in the solar spectrum while exhibiting a minimum infrared emittance. This may be accomplished by using a so-called "selective" absorber coating (Fig. lb) [2]. Such thin layers deposited on the surfaces act as filters where some photons at wavelengths of data must be able to be reflected or absorbed, while others, at different wavelengths, must be pushed or trapped. This ability to "filter" is a feature of so called "selective surfaces".

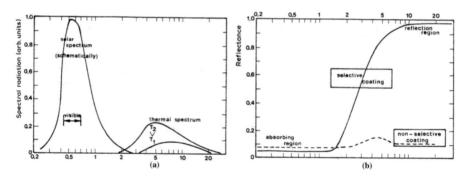

Fig. 1. Solar and thermal spectra (a) together with the typical spectrometric curves of a selective and a non-selective absorber coating (b)

2   Radiative Balance and Heat Balance of an Absorber

An absorber, with its rear faces and side perfectly isolated, exposed to solar radiation reaches an equilibrium temperature as the received power equals the power lost. The power received per square meter is the sum of two terms [3].

$$P_r = \alpha H + \alpha' \sigma T_a^4 \tag{1}$$

The power loss is also a function of two components [3]:

$$P_p = \varepsilon \sigma T^4 + P_c \tag{2}$$

The heat balance per unit area is:

$$\varepsilon \sigma T^4 + P_c = \alpha' \sigma T_E^4 + \alpha H \tag{3}$$

For a black surface $\alpha = \alpha' = \varepsilon = 1$, Eq. (3) becomes:

$$H = P_c + \sigma(T^4 + T_E^4) \tag{4}$$

In the case where part of the absorbed power is converted into heat then transmitted to the fluid, we
Should consider this power φ_u, either:

$$\varepsilon \sigma T^4 + P_c = \alpha' \sigma T_E^4 + \alpha H \tag{5}$$

Such For H = constant, we define the conversion efficiency as: η

$$\eta = \frac{\varphi_u}{H} = \alpha + \frac{\alpha' \sigma T_E^4}{H} - \frac{\varepsilon \sigma T^4}{H} - \frac{P_C}{H} \tag{6}$$

Examination of the radiation balance:

$$\eta_R = \alpha \left[1 - \frac{\varepsilon}{\alpha} \frac{\sigma \left(T^4 - T_E^4\right)}{H} \right] \tag{7}$$

By considering Eq. (7), we can say that:

1. The yield η_R is maximum when T = TE, it means, if there is no rise in temperature of the absorber from the outside temperature.
2. It is possible to increase the value of output by increasing the ratio α/ε in view Kirchhoff's law [4], we have, for the wavelength λ:

$$\alpha_\lambda = \varepsilon_\lambda = 1 - R_\lambda \tag{8}$$

Parameters α and ε are based on optical properties of the surface temperature T and wavelength [4]. To reduce the value of (α/ε), the surface must have a low coefficient of reflection for the visible (VIS), while for the infrared (IR), R_λ must be high.

3 Mathematical Model

In this work we consider an electromagnetic wave incident on a surface. This one is a multilayer s surface constitute of different materials with different refractive index n. The whole system is put on substratum of aluminum Al or copper Cu with a higher refractive index. During its trip, the electromagnetic wave crosses the different mediums and interfaces.

We choose a plane wave with intensity I_i, after crossing the system, we will have a reflected wave with intensity I_r and a transmitted wave with intensity I_t. The rules of optics give us the transmission coefficient and the reflection coefficient as:

$$T = \frac{I_t}{I_i} \quad \text{and} \quad R = \frac{I_r}{I_i} \tag{9}$$

The interfaces contribute to the division and recombination of the beams. The widths of the layers create a phase shift between the layers. Thus, considering a wave

$A(x) = a_i e^{i\left(\omega t - \frac{2\pi n_i x}{\lambda}\right)}$, when it travels in the medium i undergoes a phase shift $\varphi_i = 2\pi n_i e_i \cos\theta_i / \lambda$, with e_i the width of the layer, θ_i the refraction angle in the medium and λ the wave length. Thus, one can write:

$$A_{in}(x) = M_i A_{out}(x) \tag{10}$$

where $A_{in}(x)$ is the wave entering the medium i, $A_{out}(x)$ the outgoing wave and M_i a matrix which contains the phase shift, written as:

$$M_i = \begin{pmatrix} e^{-i\varphi_i} & 0 \\ 0 & e^{+i\varphi_i} \end{pmatrix} \tag{11}$$

By the same way, when the wave crosses an interface between a layer i and a layer $i + 1$ it undergoes a transformation due to the division and recombination processes. As a consequence, on can write:

$$A_i(x) = L_i A_{i+1}(x) \tag{12}$$

where the matrix L_i is given by the expression:

$$L_i = \begin{pmatrix} \frac{n_i + n_{i+1}}{2n_i} & \frac{n_i - n_{i+1}}{2n_i} \\ \frac{n_i - n_{i+1}}{2n_i} & \frac{n_i + n_{i+1}}{2n_i} \end{pmatrix} \tag{13}$$

To consider now the whole system, on can write:

$$A_0(x) = S A_p(x) \tag{14}$$

with $A_0(x)$ the incident wave, $A_p(x)$ the outgoing wave and S the global transformation matrix containing all the informations about the system, it reads as a matrix product of all the transformation matrices:

$$S = L_0 . M_1 . L_1 . M_2 \ldots M_p . L_p \tag{15}$$

Finally the transmission coefficient and the reflection coefficient are given with the help of the matrix elements of S:

$$T = \frac{1}{S_{22}} \quad \text{and} \quad R = \frac{S_{12}}{S_{22}} \tag{16}$$

The absorbance is given by $A = 1 - T - R$.

We applied this model to a three layers system composed of a three different kinds of solar absorbent paints put on an aluminum substratum. We first made the calculation for one layer, then for two and finally for three. We plot the transmission coefficient, the reflection coefficient and the absorbance for the three cases, as shown in Figs. 2(a), (b) and (c).

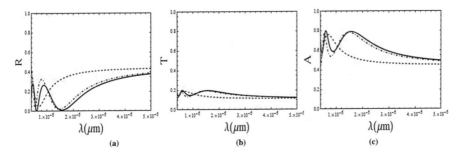

(a)   (b)   (c)

Fig. 2. (a) The reflection coefficient. (b) The transmission coefficient. (c) The absorbance. As function of the wavelength λ. Black solid line: 3 layers. Blue dot-dashed line: 2 layers. Red dotted line: one layer. The width of the layers is the same $e = 1000$ nm.

As expected, the profile of R is logarithmic with a diminution when one increases the number of layers. The transmission coefficient is nearly the same in the three cases. Whereas the absorbance increases with the number of layers, which is the desired result.

4 Experimental Setup

4.1 Filing of Selective Layers by Dip-Coating Method

We give in the table characteristics of the two black-pigmented coating used (Table 1):

Table 1. Characteristics of the two black-pigmented coating

Coatings	Based	Density	Viscosity at 20 to 25 C	Limited temperature
Silicone	Silicone	1.065 +0.05	30+10 Po (M4/V20)	500 C Without grey fosfal AM 120 C with grey fosfol AM
Polyurethane	Hydroxyl polyester/aliphatic polyisocyanate	1.251 +0.05	52+10 Po (M4/V20)	120 C

a. The first multilayer: Silicone black-pigmented/Polyurethane black-pigmented

Aluminum and copper plates of size 30 × 30 mm were mechanically polished, degreased with a solvent. The Polyurethane black-pigmented was coated on top of the absorbing layer which is composed on smooth highly reflecting aluminum or copper substrates, Dip-coating technique is used for deposits made on substrates with translational symmetry. The substrate is immersed in a solution of black paints for the two cases silicone black-pigmented coating and Polyurethane black-pigmented coating, and is removed from the solution at a constant speed and controlled so that a uniform film covers the submerged surface. This film is left to dry for a few minutes to allow the sol-gel. The withdrawal rate may vary between 1 and 50 cm/min. Dip-coater is used at the CDTA (Development Unit silicon technologies). Figure 3 shows the schematic diagram of the dip-coater that we used. The surface composed by substrate and polyurethane black-pigment was coated by the Silicone black-pigmented layer with the same sol-gel process.

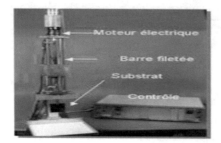 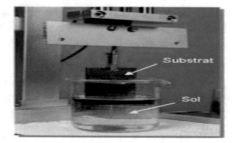

Fig. 3. Photo of dip-coater

b. The second multilayer: Polyurethane black-pigmented/SiO$_2$

We repeat The same dip coating process for the polyurethane black-pigmented. The anti-reflective films were obtained by spin-coating. A syringe containing coating solution was used to eject the liquid on top of the substrate. In a fraction of a second the substrate was fully covered with coating solution and a completely homogenous and even film was produced. Further evaporation of solvents was achieved by allowing the spinning process to continue for 30 s after the solution was ejected [5].

After the spin-coating process the samples were subjected to a heat treatment at temperatures up to 480 °C in order to form the silica.

4.2   Fitting Experimental Measure R(λ)

The experimental curves, R(λ), selective layers deposited by dip-coating technique on various substrates were measured using a spectrophotometer "Spectro 320". This spectrophotometer can scan wavelengths up to 2500 (nm).

4.3   Mounting Experimental Measure Refractive Index and Thickness

Ellipsometry has been applied in our case to the characterization of materials and deposited layers. Knowing the index of the substrate and that of the deposited layer can then deduct the value of the thickness of the layer. We opted for ellipsometry because of its sensitivity to small variations in thickness.

5   Results and Discussions

5.1   Black Paint on Aluminum

Initially, we deposited a layer of black paint identical to that applied to our first water heater, manufactured in the 1990s, on an aluminum substrate, after that carried out the characterization by spectroscopy to measure $R(\lambda)$, Fig. 4 shows the reflectance curve as a function of wavelength.

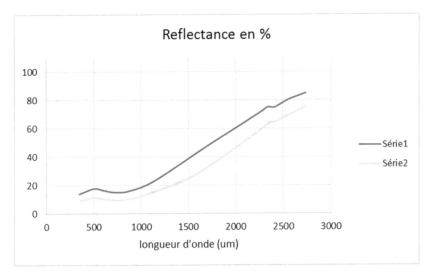

Fig. 4. Variation of reflectance as a function of wavelength for different thicknesses

We observe that the black paint produced an increase in absorption coefficient of the absorber plate and the coefficient of thermal emissivity.

5.2   Selective Painting Silicone Black-Pigmented/Polyurethane Black-Pigmented on Aluminum

In a second step, we chose a second option for selective indication Silicone black-pigmented/Polyurethane black-pigmented. After depositing the various layers from this solution on the aluminum substrate was measured and must be performed to characterize spectroscopically.

The thicknesses of the layers using ellipsometry described above. Figure 4 shows the change in reflectance function λ for different thicknesses of deposited layers 1 and 2.

There was a reduction of the reflectance gradually as the thickness of layers increases (curves 1 and 2). The thicknesses and refractive indices are shown in Table 2.

Table 2. Thicknesses and refractive indices of layers.

Layers	Serie1		Serie2	
Thickness (nm)	240	415	565	695
Refractive indices	1.54	1.58	1.60	1.61

5.3   Deposit of Polyurethane Black-Pigmented/SiO$_2$ on a Aluminum Substrate

Finally, we characterized by spectroscopy of Polyurethane black-pigmented/SiO$_2$ multi-layers deposited on aluminum substrate. The reflectance curve as a function of wavelength is shown in Fig. 5.

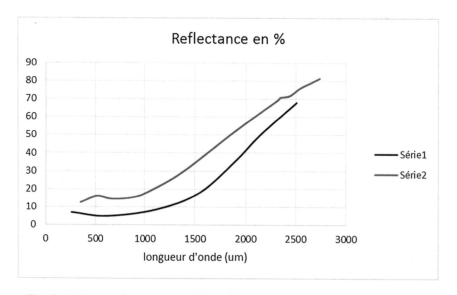

Fig. 5. Variation reflectance according to the wavelength for an optimal thicknesses.

There is a good selectivity of the deposited layers for two ranges of the spectrum (VIS) and (IR) (Table 3).

Table 3. Thicknesses and refractive indices of layers.

Layers	Serie1		Serie2	
Thickness (nm)	450	150	600	250
Refractive indices	1.52	1.38	1.56	1.61

Finally can we give comparison with the two multi-layers used (Fig. 6).

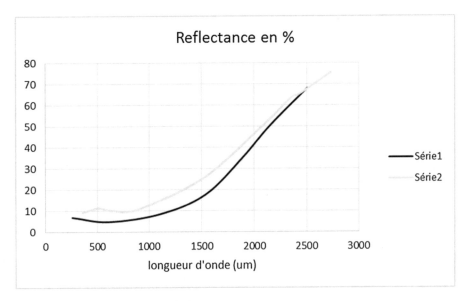

Fig. 6. Variation reflectance according to the wavelength for the two multi-layers an optimal thicknesses

6 Conclusion

These results of the deposition of one or more layers selectively allow a significant increase in conversion efficiency of the (ESC). It is in this case we can consider the realization of high performance absorbers.in this paper we exposed the process of spectrally selective paint coating with different multi-layers, we obtained a better result for selective surface with a anti-reflecting layer. Using these layers, we found an improvement in conversion efficiency of 10 to 17%. Selective layers deposited and characterized helped to highlight the effect of selectivity the two multilayers used. The choice of technologies selective deposition of these layers must be simple and economical [6].

References

1. Granqvist, G.: Solar energy materials overview and some examples. Appl. Phys. A **52**, 83–93 (1991)
2. Bogaerts, W.F., Lampert, C.M.: Review materials for photo thermal solar energy conversion. J. Mater. Sci. **18**, 2847–2875 (1983)
3. Gallet, P., Papini, F., Péri, G.: Physique des convertisseurs héliothermiques, EDISUD (1980)
4. Duffie, J.A., Beckman, W.A.: Solar Energy of Thermal Process. Wiley, New York (1974)
5. Haddad, F., Chikouche, A., Laour, M.: Simulation of the opti-physical parameters of selectives surfaces of absorber by the FDTD method. Energ. Procedia **6**, 413–421 (2011)
6. Orel, Z.C., Gunde, M.K.: Spectrally selective paint coatings: preparation and characterization. Energ. Mater. Solar Cells **68**, 337–353 (2001)

Sahara Potential and Sustainable Development of Algeria: A Thermal Experimental Study of Parabolic Trough Solar Concentrator with New Perspectives of Solar Energy

Lahlour Rafik[✉], Bellel Nadir, and Bouguetaia Nadia

Energetic Physics Laboratory, Department of Physics,
University of Constantine 1 – Algeria, Constantine, Algeria
rafik.energie@gmail.com

Abstract. In this article which presents a experimental study of parabolic trough collector concentrator. In this work, we propose the practical realization of any concentrator having a 2.88 m^2 opening and it is equipped by a solar semi-electronic tracking system with a closed circuit of the heat-transfer fluid. A study on the extensive comparison of three outdoor test methods for determining the thermal performance of parabolic trough solar collectors is presented. These test methods of vertical movement and two other case of bi-axel movement with and without the absorber tube glass envelope, in other review the principles of the work of power plants with a review of the possibility of solar energy in Algeria. There are also important conditions for power generation with economic concentrating solar power provides all the electrical grid requirements.

The experimental side has studied the influence of the prototype's orientation on the performances of training. Several trials have been carried out in order to have the possibility of attending the temperatures that can ensure the vaporization of water. These trials have been preceded in different conditions of climatic operation. The results are provided and have been discussed. They are discussed in detail and are explained in view of the on-site parabolic trough solar collectors in the real service.

Keywords: Solar energy · Experimentation · Parabolic trough
Solar collector · Realization

1 Introduction

Henceforth, the solar energy is an essential solution for the human life's development. The most basic processes surviving in the earth, such as the photosynthesis and the cycle of rain are due to the solar energy.

The recent industrial development and the environmental effects show that the solar energy is the most promising sources of energy which are not convential. The most part of available plants in the solar common trades use the cylindrical-parabolic concentrator.

© Springer International Publishing AG 2018
M. Hatti (ed.), *Artificial Intelligence in Renewable Energetic Systems*, Lecture Notes in Networks and Systems 35, https://doi.org/10.1007/978-3-319-73192-6_15

The parabolic collector contains: receiver pipe, a concentrator, power of transmission, collector's structure, and the receiver is a system on which the solar radiation is absorbed and converted to thermal energy. It contains an absorber pipe, its glass cover and the insulation at its extremity. And on top of it all Algeria is one gas reserves in the world increased by solar and wind energy.

In 1962, Algeria has established renewable energy and put the interests of the most important conditions for the change of hydrocarbons without risk to health and the environment.

The solar systems with concentration offer the ability to produce the electricity through the solar energy, the temperatures that can exceed the 500 °C and the conversion yield is generally high. By using the direct solar radiation. This one is considered as a principle resource which is very considerable in the planet; these technologies provide a real alternative to the consumption of fossil resources with low environmental impact and high potential for cost reduction, as well as the possibility of the hybridization of these installations.

Today, thousand of captors produce a power more than 674 MW in the desert of Mojave south California, this power present 90% of the solar capacity set up in the world [1].

Many researches are carried out in order to study the absorber, the different performances of pipes such as the absorber pipes without glass envelop [2] and pipes with glass envelop. It means that our study is based on two different elements in the concentrator parabolic trough solar which are: direction and absorber pipe.

2 Solar Energy

2.1 Assessment of the Potential of Concentrating Solar Power Use in Algeria

Evaluated siting parameters for centralized concentrating solar power plants are required before locating a real plant. The potential for CSP implementation in Algeria depends on identifying and analyzing these technical and economical parameters and issues which are listed in Table 1 and studied, in addition to other parameters.

Table 1. Main siting factors of concentrating solar power plant [4].

Siting factor	Requirement
Solar resource	Abundant 4(1800 kWh/m^2/year) for economical operation
Land use	20,234 m^2/MW of electricity production
Land cover	Low diversity of biological species, limited agriculture value
Site topography	Flat, slope upto 3%, 1% most economical
Infrastructure	Proximity to transmission-line corridor, natural gas pipeline
Water availability	Adequate supply, otherwise dry cooling

Due to the nature of CSP technology, only the direct normal insolation (DNI) can be used which limits the high-quality CSP sites to areas with low levels of atmospheric moisture and particulates, little or no cloud cover, and high levels of DNI around the year, deserts thus being the most typical for these conditions [3].

Further, the required solar field size for CSP is directly proportional to the level of DNI, with the solar field representing about 50% of total project cost; the DNI level will have the greatest impact on

Overall CSP system cost since the CSP systems require high DNI for cost-effective operation. Sites with excellent solar radiation can offer more attractive levelized electricity prices, and this single factor normally has the most significant impact on solar system costs [3]. It is generally assumed that concentrating solar power systems are economic only for locations with DNI above (1800 kWh/m^2/year) (circa 5 kWh/m^2/day) [3, 5]. The geographic location of Algeria, in the Sun Belt region, and the climatic conditions such as the abundant sunshine throughout the year, low humidity and precipitation, and plenty of unused flat land close to road networks and transmission grids, have several advantages for the extensive use of the solar energy as enormous potential for power generation compared to global energy demands.

According to a study of the German Aerospace Agency (DLR) based on satellite imaging (Fig. 4), Algeria, with 1,787,000 km^2, has the largest long term land potential for the concentrating solar power in the Mediterranean basin. In addition, as shown in Table 2 [6], the value of solar radiation falls between 4.66 kWh/m^2 and 7.26 kWh/m^2; this corresponds to 1700 kWh/m^2/year in the north and 2650 kWh/m^2/year in the south. The insulation time over the quasi-totality of the national territory exceeds 3000 h annually and may reach 3500 h in Sahara. With this huge quantity of sunshine per year, Algeria is one of the countries with the highest solar radiation levels in the world; this solar potential exceeds 6 billion GWh/year [6]. The economic potential for solar energy generation in Algeria has been assessed by the DLR and the CDER, mainly from satellite imaging and further processing. The derived economic potential data are gathered in the REs guide report by the MEM [5] and estimated at 169,440 TWh/year for thermal solar system.

Table 2. Solar potential in Algeria

Areas	Coastal area	High plains	Sahara	Total
Surface (%)	4	10	86	100
Area (km^2)	95,270	238,174	2,048,297	2,381,741
Mean daily sunshine duration (h)	7.26	8.22	9.59	
Average duration of sunshine (h/year)	2650	3000	3500	
Received average energy (kWh/m^2/year)	1700	1900	2650	
Solar daily energy density (kWh/m^2)	4.66	5.21	7.26	
Potential daily energy (TWh)	443.96	1240.89	14,870.63	16,555.48

2.2 Future Parabolic Trough Power Plant Projects in Algeria

Parabolic trough solar thermal power plant (PTSTPP) is one of the attractive technologies to produce electricity from thermal solar energy that use mirrors to focus sunlight onto a receiver that captures the sun's energy and converts it into heat that can run a standard turbine generator or engine. PTSTPP systems range from remote power systems as small as a few kilowatts (kW) up to grid-connected power plants of 100 s of megawatts (MW). The process of energy conversion by PTSTPP consists of two main parts:

- The concentration of solar energy and converting it into usable thermal energy.
- The conversion of heat into electricity, which is realized by a conventional steam turbine.

This paper goes on to give a review on the assessment of concentrating solar power (CSP) potential, and PTSTPP projects development in Algeria.

Three further hybrid power plant units are to be completed by 2018, with 70 MW parabolic trough solar power plants capacity for each one of them; each one will be scale-ups of Hassi R'mel, and are part of the government's plan to develop electricity production and exports from renewable energies in Algeria Table 3 lists the new PTSTPP proposed generation projects in the Algerian investment plan under MENA CSP scale-up programme, with a scheduled accumulated CSP capacity of 210 MW.

Table 3. Three new solar integrated projects under various stages of consideration [3, 9].

Solar–gas hybrid power plant	Location	Installed CSP capacity (MW)	COD
SPP II: Solar power plant	Meghaier	70	2014
SPP III: Solar power plant two	Naama	70	2016
SPP IV: Solar power plant three	Hassi R'mel	70	2018

Two options are being considered for the first project, which will be located in Meghaier, in the southeast part of Algeria. Both would include a 270–280 ha solar island using parabolic trough:

Option 1: power production only, total capacity 400 MW, of which 70 MW was generated from PTSTPP.

Option 2: integrated desalination/power production, total capacity 480 MW, of which 80 MW PTSTPP (the plant would treat local brackish water) [8].

2.3 Solar Radiation in Algeria

Algeria has a significant solar sink. From its climate, the maximum solar density in any point (clear sky, June) exceeds the 6 kW/m^2 and maximum annual received energy in Algeria is close to 2500 kW/m^2 [10]. The implantation of solar installations, dictated by the judicious choice of the preferred sites for a better collecting, is conditioned by the sweeping of all the territory (48 Provinces) which will enable us to select the sunniest cities for which the collector efficiency is optimal (Fig. 1). From its strategic importance, the area of Adrar (South-West of Algeria) is the favorable area for such projects. However, the problem of transport of energy produced could constitute a handicap which one must free oneself.

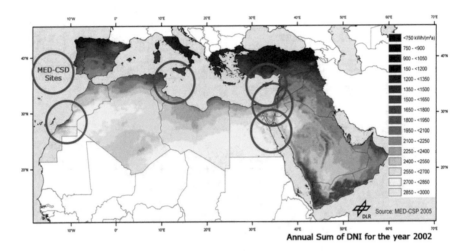

Annual Sum of DNI for the year 2002

Fig. 1. Solar thermal electricity generating potentials in Algeria [7].

The history of using solar energy in Algeria backs to 1954 with the solar furnace built by the French for ceramic fabrication purpose. Because of the geographic position, Algeria is considered one of the best places for solar energy usage.

3 Algeria Desert Irradiation

Total annual irradiation for all world deserts (31 deserts of 19 million km^2), was calculated [11, 12]. A summary of calculated results is depicted in Table 4. It can be seen easily that Algeria desert has an 11.4% share of the total irradiation of world deserts. The Algeria desert annual irradiation (395600 Mtoe/y) is more than 26 times higher the primary energy consumption of the world in 2030 (15000 Mtoe [13]) (Fig. 2).

4 Silicon (SI) Potential of Algeria Desert

VLS-PV needs a huge quantity of Silicon (Si). Is the Si present in Algeria desert sand enough to produce 60 TW PV? A calculation method [11, 12] was proposed to assess Si potential of Algeria desert. Table 4 shows a rough calculation of sand and Silicon potential of Algeria desert.

Table 4. Algerian desert irradiation compared to world energies

World region deserts	Area (km^2)	Annual irradiation (kWh/m^2)	Total Annual irradiation (PWh)	Total Annual irradiation (Mteo)	S T A I (%)
Total World deserts (31 deserts)	18,978,143	2,136	40,537	3,486,218	100
North Africa (NA)	8,600,000	2,300	19,780	1,701,080	49
Middle East (Me)	3,052,400	2,137	6,524	561,052	16
Algeria	2,000,000	2,300	4,600	395,600	11,4

S T A I: Share of total annual irradiation.

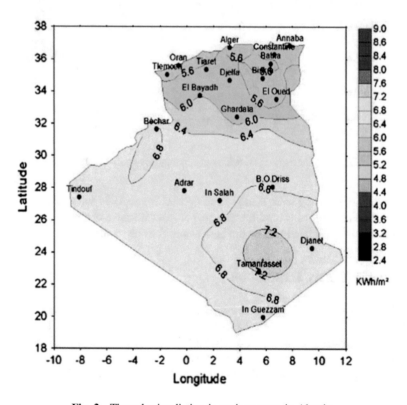

Fig. 2. The solar irradiation in various areas in Algeria.

While 1 MW PV needs 10 tons of Si [14] or 100 tons of sand, then 60 TW PV (total PV capacity, see Table 5) will be needing 60 × 108 tons of sands. The sand reserves which equal to 48,6 × 1012 tons is more than 8000 times the sand needs to produce 60 TW PV energy [15].

Table 5. PV capacity and annual energy generation for all the suitable deserts area of Algeria

PV capacity and Energy generation using VLS-PV system	PV capacity (TW)	Annual generation (PWh)	Annual generation (Mtoe)	Annual generation to Annual irradiation %
Without buffer plant	60	95	8,100	2,1
With buffer plant	40	63	5,500	1,4

5   Experimental Study

5.1   The Realization

(1)   The different steps of contribution of parabolic trough solar

These four photos show the different steps of the designing of this cylindrical parabolic concentrator. In the beginning, the structure of the galvanized support is designed to form the semi-cylindrical solar parabolic concentrator as it is shown in the photo (A), the photo (B) shows the location and the pasting of mirrors on cylindrical iron. However, for the photos (C, D), they present the final form of our cylindrical parabolic body (Fig. 3).

(a)   (b)

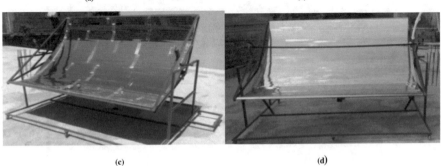

(c)   (d)

Fig. 3. The cylindrical parabolic concentrator.

(2) **Experimental Results**

It is true that the tests will be conducted in two different days. But we chose the days which have almost the same radiation [16].

(a) *Performance test with vertical orientation*

The Fig. 4 shows the results of the test which is done on July 25th 2014. There was an ambient temperature 41 °C with an occasional wind.

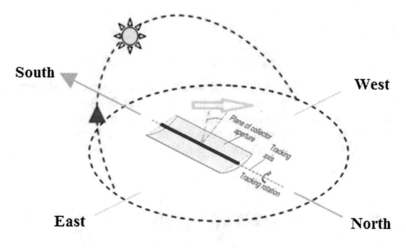

Fig. 4. GuidanceMode

In the beginning, the temperatures were very close. Then, they were increased especially in the absorber and the outgoing temperature (Fig. 5).

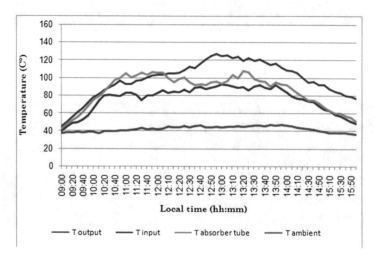

Fig. 5. Experimental results of cylindrical parabolic concentrator with vertical orientation done on July 25th 2014.

After 80 min, we observe that the temperature of the absorber has exceeded the temperature of heat-transfer fluid in the outgoing but it does not stay a long time till the fluid's temperature exceeds the temperature of the absorber in the outgoing, it attended its maximum 127 °C at 12 h 55 min. Then, it is decreased to 77 °C at 16 h in the end of the experimental session.

We notice quite changes of heat in the absorber pipe under the influence of the variety and the speed of the wind. The following table indicates the maximum temperature in the different parts of the cylindrical parabolic concentrator (Table 6):

Table 6. The maximum temperature in the different parts of concentrator at July 25, 2014

	T °C	Hour	Date
Ts max	127	12:55	July 25th 2014
Te max	93	13:00	July 25th 2014
Tabs max	108	13:20	July 25th 2014
Tamb max	47.1	13:55	July 25th 2014

(b) *Performance test and a bi-axial orientation (slanting)*

The tests are done on July **29th 2014** from **9 h 00** till **16 h**, it was a clear day but there was some wind at the beginning with an ambient temperature **41 °C**.

After the method's change of the bi-axial orientation of cylindrical parabolic concentrator, we have noticed an increase in the fluid's temperature at its outgoing, in this case, the capitation of solar rays is very important where the huge use of solar energy is insured.

We noted a temperature 137 °C which have attended only 127 °C during the last experience that proves the performance of this case.

The next table proves the experimental results (Table 7):

Table 7. The maximum temperature in the different parts of concentrator at July 29, 2014

	T °C	Hour	Date
Ts max	137	13:20	July 29th 2014
Te max	117	13:30	July 29th 2014
Tabs max	122	13:30	July 29th 2014
Tamb max	47.3	13:20	July 29th 2014

(c) *The performance test with bi-axial orientation* **5 cm**

The Fig. 6 indicates the results of the test which is done on August 13th 2014 with an ambient temperature 41 °C.

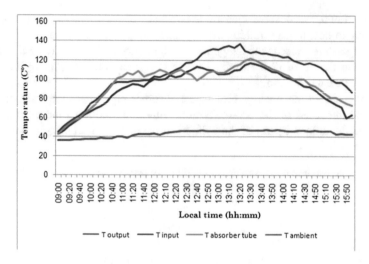

Fig. 6. Experimental results of cylindrical parabolic concentrator with a bi-axial orientation done on July 29th 2014.

In the third experiment, we have covered the absorber pipe by **05 cm** diameter of glass. In this case, the obtained results are more effective than they were in the second one (Fig. 7).

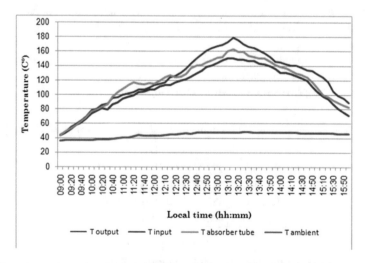

Fig. 7. Experimental results of cylindrical parabolic concentrator with bi-axial orientation and the absorber pipe covered by a glass envelope which is done on July 25th 2014.

In the temperature increase process, this one became fast and attended 100 °C at 11 h o'clock, where it does not increase in the last experiments till 11 h 40 min. The glass cover of the absorber pipe has clearly contributed in the increase of the outgoing temperature which has attended the maximum (179 °C) at 13 h 15 min (Table 8).

Table 8. The maximum temperature in the different parts of concentrator at August 13, 2014

	T °C	Hour	Date
Ts max	179	13:15	August 13th 2014
Te max	151	13:15	August 13th 2014
Tamb max	163	13:15	August 13th 2014
Tamb max	48.3	13:30	August 13th 2014

6   Conclusion

This work suggests an analytical study for evaluating future projects of solar energy in Algeria. While being interested by the average temperatures [200 °C, 500 °C], the bibliographical analysis shows that the solar concentrators meet this aim rather well. And experimental study of cylindrical parabolic concentrator. The theoretical study is based on the solar rays, the thermal balance and the equation system, and then we have realized a cylindrical parabolic concentrator with a system, which allows the absorption using the best method of solar energy.

After that, we have applied tests on this concentrator in several cases that are related either to the meaning or to the absorber pipe.

We have remarked that the temperature of the outgoing heat-transfer fluid has exceeded **100** °C with a difference in time according to the applied method. Considering these experiments, we deduce that the bi-axial orientation of solar concentrator, which uses an absorber pipe covered by glass envelop, is the most efficient method. Out of the other experiments.

References

1. Guerraiche, D., Benderradji, A., Benmoussa, H.: Facteurs optiques et géométriques caractérisant un concentrateur cylindro-parabolique. Rev. Energ. Renouv. **14**(2), 229–238 (2011)
2. Roesle, M.: Numerical analysis of heat loss from a parabolic trough absorber tube with active vacuum system. J. Sol. Energ. Eng. **133**, 031015 (2011)
3. Cohen, G., Skowronski, M., Cable, R., Morse, F., Jaehne, C.H., Kearney, D., et al.: Solar thermal parabolic trough electric power plants for electric utilities in California PIER final project report. California Energy Commission. http://www.energy.ca.gov/2005publications/CEC-500-2005-175/CEC-500-2005-175.pdfs. Accessed 2012
4. Kaygusuz, K.: Prospect of concentrating solar power in Turkey: the sustainable future. Renew. Sustain. Energ. Rev. **15**, 808–814 (2011)
5. Steinhagen, H.M., Trieb, F.: Concentrating solar power—a review of the Technology. http://www.dlr.de/Portaldata/41/Resources/dokumente/institut/system/publications/Concentrating_Solar_Power_Part_1.pdfs. Accessed 2012
6. Stambouli, A.B., Khiat, Z., Flazi, S., Kitamura, Y.: A review on the renewable energy development in Algeria: current perspective, energy scenario and sustainability issues. Renew. Sustain. Energ. Rev. **16**, 4445–4460 (2012)
7. German Aerospace Center (DLR). Concentrating solar power for the Mediterranean region final report. http://www.dlr.de/Portal

8. Ministry of Energy and Mines. Renewable energy and energy efficiency program. http://www.mem-algeria.org. Accessed 2011
9. Sonelgaz. http://www.sonelgaz.dz. Accessed 2012
10. Bernard, J.: Energie Solaire, Calculs et Optimisation, Edit. Ellipses (2004)
11. Flazi, S., Stambouli, A.B., Khiat, Z.: Sahara solar potentials: energetic, socio-economic and sand reserve. In: 2AASE Forum and 4SSB Workshop, USTO/ORAN, 15–16 May 2012
12. Flazi, S., Stambouli, A.B., Khiat, Z.: Sahara photovoltaic potential and silicon reserve. In: 3AASEF & 5SSBWS, Hirosaki, Japan, 6–8 May 2013
13. BP Statistical Review of world Energy, June 2012
14. Koinuma, H., Kurokawa, K., Stambouli, A.: Shift the global energy paradigm from the Sahara. Sahara solar breeder (SSB) Plan: SCJ proposal in G8+5 Academies
15. Flazi, S., Stambouli, A.B., Siali, M.: Sahara potentials and sustainable development of Algeria
16. Gama, A., et al.: Review of the renewable energies **11**(3), 473–451 (2008)

Ensemble of Support Vector Methods to Estimate Global Solar Radiation in Algeria

Nahed Zemouri[1(✉)] and Hassen Bouzgou[2]

[1] Department of Electronics, Faculty of Technology,
University of Mohamed Boudiaf, Msila, Algeria
zemouri_nahed@yahoo.fr
[2] Department of Industrial Engineering, Faculty of Technology,
University of Batna 2, Batna, Algeria
bouzgou@gmail.com

Abstract. In this paper, we propose a set of times series forecasting techniques based on the combination of Support Vector Regression methods to predict global horizontal solar radiation in Algeria. The models were constructed and tested using different architectures of Support Vector Machine (SVM), namely, (RBF kernel, Polinomial kernel and Linear kernel). We use individual time series models and linear combination techniques to predict global solar radiation indifferent sites in Algeria. For this aim, the recorded data of 4 stations spread over Algeria were used to build different combination schemes for the different times series algorithms. The efficiency of the different models was calculated using a number of statistical indicators: the Mean Absolute Percentage Error (MAPE), the Mean Squared Error (RMSE), Mean Bias Error (MABE) and the Coefficient of Determination (R^2). The results obtained from these models were compared with the measured data.

Keywords: Support vector regression · Global horizontal irradiance
Combining forecasts · Algeria

1 Introduction

The precise information of solar radiation at any specific locations is necessary for different areas such as hydrological, agricultural and ecological in addition to solar energy applications. It can be seen that the plentiful potential of solar energy can play a major role to meet the increasing world energy demand [1]. Among different types of renewable resources, solar energy has enticed enormous consideration since not only it is worldwide abundant, but also it is sustainable and environmental friendly [2].

Solar energy is the portion of the sun's energy available at the earth's surface. It is known as an ancient clean source, and it is the basic ingredient of almost all renewable types of energy on earth. This abundant source of energy is accessible in most places and its utilization is important. Unfortunately, for a lot of developing countries, solar radiation measurements are not easily accessible. Hence, it is indispensable to develop techniques to predict the solar radiation on the basis of the more readily available meteorological data [3].

© Springer International Publishing AG 2018
M. Hatti (ed.), *Artificial Intelligence in Renewable Energetic Systems*, Lecture Notes in Networks and Systems 35, https://doi.org/10.1007/978-3-319-73192-6_16

Several models have been developed to forecast the amount of global solar radiation on horizontal surfaces using various climatic parameters, such as sunshine duration, cloud cover, humidity, maximum and minimum ambient temperatures, wind speed,… etc. [5–7].

The paper is organized as follow: after a short presentation of the general framework of time series methods for prediction global solar radiation, the Sect. 2 describes the statistical methods used. Firstly, those based on individual methods and then the different linear combination methods to combine the individual ones. Section 3 presents the experimental results obtained on real data sets and compares with that of single prediction methods. Finally, Sect. 4 draws the conclusion of this work.

2 Materials and Methods

2.1 Descriptions of Data Sets

Algeria situated in the center of North Africa along the Mediterranean coastline, between latitudes 191 and 38 North and longitudes 81 west and 121 east, has an area of 2,381,741 km^2. This geographic location of Algeria signifies that it is in a position to play an important strategic role in the implementation of solar energy technology in the north of Africa [8].

Algeria has an enormous potential of solar energy. As shown in Table 1, the potential of daily solar energy is significant; it varies from a low average of 4.66 kwh/m^2 in the north to a mean value of 7.26 kwh/m^2 in the south. The mean yearly sunshine duration varies from a low of 2650 h on the coastal line to 3500 h in the south. With this huge quantity of sunshine per year, Algeria is one of the countries with the highest solar radiation potential in the world [8].

Table 1. Daily solar energy and Sunshine duration in Algeria [14].

Parameters	Coastal line	High plains	Sahara
Area (km^2)	95.271	238.174	2.048.296
Mean daily sunshine duration (h)	7.26	8.22	9.59
Solar daily density kwh/m^2	4.66	5.21	7.26
Potential daily energy 10^{12}wh	443.96	1.240.89	14.870.63

The different models were constructed and tested for four Algerian sites with three different climates. Geographical details of different sites are given in Table 2.

Table 2. Summary of geographical data for the locations under study.

Location	Latitude (°N)	Longitude (°E)	Elevation (m)
Alger	36.760	3.050	66
Msila	35.700	4.540	468
Batna	35.550	6.170	1042
Bechar	31.610	−2.220	767

2.2   Support Vector Regression Algorithm

In the early 1990s, a support-vector machine (SVM) was inspired from the statistical learning theory [9, 10]. SVM can be classified into a classification support-vector machine and a regression support vector machine. The first is principally used for classification problems, and the last is principally used for continuous problems prediction, which is frequently known as a support vector machine regression (SVR). SVM is recognized for its better generalization capability and its capacity to deal with nonlinear problems.

SVM uses the margin-based loss functions (e.g. Fig. 1c) and maps the learning data (linearly or non-linearly) into a higher dimensional space (e.g. Fig. 1) and then It seeks to determine the best decision that gives a good generalization. The final SVR model can be written as:

$$Y = f(x) = \sum_{i=1}^{N} (\alpha_i - \alpha_i^*)k(x_i, x) + b = \sum_{i=1}^{N} W_i k(x_i, x) + b \qquad (1)$$

Where $k(x_i, x)$ is the kernel function α_i, α_i^* are the Lagrangian multipliers and b is a constant [11].

2.3   Kernel Function

Kernel functions $k(x_i, x)$ are in fact a mapping function. The function projects the original feature space into a high dimensional feature space, where the mapped data can be represented linearly [10]. A kernel function must satisfy the Mercer's conditions [10]. The frequently used kernel functions are as follow:

(1) **Linear kernel function:**

$$k(x_i, x) = x_i.x \qquad (2)$$

(2) **Polynomial kernel function:**

$$k(x_i, x) = (x_i.x + 1)^d \qquad (3)$$

(2) **RBF kernel function:**

$$k(x_i, x) = \exp(-\gamma \|x_i - x^2\|) \qquad (4)$$

These kernels can be classified into two forms: the global kernels and local the kernels [11]. The global kernels permit to learning points that are distant from each other to influence the kernel value. The polynomial kernel in Eq. (3) is an example of global kernel. By increasing the polynomial order we can achieve better interpolation ability. In contrast, by decreasing the polynomial kernel order we can get a better extrapolation. Concerning local kernels, the data that are close can have an influence on the kernel values. The RBF in Eq. (4) is a typical local kernel function. The smaller value of γ is, the worse its interpolation ability is, and vice versa.

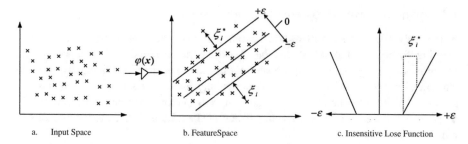

a.   Input Space   b. FeatureSpace   c. Insensitive Lose Function

Fig. 1. Mapping transformation of a SVR model, (a) nonlinear mapping, (b) support vectors points; (c) the ε-insensitive loss function where the slope is determined by C.

2.4   Linear Combination Methods

The linear combination of forecasts is a process that aims to enhance the forecasting precision for diverse time series. We propose here to use three linear strategies to identify numerical weights to combine the outcome of the support vector forecasts. Combining the forecasts yielded by different methods usually enhances the forecasting precision [12, 13]. In the linear combination of forecasts, a weight is associated to each single method, and the combined forecast could be expressed as the weighted sum of the forecasts separately yielded by the methods. The linear combination of K methods can be illustrated as follows. Let $Y = [y_1, y_2, \ldots, y_M]^T$ be the actual times series values whose forecasts are offered by n distinct models $\hat{Y}^j = [\hat{y}_1^j, \hat{y}_2^j, \ldots, \hat{y}_M^j]$ and $j = 1, 2, \ldots, n$. The forecasts of linear combination are given by:

$$\hat{y}_t = w_1\hat{y}_t^1 + w_2\hat{y}_t^2 + \ldots + w_n\hat{y}_t^n = \sum_{j=1}^{n} w_j\hat{y}_t^j \tag{5}$$

The combining weights $w_j (j = 1, 2, \ldots, n)$ are positive numerical values that designate the contribution part of each individual method in the combined forecasts and $\sum_{j=1}^{n} w_j = 1$ the structural design of the linear ensemble mechanism is illustrated in (e.g. Fig. 2). In the present work, we use three linear methods to combine the outcomes of the different forecasts.

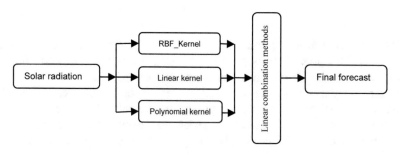

Fig. 2. Block diagram of the proposed system.

(1) Simple *Average*

Is very straightforward approach and easiest combination method that attribute equal weights to all constituent forecasts ($w_j = 1/n$). The *SA* method has proved to be robust in the forecasting of several series [14].

(2) *The Least Square Regression (LSR)*

This method assigns the weights of a linear combination by minimizing the Sum of Squared Error (*SSE*), derived from the measured and forecasted data. The typical formulation of the linear combination (7) can be expressed as:

$$\hat{Y} = UW \tag{6}$$

where,

$$U = \left[\hat{y}_1^{(1)} \hat{y}_1^{(2)} \ldots \hat{y}_1^{(n)}; \hat{y}_2^{(1)} \hat{y}_2^{(2)} \ldots \hat{y}_2^{(n)}; \ldots; \hat{y}_M^{(1)} \hat{y}_M^{(2)} \ldots \hat{y}_M^{(n)} \right]^T \tag{7}$$
$$W = [w\_1, w\_2, \ldots, w\_n]^T$$

Then the forecast SSE is expressed as:

$$\begin{aligned}
SSE &= \sum_{t=1}^{M} (y_t - \hat{y}_t)^2 \\
&= (Y - UW)^T (Y - UW) \\
&= Y^T Y - 2W^T U^T Y + W^T U^T UW
\end{aligned} \tag{8}$$

By minimizing SSE with regard to W, the weight vector is acquired as:

$$W = (U^T U)^{-1} U^T Y \tag{10}$$

(3) *The Error-Based Method*

In this method, the time series to be forecasted is subdivided into two paired subsets, the training and the validation sets. The constituent models are trained on the training set and their obtained forecast errors on the validation set are recorded. The weight of each individual model is then taken to be inversely proportional to the forecast error of the corresponding model [14].

$$w_j = e_j^{-1} \Big/ \sum_{j=1}^{n} e_j^{-1} \tag{11}$$

Were e_j indicates the error of the j forecasting model.

Four evaluation measures are used in this paper. which are: the Mean Absolute Percentage Error (*MAPE*), the Mean Squared Error (*MSE*), Mean Absolut Bias Error (*MABE*) and the Coefficient of Determination (R^2), which are defined as follow:

$$MAPE = \frac{1}{M} \sum_{t=1}^{M} \frac{|y_t - \hat{y}_t|}{|y_t|} \times 100 \qquad (12)$$

$$RMSE = \sqrt{\frac{1}{M} \sum_{t=1}^{M} (y_t - \hat{y}_t)^2} \qquad (13)$$

$$MABE = \frac{1}{M} \sum_{t=1}^{M} |y_t - \hat{y}_t| \qquad (14)$$

$$R^2 = 1 - \frac{var(y_t - \hat{y}_t)}{var(y_t)} \qquad (15)$$

Here, y_t and $\hat{y}$ are the actual and forecasted values respectively and M is the number of forecasted data points.

3 Results and Discussion

This section presents the experimental assessement, carried out to study the efficiency of the proposed linear combination methods. The six models established and evaluated for four sites by four performance indicators where used to compare the different proposed kernel models. These statistical indicators are most broadly used by researchers to assess the performance of solar radiation regression models. The available real data cover a period of 2 years between 2004 and 2005.

A soft computing program developed under Matlab has been used to train the SVM model. A set of data samples has been divided into three subsets: For the experiments, 50% (5000 data points) for the period 2004–2005 were used for training and the remaining 50% (2500 data points) are used for validation while (2500 data points) are used for testing. It should be noted that nighttime values are removed from the data sets used in this study.

3.1 Performance Analysis

In order to evaluate the performance of the deferent models, experimental work was carried out to determine the importance of each independent input variable for the output. Root-mean-square error (*RMSE*), coefficient of determination (R^2), mean absolute percentage error (*MAPE*) and Mean Absolut Bias Error (*MABE*) served to evaluate the differences between the predicted and actual values for both single and combined SVMs models. Tables 3–4 compares the single SVM–based models with that of different combinations.

3.2   Site Dependent Models

(1) *Alger*

For Alger, the best result was obtained for the *RBF_kernel* model with $R^2 = 0.9998$, $MAPE = 0.3734$, $MABE = 1.5546$ and $RMSE = 3.9160$. While the worst performance in the same site was obtained with the *Lin_kernel* model with a coefficient of determination $R^2 = 0.9942$. *LSR* obtained the best result in combination methods with $R^2 = 0.9998$. *MAPE, MABE* and *RMSE* are 0.3547, 1.4768 and 3.7808 respectively (see Table 3, e.g. Fig. 3).

(2) *Msila*

As shown in Table 4, e.g. Fig. 4, the best result was obtained with the *RBF_kernel* model with $R^2 = 0.9998$, $MAPE = 0.3462$, $MABE = 1.4215$, and RMSE = 1.4215. While the worst result in the same site was obtained from the *Lin_kernel* model with a coefficient of determination $R^2 = 0.9952$. LSR has got the best result in combination with $R^2 = 0.9999$ and *MAPE, MABE* and *RMSE* equal to 0.3159, 1.2970 and 3.4515 respectively.

Table 3. The obtained forecasting results of all methods for Alger and Msila site.

	SITE	RBF_kernel	Poly_kernel	Lin_kernel	SA	LSR	EB
RMSE	*Alger*	**3.9160**	4.4112	21.7593	8.1733	**3.7808**	4.2887
	Msila	**3.803**	3.8175	19.6098	7.5502	**3.4515**	3.7317
MAPE	*Alger*	**0.3734**	0.4248	4.0484	1.4334	0.3642	0.3936
	Msila	**0.3462**	0.4118	3.7860	5.5274	**0.3159**	0.3448
MABE	*Alger*	**1.5546**	1.7687	16.8555	5.9682	1.5165	1.5222
	Msila	**1.4215**	1.6907	15.5441	1.3463	**1.2970**	1.3979
R^2	*Alger*	**0.9998**	0.9998	0.9942	0.9992	**0.9998**	0.9998
	Msila	**0.9998**	0.9998	0.9952	0.9993	**0.9999**	0.9998

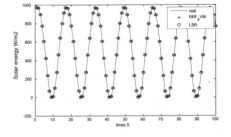 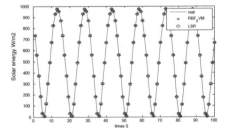

Fig. 3. Comparison between the measured and predicted global solar radiation in Alger.

Fig. 4. Comparison between the measured and predicted global solar radiation in Msila.

(3) *Batna*

As shown in (Table 4, e.g. Fig. 5), and for individual models, the highest R^2 calculated at Batna station is 0.9999 obtained with *RB_kernel* model, while the lowest value of R^2 acquired by *Lin_kernel* was 0.9957.

(4) *Bechar*

For Bechar site, and for individual models, the highest R^2 calculated is obtained as 0.9999 with *RBF_kernel* model, while the lowest value of R^2 acquired by *Lin_kernel* as 0.9960. (see Table 4, e.g. Fig. 6). We can also clearly observe that for each times series, the combination methods achieve the lowest errors and hence the best accuracies in terms error measures. So we see that the combination methods have an overall better forecasting accuracies compared to the individual methods.

Table 4. The obtained forecasting results of all methods for Batna site.

	SITE	RBF_kernel	Poly_kernel	Lin_kernel	SA	LSR	EB
RMSE	*Batna*	**3.5793**	3.8812	20.2637	7.9983	**3.4201**	3.4888
	Bechar	**3.0060**	3.5600	18.5022	6.8842	**2.9952**	3.0002
MAPE	*Batna*	**0.2621**	0.4039	3.5329	1.3365	**0.2503**	0.2505
	Bechar	**0.2288**	0.2924	3.2088	1.1266	**0.2751**	0.2190
MABE	*Batna*	**1.1440**	1.7629	15.9829	**5.8330**	**1.0850**	1.1297
	Bechar	**1.0924**	1.2682	13.9180	4.8865	**0.9931**	**1.0667**
R^2	*Batna*	**0.9999**	0.9998	0.9957	0.9993	**0.9999**	0.9999
	Bechar	**0.9999**	0.9999	0.9960	0.9994	**0.9999**	**0.9999**

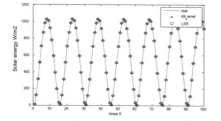

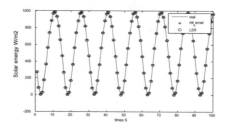

Fig. 5. Comparison between the measured and predicted global solar radiation in Batna

Fig. 6. Comparison between the measured and predicted global solar radiation in Bechar.

4 Conclusion

In this study, 6 different models classified into 2 categories (3 individual models and 3 combinations methods) were proposed. The results obtained in this study, with regard to individual models, demonstrate that the *RBF_Kernel* model (with coefficients of determination varied from 0.9763 to the highest one with 0.999 for the four stations) achieves the best performance. With respect to the whole models, the best performances for the other error criteria were obtained from the linear combination method based on LSR method. Consequently, the proposed combination strategies of the autoregressive models can be used for predicting the global solar radiation in Algerian locations with the lack of measured solar radiation data.

References

1. Ming, T., Liu, W., Caillol, S.: Fighting global warming by climate engineering: is the Earth radiation management and the solar radiation management any option for fighting climate change? Renew. Sustain. Energy Rev. **31**, 792–834 (2014)
2. Himri, Y., BoudgheneStambouli, A., Draoui, B., Himri, S.: Review of wind energy use in Algeria. Renew. Sustain. Energy Rev. **13**, 910–914 (2009)
3. Bouzgou, H.: A fast and accurate model for forecasting wind speed and solar radiation time series based on extreme learning machines and principal components analysis. J. Renew. Sustain. Energy **6**(1), 013114 (2014)
4. Chakhchoukh, Y., Panciatici, P., Mili, L.: Electric Load Forecasting Based on Statistical Robust Methods. IEEE Trans. Power Syst. **26**, 982–991 (2011)
5. Wu, G., Liu, Y., Wang, T.J.: Methods and strategy for modeling daily global solar radiation with measured meteorological data: a case study in Nanchang station, China. Energy convers. manage. **48**, 2447–2452 (2007)
6. Sen, Z.: Simple nonlinear solar irradiation estimation model. Renew. Energy **32**, 342–350 (2007)
7. Salmi, M., Bouzgou, H., Al-Douri, Y., Boursas, A.: Evaluation of the hourly global solar radiation on a horizontal plane for two sites in Algeria. Adv. Mater. Res. **925**, 641–645 (2014). Trans Tech Publications
8. Stambouli, A.B., Khiat, Z., Flazi, S., Kitamura, Y.: A review on the renewable energy development in Algeria: current perspective, energy scenario and sustainability issues. Renew. Sustain. Energy Rev. **16**(7), 4445–4460 (2012)
9. Vapnik, V.N., Vapnik, V.: Statistical Learning Theory, vol. 1. Wiley, New York (1998)
10. Smola, A.J., Schölkopf, B.: A tutorial on support vector regression. Stat. Comput. **14**(3), 199–222 (2004)
11. Scholköpf, B., Smola, A.J.: Learning with Kernels. MIT Press, Cambridge (2002)
12. Hibon, M., Evgeniou, T.: To combine or not to combine: selecting among forecasts and their combinations. Int. J. Forecast. **21**(1), 15–24 (2005)
13. Bouzgou, H., Benoudjit, N.: Multiple architecture system for wind speed prediction. Appl. Energy **88**(7), 2463–2471 (2011)
14. De Menezes, L.M., Bunn, D.W., Taylor, J.W.: Review of guidelines for the use of combined forecasts. Eur. J. Oper. Res. **120**(1), 190–204 (2000)

Fuzzy-Direct Power Control of a Grid Connected Photovoltaic System Associate with Shunt Active Power Filter

Sabir Ouchen[1(✉)], Achour Betka[1], Jean Paul Gaubert[2],
Sabrina Abdeddaim[1], and Farida Mazouz[3]

[1] Electrical Engineering Laboratory (LGEB),
University of Mohammed Khider Biskra, Biskra, Algeria
Ouchen_sabir@yahoo.fr, betkaachour@gmail.com,
s_abdeddaim@yahoo.fr
[2] Laboratory of Computer Science and Automatic Control for Systems
(LIAS-ENSIP), University of Poitiers, Poitiers, France
jean.paul.gaubert@univ-poitiers.fr
[3] Electrical Engineering Laboratory LEB, University of Batna 2, Batna, Algeria
fari_maz@yahoo.fr

Abstract. The present paper suggests a simulation study for a grid connected photovoltaic system, associated to a shunt active power filter (SAPF). On the one hand, to extract the maximum power from the PV generator a fuzzy logic Maximum power taking point (MPPT) is proposed in order to track permanently the optimum point. On the other hand, direct power control (DPC) is used to control the shunt active power filter in order to compensate the undesirable harmonic. The simulation of the system under the Matlab/Simulink™ environment demonstrates the robustness of the proposed controls that simultaneously guarantee harmonic current compensation, unit power factor operation and injection of solar power to the power grid.

Keywords: PV grid · Fuzzy logic MPPT · Shunt active power filter (SAPF)
Direct power control (DPC)

1 Introduction

The progress of energy request in the world is progressively increasing and new forms of energy sources must be developed and discovered in order to cover the future demands. Recently, the grid-connected PV systems are extensively used to support conventional power generation and can be proposed a solution to the depletion of fossil energy resources. The grid-connected PV system injects the active power to the electrical grid by using maximum power point tracking algorithms to extract the maximum power from the PV system.

Numerous maximum power point tracking (MPPT) methods are proposed in literature to reach the MPP point such as hill climbing (HC) [1], Incremental Conductance (IC) and perturb and observe P&O [2]. The principle of these methods consists on perturbing the system with a constant voltage/duty cycle, and checking its behavior. They are the three most used methods, because they have the advantage of easy to

© Springer International Publishing AG 2018
M. Hatti (ed.), *Artificial Intelligence in Renewable Energetic Systems*, Lecture Notes
in Networks and Systems 35, https://doi.org/10.1007/978-3-319-73192-6_17

implement. These algorithms suffer from a notable incompatibility under quick irradiance variations. Although intelligent MPPT algorithms are suggested to be the alternatives, in term of rapidity and accuracy such as: fuzzy logic, [3], neural network, particle swarm optimization (PSO) [4]. On the other hand, due to the growth of development in the fields of power electronics, intensive use of non-linear loads caused serious disturbances, such as harmonics, unbalanced currents, etc., injected into the network electric. Harmonics reduce efficiency and power factor, increasing losses and causing electromagnetic interference with neighboring communication lines and other adverse consequences. Improving the quality of energy has become a major research topic in the modern electricity distribution system. Nearly 20 years ago, most of the loads used by industries and consumers were passive and linear, with fewer non-linear loads having less impact on the feed system [5]. With the arrival of semiconductors and power electronics devices and their ease of control have caused the wide use of non-linear loads, such as chopper, inverter and rectifier. The use of these devices is the major cause of harmonic and reactive power disturbances. These induce overheating of the transformers, distortion of the supply voltage, the low power factor and the malfunction of the sensitive equipment [6, 7].

To reduce the impact of these harmonics on the power system, the active power filters are installed at the common connection point PCC. The SAPF injects the compensation current to the PCC to cancel the harmonics and to make the sinusoidal source current. By installing the SAPF, the harmonic pollution as well as the low power factor in the supply system can be improved [8]. Direct power control (DPC) is the most control strategy used for the control of SAPF. It was first proposed by Ohnishi [9] for PWM rectifier. The DPC control comes from the direct torque control (DTC) proposed by Takahashi [10] used in the control of electrical machines. The principle of the DPC control is based on the calculation of the active and reactive powers through the current and input voltage measurements and instantly performs the power control using a switching table and tow hysteresis comparators [11].

The present study is focused in two sections:

1. The extraction of the maximum PV-power through a fuzzy logic MPPT algorithm.
2. Direct power control for the SAPF in order to eliminate the undesirable harmonics on the grid side.

The rest of the paper is planned as follows: In Sect. 2, explicit description of the system is specified, while in Sect. 3, the various proposed control techniques are proposed. To test the effectiveness of these approaches, Sect. 4 displays and comments the obtained results. Finally, Sect. 5 give to this study.

2 Système Description

The proposed control scheme of the studied system is illustrated in Fig. 1. The first conversion chain stage is composed of a PV system connected with a boost converter. The main role of this DC/DC converter is to make an impedance matching between the PV system and the grid, so that the PV system offers progressively its maximum power, by using a fuzzy logic MPPT algorithm. The second stage is comprised of three phase two level inverter, connected to the grid via an inductive filter. This DC/AC converter acts as a shunt active filter, to eliminate the harmonics caused by the nonlinear load at the AC main, and the reactive power also to properly ensure a power distribution of the load request under various irradiance levels between the PV array and the grid, a direct power control (DPC) approach is proposed as a control strategy of the DC/AC converter.

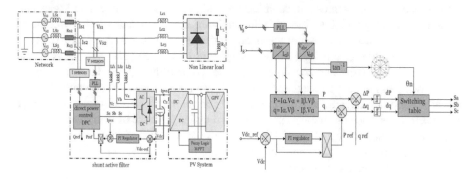

Fig. 1. Block diagram of the configuration studied   **Fig. 2.** Block scheme of the principle of DPC

3   Control Approaches

3.1   Direct Power Control Strategy (DPC)

Direct power control strategy has been proposed by Noguchi. It uses a switching table to determine the switching states there is no PWM block and internal current control loop. The instantaneous active and reactive powers are compared to a reference value. The reference active power is calculated from the output of the DC bus voltage regulator Vdc and the reference of the reactive power is set to zero to ensure a unit power factor as shown in Fig. 2 [12].

It is known that the calculation of the active power P is a scalar product of the voltages and the currents grid, although the reactive power q is calculated by a vector product between the voltages and the currents grid. Active and reactive power are given by the following expressions:

$$\begin{cases} p = i_\alpha.v_\alpha + i_\beta.v_\beta \\ q = i_\alpha.v_\beta - i_\beta.v_\alpha \end{cases} \tag{1}$$

The active power error ΔP and reactive power error Δq are assumed by the following expression

$$\Delta P = Pref - P \tag{2}$$

$$\Delta q = qref - q \tag{3}$$

The phase of the voltage vector of the grid is converted into a digitized signal θn (angular position). The calculation of this position requires knowledge of the components $v\alpha$ and $v\beta$, which can be calculated from the transformation of the voltages sources from the three-phase plane abc to the stationary plane α–β:

$$\theta_n = \arctan\left(\frac{v_\alpha}{v_\beta}\right), \ (n-2).\frac{\pi}{6} \le (\theta_n - 1) \le (n-1).\frac{\pi}{6}, n = 1, 2, 3, \ldots, 12 \quad (4)$$

Depending on the instant errors in active and reactive power Δp, Δq, and the phase angle θn position. The switching state Sa, Sb, and Sc of the switches are selected through twelve sectors switching table, which may be considered the heart of the direct power control, as mentioned in Table 1:

Table 1. DPC switching table

dp	dq	$\theta1$	$\theta2$	$\theta3$	$\theta4$	$\theta5$	$\theta6$	$\theta7$	$\theta8$	$\theta9$	$\theta10$	$\theta11$	$\theta12$
1	0	v6	v7	v1	v0	v2	v7	v3	v0	v4	v7	v5	v0
	1	v7	v7	v0	v0	v7	v7	v0	v0	v7	v7	v0	v0
0	0	v6	v1	v1	v2	v2	v3	v3	v4	v4	v5	v5	v6
	1	v1	v2	v2	v3	v3	v4	v4	v5	v5	v6	v6	v1

v1(100), v2(110), v3(010), v4(011), v5(001), v6(101), v0(000), v7(111)

3.2 DC Bus Regulator

To reduce the instability and variations in the DC bus voltage, a proportional-integral (PI) controller with anti-windup compensation is proposed for DC bus voltage regulation as shown in Fig. 3 [3].

From the simplified diagram in Fig. 4, the transfer function of the closed loop system is:

$$\frac{V_{dc}}{V_{dc-ref}} = \frac{K_p.s + K_i}{C.s^2 + K_p.s + K_i} = \frac{K_p/C(K_i/K_p)}{s^2 + K_p/C.s + K_i/C} \quad (5)$$

From the equation, the coefficients Kp and Ki are identified by the following relations:

$$K_p = 2.\xi.\omega_n.C \quad (6)$$

$$K_i = C.\omega_n^2 \quad (7)$$

Therefore, the reference of active power quantity is deduced as:

$$P_{ref} = V_{dc} * I_{sm} \quad (8)$$

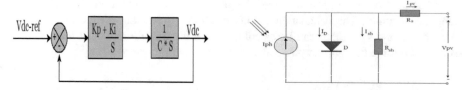

Fig. 3. DC link voltage control   **Fig. 4.** Model of a photovoltaic panel.

3.3 Photovoltaic Cell Modeling

The equivalent circuit of a photovoltaic solar cell (model with a diode) is illustrated in the Fig. 5 below, it comprises a current source Iph, which models the photocurrent associated with a parallel diode which models the P-N junction, two resistors Rsh and Rs which model the defects of the cell. The operation of the photovoltaic cell can be formalized by a system of equations resulting from Kirchhoff's laws.

$$I_{PV} = I_{PH} - I_0\left[exp(q\frac{(V_{PV} + R_S I_{PV})}{nKT}) - 1\right] + \frac{V_{PV} + R_S I_{PV}}{R_{SH}} \tag{9}$$

The operation of the photovoltaic cell can be formalized by a system of equations resulting from Kirchhoff's laws.

$$I_{PV} = I_{PH} - I_0\left[exp(q\frac{(V_{PV} + R_S I_{PV})}{nKT}) - 1\right] + \frac{V_{PV} + R_S I_{PV}}{R_{SH}} \tag{10}$$

3.4 Fuzzy Logic MPPT Controller Design

Recently, the control strategies based on fuzzy logic have been utilized in the tracking systems of the maximum power point, this command give the advantage of a robust control and that require no precise mathematical model of the system. The operation of this algorithm is based on three steps: the fuzzification, inference and defuzzification as illustrated in Fig. 5 [3, 13].

Fuzzification, that allows the conversion of the input physical variables in fuzzy sets. In our case, we have two inputs the error "e" and the variation of the error "Δe" defined as follows:

$$e = \frac{P_{pv}(n) - P_{pv}(n-1)}{V_{pv}(n) - V_{pv}(n-1)} \tag{11}$$

$$\Delta e = e(n) - e(n-1) \tag{12}$$

In the inference step, we establish logical relationships between input and output while defining the rules of belonging. Thereafter, the table is erected inference rules (Table 2).

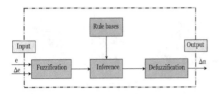

Fig. 5. Principal of fuzzy logic MPPT controller

Table 2. Rule base for the FLC controller

$e \ \Delta e$	NB	NM	NS	ZE	PS	PM	PB
NB	NB	NB	NB	NB	NM	NS	ZE
NM	NB	NB	NB	NM	NS	ZE	PS
NS	NB	NB	NM	NS	ZE	PS	PM
ZE	NB	NM	NS	ZE	PS	PM	PB
PS	NM	NS	ZE	PS	PM	PB	PB
PM	NS	ZE	PS	PM	PB	PB	PB
PB	ZE	PS	PM	PB	PB	PB	PB

Finally, defuzzification, which convert fuzzy output member ship into a digital value and computes the incremental duty cycle, using the center of gravity method.

$$\Delta\alpha = \left(\sum_{k=1}^{m} c(k) * W_k \right) / \left(\sum_{k=1}^{n} W_k \right) \tag{13}$$

4 Simulation Results

The simulations of the photovoltaic (PV) system connected to the network associate with a shunt active power filter controlled by the fuzzy logic and direct power control were extracted under the MATLAB/SimulinkTM environment.

To test the robustness of the control and the maximum power point tracking algorithm, an irradiance profile is established to scan all the operating modes of our system as show in Fig. 6.

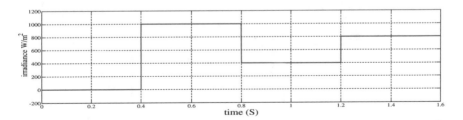

Fig. 6. Irradiance profile

The present study is separated into two modes: the first mode contains just the operation of the active filter APF between t = 0.1 s and t = 0.4 s, before the second mode contains all of the active filter APF associated with the PV-grid system between 0.4 s and 1.6 s.

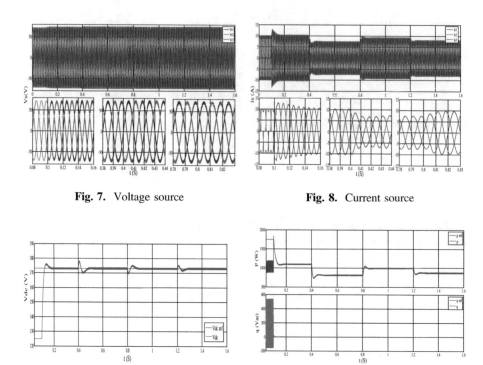

Fig. 7. Voltage source   **Fig. 8.** Current source

Fig. 9. DC link voltage   **Fig. 10.** Active and reactive power

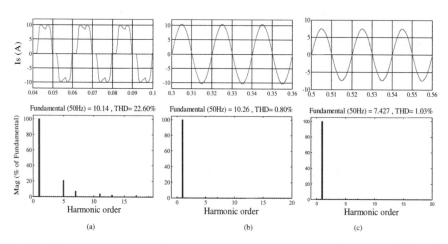

Fundamental (50Hz) = 10.14 , THD= 22.60%   Fundamental (50Hz) = 10.26 , THD= 0.80%   Fundamental (50Hz) = 7.427 , THD= 1.03%

(a)   (b)   (c)

Fig. 11. Grid current and its spectrum before (a), after (b) filtering and after introducing PV at 1000 W/m2

First mode t[0.1–0.4]

After the launch of SAPF at t = 0.1 s, the current source becomes purely sinusoidal with a total harmonic distortion which varies from THDi = 22.60% to THDi = 0.80% (Fig. 11a, b). With a transient of ΔT = 40 ms before reaching the steady state. The voltage at the DC bus joins its fixed reference Vdc_ref = 173 V after a time Δt = 60 ms, which gives a good dynamic to the power control (Fig. 9). The active power is constant and closely follows its reference value. The reactive power is zero on average, thus ensuring a unit power factor operation (Fig. 10).

Second mode t[0.4–1.6]

Figure 7, illustrate the voltage source, which keeps its form and amplitude during all the changes of solar irradiance.

Figure 8, present the evolution of the source current during the change of solar irradiance. the source current is quasi sinusoidal with changes in amplitude according to each variation of solar irradiance.

Figure 9, show that Vdc follows impeccably its reference (173 V). add to that, a transition state is observed in t = 0.8 and t = 1.2 resulting to the change in solar irradiance.

In Fig. 10, active and reactive powers track the trajectories imposed by the control system. Between 0 s and 0.4 s the voltage source provides all the power to the load (when the irradiance is null). After that from 0.4 s to 1.6 s the PV system and the voltage source inject the power to the non-linear load at the same time. Through all these tests the reactive power is zero, which insure a unit power factor operation.

From the Fig. 11, the total harmonic distortion (THD) is about 0.80%, meaning compliance with IEEE 519-1992 norm. Additionally, it is remarked, that the adding of the PV system increases slightly the THD factor (1.3%) but it is kept within the tolerable range (Fig. 11c).

5 Conclusion

In this paper, a simulation study on a photovoltaic system connected to the grid associated with an active parallel filter is achieved. The results obtained confirm the feasibility of the proposed configuration and validates various features affected to the system controlled by fuzzy logic controls for the control of MPPT and the DPC for the active filter. The control strategies proved a permanent flowing of the maximum PV power amounts, while the P-DPC strategy allowed to get a suitable injection of this power into the grid and a compensation of undesirable harmonic current, under an acceptable THD and unite power factor.

References

1. Ika, R., Wibowo, S., Rifa, M.: Maximum power point tracking for photovoltaic using incremental conductance method. Energ. Procedia **68**, 22–30 (2015)
2. Manickam, C., Raman, G.P., Raman, G.R., Ganesan, S.I., Chilakapati, N.: Fireworks enriched P&O algorithm for GMPPT and detection of partial shading in PV systems. IEEE Trans. Power Electron. **8993**(c), 1 (2016)
3. Ouchen, S., Betka, A., Abdeddaim, S., Menadi, A.: Fuzzy-predictive direct power control implementation of a grid connected photovoltaic system, associated with an active power filter. Energ. Convers. Manag. **122**, 515–525 (2016)
4. Durand, F.R., Bacon, V.D., Oliveira da Silva, S.A., Sampaio, L.P., de Oliveira, F.M., Campanhol, L.B.G.: Grid-tied photovoltaic system based on PSO MPPT technique with active power line conditioning. IET Power Electron. **9**(6), 1180–1191 (2016)
5. Ouchen, S., Betka, A., Gaubert, J., Abdeddaim, S.: Simulation modelling practice and theory simulation and real time implementation of predictive direct power control for three phase shunt active power filter using robust phase-locked loop. Simul. Model. Pract. Theor. **78**, 1–17 (2017)
6. Ouchen, S., Abdeddaim, S., Betka, A., Menadi, A.: Experimental validation of sliding mode-predictive direct power control of a grid connected photovoltaic system, feeding a nonlinear load. Sol. Energ. **137**, 328–336 (2016)
7. Afghoul, H., Krim, F., Chikouche, D., Beddar, A.: Design and real time implementation of fuzzy switched controller for single phase active power filter. ISA Trans. **58**, 614–621 (2015)
8. Tareen, W.U., Mekhilef, S., Seyedmahmoudian, M., Horan, B.: Active power filter (APF) for mitigation of power quality issues in grid integration of wind and photovoltaic energy conversion system. Renew. Sustain. Energ. Rev. **70**, 635–655 (2017). November 2016
9. Ohnishi, T.: Three phase PWM converter/inverter by means of instantaneous active and reactive power control. In: Proceedings of 1991 International Conference on Industrial Electronics, Control and Instrumentation, IECON 1991, pp. 819–824 (1991)
10. Takahashi, I., Noguchi, T.: A new quick-response and high-efficiency control strategy of an induction motor. Ind. Appl. IEEE Trans. **IA-22**(5), 820–827 (1986)
11. Ouchen, S., Betka, A., Gaubert, J.P., Abdeddaim, S.: Simulation and practical implementation of direct power control applied on PWM rectifier. In: 6th International Conference on Systems and Control. University of Batna 2, Batna (2017)
12. Ouchen, S., Betka, A., Abdeddaim, S., Mechouma, R.: Design and experimental validation study on direct power control applied on active power filter. In: 2nd International Conference on Intelligent Energy Power Systems, pp. 1–5, June 2016
13. Rezk, H., Eltamaly, A.M.: A comprehensive comparison of different MPPT techniques for photovoltaic systems. Sol. Energ. **112**, 1–11 (2015)

Fuzzy Control of a Wind System Based on the DFIG

Farida Mazouz[1(✉)], Sebti Belkacem[1], Sabir Ouchen[2],
Youcef Harbouche[1], and Rachid Abdessemed[1]

[1] Electrical Engineering Department Laboratory LEB,
University of Batna 2, Batna, Algeria
Fari_maz@yahoo.fr, belkacem_sebti@yahoo.fr,
harbouch_youcef@yahoo.fr, rachid.abdessemed@gmail.com
[2] Electrical Engineering Department Laboratory LGEB,
University of Biskra, Biskra, Algeria
ouchen_sabir@yahoo.fr

Abstract. This paper proposes the modeling and control of Wind Energy Conversion Systems (WECS) based on Doubly Fed Induction Generator (DFIG). The fuzzy logic control is used to improve the extracted wind power at given wind velocity; the mechanical power available from a wind turbine is a function of its shafts speed. Then, the rotor side converter (RSC) is controlled in the aim to follow the optimal torque for given maximal wind power. The effectiveness of the proposed control strategy is validated by theoretical analysis and simulation carried out in Matlab/Simulink environment.

Keywords: Wind energy conversion system · WECS · DFIG
Fuzzy logic control · FLC

1 Introduction

Wind energy conversion systems (WECS). Among various other techniques of wind power generation, the doubly fed induction generator (DFIG) has been popular because of its higher energy transfer capability, low investment and flexible control [1–3]. Double-fed induction generator extracts has two main parts stator winding which is directly connected to the grid where as rotor winding is connected to the grid via coupling transformer and PWM converter to operate at variable frequencies. This helps the doubly fed induction generator to operate in both below and above synchronous speeds. The DFIG wind turbine is used in varying speeds and variable pitch control application in a finite range around the synchronous speed, for example ±30% due to decrease in the power rating of the frequency converter. Now a day's DFIG wind turbine are the common variable speed wind turbines [4, 5]. Also, it has the capability of capturing more energy and less noise compared to other machines. It has better capability to control active and reactive power for grid integration. There are various control algorithms and method for controlling of power conversion. The prominent conventional control method for DFIG is vector control technique uses rotor currents which are decoupled into stator active power. The DFIG control strategy is based on

© Springer International Publishing AG 2018
M. Hatti (ed.), *Artificial Intelligence in Renewable Energetic Systems*, Lecture Notes in Networks and Systems 35, https://doi.org/10.1007/978-3-319-73192-6_18

conventional Proportional Integral (PI) technique which is well accepted in the industry, [6]. The decoupled control of DFIG has following controllers namely reference power (Pref), DC reference voltage (Vdcref) and reference power qcref. However, the intelligent controllers like fuzzy and neural network controllers, capturing the system operators' experience, outperform the conventional PI controllers, [6]. The fuzzy logic approach provides the design of a non-linear, model free controller and hence, can be used for the coordinated control of RSC and GSC in the DFIG system.

Using fuzzy control, we can use linguistic variables and rule base or fuzzy sets are easily altered to produce controller outputs more predictable. It allows for accelerated prototyping because the system designer doesn't need to know everything about the system before starting. The fuzzy logic control has the capability that it can operate accurately without using an exact mathematical model of the system. The operation of the system can be varied by using the observation of the system operation and performance. The adjustment of parameters is varied very comfortable [4].

In this paper, a fuzzy control algorithm is proposed for power regulation of a wind energy conversion system (WECS) based on a doubly fed induction generator (DFIG). We started by presenting the modeling of both the wind turbine and the generator. Then technique fuzzy logic control (FLC) the power generator system will be presented. Finally, simulation results concerning performances of the conversion system will be presented and discussed.

2   System Modeling

The system consists of an aero turbine, which converts wind energy into mechanical energy, a gearbox, which serves to increase the speed and decrease the torque and a generator to convert mechanical energy into electrical energy, [7] (Fig. 1).

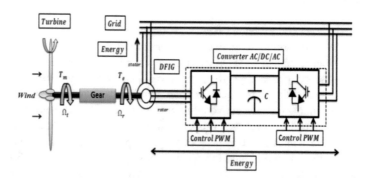

Fig. 1. Wind energy conversion system's structure.

2.1   Turbine Model

The aerodynamic power developed by a wind turbine is given by the following expression [2, 8–10]:

$$P_{aer} = C_p.P_v = \frac{1}{2}C_p(\beta, \lambda).\rho.S.v^3 \tag{1}$$

$$\lambda = \frac{R.\Omega_{turbine}}{v} \tag{2}$$

Where:

ρ is the air density; v is the wind speed; C_p is the power coefficient; β is the blade pitch angle; λ is the tip-speed ratio; R is rotor radius and $\Omega_{turbine}$ is the turbine rotor speed.

The power coefficient Cp defines the aerodynamic efficiency of the wind turbine rotor. It is represented by various approximation expressions. In this paper, Cp is expressed by (3), [2, 11]:

$$C_p(\lambda, \beta) = (0.5 - 0.0167.(\beta - 2)). \sin\left[\frac{\pi.(\lambda + 0.1)}{18.5 - 0.3.(\beta - 2)}\right]$$
$$- 0.00184.(\lambda - 3).(\beta - 2) \tag{3}$$

The shaft mechanical equation including both the turbine and the generator masses is given by:

$$J_t.\frac{d\Omega}{dt} = T_g - T_{em} - f_t.\Omega \tag{4}$$

Where J_t and ft are totals moment of inertia and viscous friction coefficient appearing at the generator side, T_g is the gearbox torque, T_{em} is the generator torque and Ω is the mechanical generator speed.

2.2 DFIG Model

The DFIG model is a multivariable,nonlinear and strong coupling system under three phase stationary coordinate framework. In order to adjust active and reactive power independently, [10], the generator model under the two phase rotating coordinate is deduced [7, 12–14].

The voltages equations and the flux equations are given by:

$$\begin{cases} V_{ds} = R_s I_{ds} + \frac{d\varphi_{ds}}{dt} - \omega_s \varphi_{qs} \\ V_{qs} = R_s I_{qs} + \omega_s \varphi_{ds} + \omega_s \varphi_{qs} \\ V_{dr} = R_r I_{dr} + \frac{d\varphi_{dr}}{dt} - (\omega_s - \omega)\varphi_{qr} \\ V_{qr} = R_r I_{qr} + \frac{d\varphi_{qr}}{dt} + (\omega_s - \omega)\varphi_{dr} \end{cases} \tag{5}$$

$$\begin{cases} \varphi_{ds} = L_s I_{ds} + M I_{dr} \\ \varphi_{qs} = L_s I_{qs} + M I_{qr} \\ \varphi_{dr} = L_r I_{dr} + M I_{dr} \\ \varphi_{qr} = L_r I_{qr} + M I_{qr} \end{cases} \tag{6}$$

The arrangement of the Eqs. (5) and (6) gives the expression of the rotor voltages according to the rotor currents by:

$$
\begin{aligned}
V_{dr} &= R_r I_{dr} + \left(L_r - \frac{M^2}{L_s}\right)\frac{dI_{dr}}{dt} - g\omega_s\left(L_r - \frac{M^2}{L_s}\right)I_{qr} \\
V_{qr} &= R_r I_{qr} + \left(L_r - \frac{M^2}{L_s}\right)\frac{dI_{qr}}{dt} + g\omega_s\left(L_r - \frac{M^2}{L_s}\right)I_{dr} + g\frac{MV_s}{L_s}
\end{aligned}
\tag{7}
$$

The torque equation and the supplied active and reactive power is defined as follows:

$$
C_{em} = p\frac{M}{L_s}\left(I_{qr}\varphi_{ds} - I_{dr}\varphi_{qs}\right)
\tag{8}
$$

$$
\begin{aligned}
P_s &= V_{ds}I_{ds} + V_{qs}I_{qs} \\
Q_s &= V_{qs}I_{ds} - V_{ds}I_{qs}
\end{aligned}
\tag{9}
$$

Adopting the assumption of a negligible stator resistance Rs and the stator flux is constant and oriented along the axis, we deduce: $\varphi_{ds} = \varphi_s$ and $\varphi_{qs} = 0$

$$
\begin{cases}
V_{ds} = \frac{d\varphi_s}{dt} = 0 \\
V_{qs} = V_s = \omega_s\varphi_s
\end{cases}
\tag{10}
$$

The stator active and reactive power can be expressed as the rotor currents as follows:

$$
\begin{aligned}
P_s &= -V_s\frac{M}{L_s}I_{qr} \\
Q_s &= \frac{V_s^2}{L_s\omega_s} - V_s\frac{M}{L_s}I_{dr}
\end{aligned}
\tag{11}
$$

From the Presidents equations we can construct the scheme of DFIG illustrated in Fig. 2:

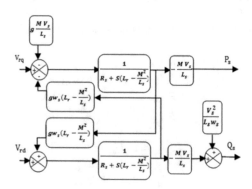

Fig. 2. DFIG block diagram representation.

3   Control Strategy Fuzzy Logic of the WECS

In this work the developing a novel control strategy using FLC. The benefits of FLC over the conventional controller are that FLC even works without a perfect mathematical model, [15, 16]. To achieve a stator active and reactive power vector control, d_q reference frame synchronized with the stator flux has been chosen [7]. By setting the stator flux vector aligned with d axis (Eq. 10). To regulate the DFIG wind turbine, fuzzy logic controllers are used because of the nonlinearity of the system, [1, 7]. The fuzzy controller includes three parts: fuzzification, fuzzy rule base and defuzzification. FLC is one of the most successful operations of fuzzy set theory. Its chief aspects are the exploitation of linguistic variables rather than numerical variables. FL control technique relies on human potential to figure out the systems behavior and is constructed on quality control rules. FL affords a simple way to arrive at a definite conclusion based upon blurred, ambiguous, imprecise, noisy, or missing input data, [15, 16]. The basic structure of an FLC is represented in Fig. 3.

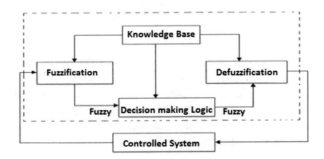

Fig. 3. Basic structure of fuzzy logic controller.

A Fuzzification interface alters input data into suitable linguistic values, [15, 16].

- A Knowledge Base which comprises of a data base along with the essential linguistic definitions and control rule set.
- A Decision Making Logic which collects the fuzzy control action from the information of the control rules and the linguistic variable descriptions.
- A Defuzzification interface which surrenders a non fuzzy control action from an inferred fuzzy control action.

There are two input signals to the FLC, the speeds error e and the change of the error ce. The error and change of error are converted by gain factors. To obtain the output of the FLC, the defuzzification used is based on the center of gravity method, and the triangular membership with overlap is used for the inputs and for the output of each FLC. G1, G2 and G3 are the gain factors. Before you begin to format your paper, first write and save the content as a separate text file. Keep your text and graphic files separate, [15, 16] (Fig. 4 and Table 1).

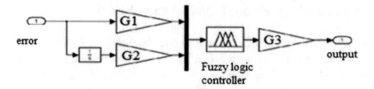

Fig. 4. Structure of the fuzzy logic controller

Table 1. Table of decision rules for the power controller.

e	de				
	NB	NS	Z	PS	PB
NB	PB	PS	NS	NS	NB
NS	PS	PS	NS	PB	NB
Z	NB	NB	NS	PS	PB
PS	NS	NS	PB	NB	PS
PB	NS	NS	PB	PB	PB

Figure 5 shows the overall vector control scheme of the RSC, in which the independent control of the rotor active power Ps and reactive power Qs is achieved by means of rotor current regulation. The control of RSC consists of regulating the stator active power Ps and reactive power Qs independently. In this paper, we assume that both frequency and voltage of grid are constant, and then in our control system, there is one control loop as shown in Fig. 5. Fuzzy logic current controllers are used to generate the reference signals of the direct and quadratic axis current components.

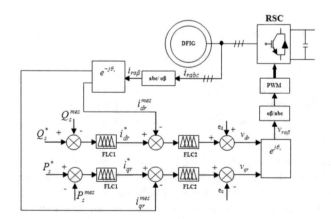

Fig. 5. Overall vector control of the RSC with fuzzy logic controller.

4 Simulation Results

The following figures show the simulation results of the fuzzy logic control applied to the WECS. The reference in reactive power is adjusted to zero so that the power factor to the stator is kept unitary, Thereafter, We have given different values of reactive power a t = 0.3 s and 0.7 s, in order to study the effectiveness of the control technique applied to the wind system. The reference active power is generated by the wind turbine.

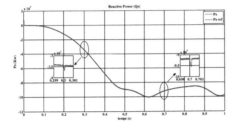

Fig. 6. Active power "Ps".

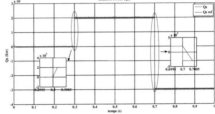

Fig. 7. Reactive power "Qs".

The study of the principles of the structure of the DPC control in the generating case has been developed from ideal operating conditions where neglecting the effect of stator resistance. In addition, the switching frequency is variable and difficult to control. The study of the principles of the structure of the DPC control in the generating case has been developed from ideal operating. The figures (Figs. 6 and 7), Showing the active and reactive powers, the two powers are well adapted to the variations in applied reference, At t = 1.3 s and t = 0.7 s, we notice the influence of the sudden change of the active power on the stator and rotor currents (Figs. 8 and 9), the two currents are well adapted to the variations of the active power with sinusoidal shapes and almost without harmonics. The variation of the reactive power directly influences on the active power with peaks which are seen in Fig. 6. From the results obtained, it can be seen that the results are well followed with the variation of the references, even on a relatively short time scale.

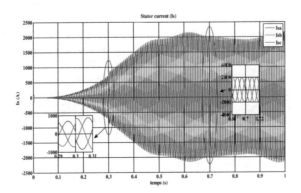

Fig. 8. Stator current "Is".

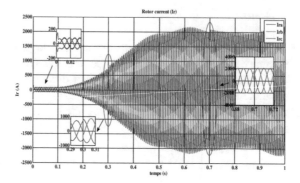

Fig. 9. Rotor current "Ir".

5   Conclusion

The global control scheme of the WECS based on the DFIG is studied in this article. To obtain the maximum wind energy generation, a new strategy using a fuzzy logic controller (FLC) is presented. This strategy has many advantages. The simulation in the Matlab/Simulink environment shows that the results are well adapted to the variations applied. The results of the simulation show that the entire control strategy gives good performance.

References

1. Belmokhtar, K., Doumbia, M.L., Agbossou, K.: Modelling and fuzzy logic control of DFIG based wind energy conversion systems. In: IEEE International Symposium on Industrial Electronics, pp. 1888–1893 (2012)
2. Boulâam, K., Boukhelifa, A.: Fuzzy sliding mode control of DFIG power for a wind conversion system. In: 16th International Power Electronics and Motion Control, Antalya, Turkey, 21–24 September 2014, pp. 353–358 (2014)
3. Boulâam, K., Boukhelifa, A.: A fuzzy sliding mode control for DFIG-based wind turbine power maximisation. In: 7th IET International Conference on Power Electronics, Machines and Drives (PEMD 2014), pp. 1–6 (2014)
4. Lalitha, M.P., Janardhan, T., Mohan, R.M.: Fuzzy logic based wind energy conversion system with solid oxide fuel cell. In: International Conference on High Performance Computing and Applications (ICHPCA), pp. 1–6 (2014)
5. Yunus, A.M.S., Abu-Siada, A., Masoum, M.A.S.: Improving dynamic performance of wind energy conversion systems using fuzzy-based hysteresis current-controlled superconducting magnetic energy storage. IET Power Electron. **5**(8), 1305–1314 (2012)
6. Mishra, S., Mishra, Y., Li, F., Dong, Z.Y.: TS-fuzzy controlled DFIG based wind energy conversion systems. In: IEEE Power Energy Society General Meeting, pp. 1–7 (2009)
7. Hamane, B., Doumbia, M.L., Bouhamida, M., Benghanem, M.: Control of wind turbine based on DFIG using Fuzzy-PI and sliding mode controllers. In: Ninth International Conference on Ecological Vehicles and Renewable Energies (EVER), pp. 1–8 (2014)

8. Zhang, S., Mishra, Y., Shahidehpour, M.: Fuzzy-logic based frequency controller for wind farms augmented with energy storage systems. IEEE Trans. Power Syst. **31**(2), 1595–1603 (2016)
9. Ashouri-Zadeh, A., Toulabi, M., Ranjbar, A.M.: Coordinated design of fuzzy-based speed controller and auxiliary controllers in a variable speed wind turbine to enhance frequency control. IET Renew. Power Gener. **10**(9), 1298–1308 (2016)
10. Liu, X., Han, Y.: Sliding mode control for DFIG-based wind energy conversion optimization with switching gain adjustment. In: Proceeding of the 11th World Congress on Intelligent Control and Automation, Shenyang, China, 29 June–4 July, 2014, pp. 1213–1218 (2014)
11. Mazouz, F., Belkacem, S., Harbouche, Y., Abdessemed, R., Ouchen, S.: Active and reactive power control of a DFIG for variable speed wind energy conversion. In: 6th International Conference on Systems and Control (ICSC 2017), Batna, Algeria (2017)
12. de Santana, M.P., de A.M. José Roberto, B., de P. Gey Versori, T., de A. Thaies, E.P., de A. P. William, C., Carlos, O.: Fuzzy logic for stator current harmonic control in doubly fed induction generator. In: 40th Annual Conference of the IEEE Industrial Electronics Society (IECON), pp. 2109–2115 (2014)
13. Vrionis, T.D., Koutiva, X.I., Vovos, N.A.: A genetic algorithm-based low voltage ride-through control strategy for grid connected doubly fed induction wind generators. IEEE Trans. Power Syst. **29**(3), 1325–1334 (2014)
14. Duong, M.Q., Grimaccia, F., Leva, S., Mussetta, M., Zich, R.: Improving LVRT characteristics in variable-speed wind power generation by means of fuzzy logic. In: 2014 IEEE International Conference on Fuzzy Systems (FUZZ-IEEE), 6–11 July, Beijing, China, pp. 332–337 (2014)
15. Bhavani, R., Prabha, N.R., Kanmani, C.: Fuzzy controlled UPQC for power quality enhancement in a DFIG based grid connected wind power system. In: 2015 International Conference on Circuits, Power and Computing Technologies (ICCPCT 2015), pp. 1–7 (2015)
16. Nam, T., Pardo, T.A.: Conceptualizing smart city with dimensions of technology, people, and institutions. In: Proceedings of the 12th Annual International Digital Government Research Conference: Digital Government Innovation in Challenging Times, pp. 282–291 (2011)

Intelligent Maximum Power Point Tracking

MPPT Technique for Standalone Hybrid PV-Wind Using Fuzzy Controller

Saidi Ahmed[✉], Cherif Benoudjafer, and Chellali Benachaiba

Department of Electrical Engineering, Faculty of Technology,
Tahri Mohammed University, Route de Kenadsa, BP 417, 08000 Béchar, Algeria
ahmedsaidi@outlook.com,
benoudjafer.cherif@mail.univ-bechar.dz,
chellali99@yahoo.fr

Abstract. This paper present an intelligent technique extraction of Maximum Power Point Tracking (MPPT) by using fuzzy logic from a standalone hybrid generation power system comprising a permanent magnet synchronous generator branch (PMSG) based wind turbine and photovoltaic generator branch compared to the conventional hill climb search (HCS) and Perturb and Observe (P&O) algorithms respectively. The fuzzy controller for solar wind MPPT scheme shows a high precision in current transition and keeps the voltage without any changes, in variable-load case, represented in small steady state error and small overshoot. The proposed scheme ensures optimal use of the photovoltaic (PV) array and PMSG wind proves its efficacy in variable load conditions, unity and lagging power factor at the inverter output (load) side.

Keywords: PV · PMSG wind · MPPT · FLC · P&O · HCS

1 Introduction

The demand for electricity is increasing exponentially because of the industrially revolution, which cannot be fulfilled by non-renewable energy sources alone. Renewable energy sources such as solar and wind are omnipresent and environmental friendly [1]. The renewable energy provides a profound public assessment of the environmental impacts of using fossil fuels to generate electricity.

With their advantages of being abundant in nature and nearly nonpollutant, renewable energy sources have attracted wide attention [2]. By 2050. The International Energy Agency (IEA) projects that as high as 16% of global electricity will be generated from solar photovoltaic (PV) with a wind contributing about 15–18% [3, 4].

A standalone wind/solar hybrid power system, making full use of the nature complementarily between wind and solar energy, has an extensive application prospect among various newly developed energy technologies the capacity of the hybrid power system needs to be optimised in order to make a tradeoff between power reliability and cost. [5] It is probable to endorse that hybrid stand-alone electricity generation systems are usually more reliable and less costly than systems that depend on a single source of energy [6].

© Springer International Publishing AG 2018
M. Hatti (ed.), *Artificial Intelligence in Renewable Energetic Systems*, Lecture Notes in Networks and Systems 35, https://doi.org/10.1007/978-3-319-73192-6_19

On other hand, you can choose the most profitable type of RES than other. For example in ALGERIA, Photovoltaic (PV) system is ideal for south, having more solar illumination levels and wind power system is ideal for locations as Tiaret and Elkheiter and other who having better wind flow conditions and there are the locations having the both more solar illumination levels and better wind flow at the same time as Adrar, Béchar and Timimoun.

Much research effort has been made towards modeling of renewable energy resources, solar standalone, wind standalone, or hybrid solar wind hybrid energy systems for reliability assessment and economic viability [7, 8].

Photovoltaic energy generation is ever more important as a renewable resource since it does not cause in fuel costs, pollution, maintenance, and emitting noise compared to other renewable resources as more accessibility of solar irradiation [9].

Permanent Magnet Synchronous Generator (PMSG) is most preferred wind generator due to its reliability and size for standalone wind energy conversion system [10, 11].

In recent years, there are many publications to connect multiple renewable power energy sources of different type (e.g. solar, wind, supercapacitor and fuel cell) and capacities to power grid or load [12, 13].

2   System Modelisation and Configuration

Figure 1 shows the configuration of the proposed PMSG wind-PV hybrid generation system with battery bank. This hybrid power generation system is constituted by a PV Power generation branch, PMSG wind power generation branch and storage power system. The PV power generation branch is composed of an array 2×2 PV panel, MPPT controller, a DC-DC converter, a shared battery bank and a shared inverter. On the other hand the wind power generation branch is composed of a PMSG machine, MPPT controller, a shared battery bank and a shared inverter with the P&O algorithm and Hill climb search respectively performs the MPPT for the PV panel and PMSG wind. The DC-DC converter functions to step up the output voltage so that the same battery bank is shared by the both wind and Photovoltaic power branch.

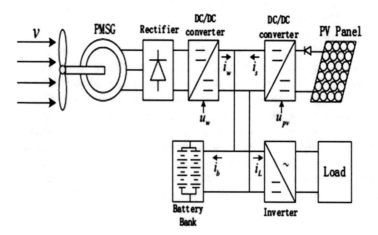

Fig. 1. Block diagram of PV-wind hybrid system proposed

The inverter is used to convert the DC voltage to the AC voltage so that a typical household can freely utilize electricity from the AC grid. Synchronization among inverters needs to be taken. Since the additional inverter will inevitably cause efficiency loss, the concept of dc microgrid is under active research.

The proposed wind–PV hybrid generation system possesses the following features.

1. The wind power generation branch and the PV power generation branch of the hybrid generation system can independently be controlled, hence achieving the MPPT simultaneously.
2. Compared with an individual generation system (the wind power or the PV power), the wind–PV hybrid generation system cannot only harness more energy from nature but also allow the wind power and the PV power to complement one another to some extent between day and night.

This paper analyses the maximum power point tracking control approach for standalone WECS. The performance of the conventional PI controller, P&O controller and FLC under wind speed is evaluated. The proposed control strategy possess the improved capability of capturing the maximum power from wind. Comparative efficiency of the controllers is analysed from the output power of the converter.

3 System Models

3.1 MPPT For Photovoltaic (PV)

The DC current generated by the PV cell is expressed as follows

$$I = I_{PV,Cell} - I_{s,Cell}\left(e^{\frac{V}{aV_t}} - 1\right) \tag{1}$$

The first term in Eq. (1), that is Ipv, cell, is proportional to the irradiance intensity whereas the second term, the diode current, expresses the non-linear relationship between the PV cell current and voltage. A practical PV cell, shown in Fig. 2, includes series and parallel resistances [14]. The series resistance represents the contact resistance of the elements constituting the PV cell while the parallel resistance models the leakage current of the P–N junction [9].

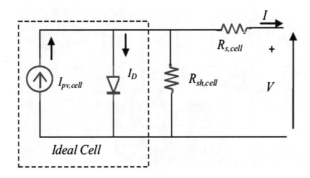

Fig. 2. Equivalent circuit of an ideal and practical PV cell.

The equation of the I–V characteristic of the PV module is obtained from Eq. (1) by including the equivalent module series resistance, shunt resistance and the number of cells connected in series and in parallel.

$$I = N_p \left(I_{PV} - I_s \left(e^{\frac{q(V + I.R_s)}{aN_sKT}} - 1 \right) \right) - \frac{(V + I.R_s)}{R_{sh}} \tag{2}$$

Where Vt the PV cell thermal voltage in Eq. (1) is substituted by that of the module thermal voltage given by $V_t = \frac{N_sKT}{q}$ and Ns and Np are respectively the number of cells connected in series and in parallel forming the PV module.

The constant a expressing the degree of ideality of the diode may be arbitrary chosen from the interval (1, 1.5) [15]. The light generated current of PV cell depends linearly on the irradiance and is also influenced by the temperature:

$$I_{PV} = \left(\frac{G}{G_{STC}} \right) (I_{PVn} + K_i(T - T_{STC})) \tag{3}$$

Ipvn is the nominal light-generated current provided at GSTC, TSTC which refer to the values at nominal or Standard Test Conditions (1 kW/m², 25 °C).

To achieve the maximum power point although different MPPT methods have been proposed in the literature for PV power generation, The Perturb and observe one is still the most common method in practice due to its simplicity and system independence.

Figure 3 shows the circuit topology of the P&O method for the PV power generation branch.

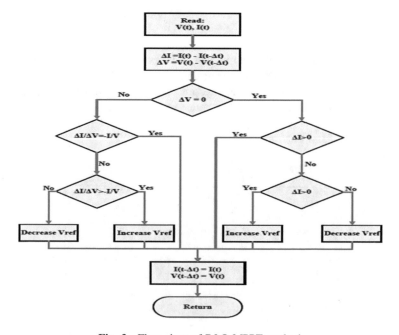

Fig. 3. Flow chart of P&O MPPT method

In the Fuzzy Implementation of MPPT algorithm, a Mamdani FLC is selected and the inputs of the FLC are error and change in error. They are computed by considering PV input current and voltage, and duty cycle has been calculated as output [15]. Figure 4 indicates the sub-system implementation of generating error (E) and change in error (CE) using FLC for MPPT control of PV based generation system.

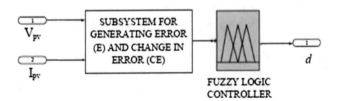

Fig. 4. FLC implementation of MPPT control

The rules for the Mamdani FLC implementation is tabulated as shown in Table 1.

Table 1. Rule base for FLV implementation

DE/E	NG	NM	NP	ZR	PP	PM	PG
NG	NG	NG	NG	NM	NM	NP	ZR
NM	NG	NG	NM	NM	NP	ZR	PP
NP	NG	NM	NM	NP	ZR	PP	PM
ZR	NM	NM	NP	ZR	NP	PM	PM
PP	NM	NP	ZR	PP	NP	PM	PG
PM	NP	ZR	PP	PM	NP	PG	PG
PG	ZR	PP	PM	PG	ZR	PG	PG

3.2 MPPT For PMSG Wind

Depending on the aerodynamic characteristics, the wind power captured by the wind turbine can be expressed

$$p = \frac{1}{2}C_p(\lambda, \beta)\rho\pi R^2 V^3 \tag{1}$$

Where $C_p(\lambda, \beta)$ is the wind turbine power coefficient which is a function of λ and β, ρ is the air density, R is the radius of wind turbine blade, V is the wind speed, β is the blade pitch angle, and λ is the tip speed ratio:

$$\lambda = \frac{wR}{V} \tag{2}$$

Where w is the wind turbine rotational speed. There exits an optimal tip speed ratio λ_{opt} that can maximize Cp and P.

Then, the maximum wind power Pmax captured by wind turbine can be described as

$$P_{max} = \frac{1}{2} \rho \pi R^5 \frac{C_{p,max}}{\lambda_{opt}^3} w^3 \tag{3}$$

Figure 5 shows a hill-climbing method for MPPT. The wind turbine works under wind speed of v1, the generator's initial operating point is A. The system measured power is PA, through checking the look-up stable, PA's corresponding maximum power point is M1, update the generator side's speed command to ωM1, when the turbine speed is stable at ωM1, and use the hill-climbing method to do vernier regulation. Maybe the generator's operating point would change into B, then just begin a new round of adjustment according to the measured power at that time, and so forth, usually only 2 or 3 times adjust is good enough to get close to the maximum power point C under the wind speed of v1.

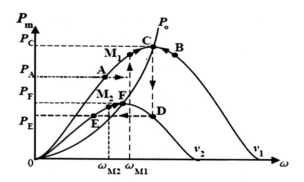

Fig. 5. A hill-climbing method for MPPT

The inputs to the controller are voltage, current and speed of PMSG. Using the speed and voltage samples the reference current is calculated. It is compared with the current measured and the error is utilized to compute the duty cycle of the power electronic switch in boost converter which controls the operation of wind power generation at MPP.

A Sub-system implementation of HCS algorithm using FLC as MPPT control for wind power generation system is shown in Fig. 6. The error and change in error are computed using the current samples measured at the input of the boost converter and speed of PMSG and based on the Mamdani rule base duty cycle is generated.

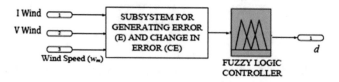

Fig. 6. FLC implementation of MPPT control using HCS Algorithm

The Rule base for FLC implementation of HCS algorithm is tabulated as shown in Table 2. The generated duty cycle is to control the power electronic switch in the boost converter which tracks the MPP.

Table 2. Rule base for FLC HCS implementation

DE/E	NL	NM	NS	Z	PS	PM	PL
NL	PL	PL	PL	PL	Z	Z	Z
NM	PL	PL	PM	PM	Z	Z	Z
NS	PL	PM	PS	PS	NM	NS	NM
Z	PL	PL	PS	Z	NS	NM	NL
PS	NM	PS	PS	NS	NL	NL	NL
PM	Z	Z	Z	NM	NM	NL	NL
PL	Z	Z	Z	NL	NL	NL	NL

3.3 Storage Power System

The controlled voltage source is given by the following expression [16]:

$$E = E_o - \frac{V_P Q_b}{Q_b - \int i_b dt} + \widetilde{A} \exp(-B_t \int i_b dt) \tag{9}$$

Where Eo is the constant voltage of battery (V), Vp is the polarization voltage (V), Qb is the battery capacity (AH), ib is the battery current (A), A is exponential zone amplitude (V), Bt is exponential zone time constant inverse (AK') [17].

4 Results

In order to verify the concept and simulation results of the solar wind hybrid power systems. Dynamic performance of the complete control scheme for various stages of the proposed hybrid PV/PMSG wind system is evaluated in this section. The complete system described in Sect. 2 is modeled under MATLAB-Simulink.

First, the PV power generation branch characteristics are analyzed. Figures 7 and 8 show the characteristics of the PV output voltage and power under stable irradiance level of 1000 W/m². It verifies that the PV power generation branch can readily perform the MPPT and achieve the maximum output power at a given irradiance.

The improvement of FLC algorithm regarding ripple is undeniably clear in Fig. 7, The overshoot due to irradiation level 1000 W/m² is less important in case of conventional P&O algorithm (4.2%) than in case of FLC algorithm (4.8%). This is considered using only the pic value that is the maximum, knowing that the duration of this peak is less in the case of FLC proposed algorithm compared to the conventional P&O algorithm (Fig. 7).

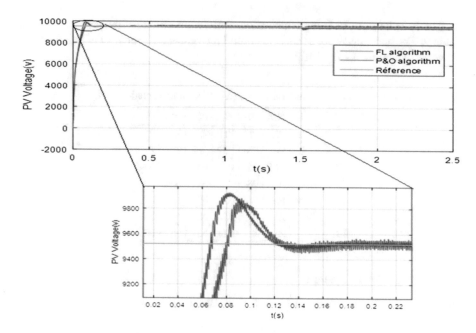

Fig. 7. Voltage of PV for different controllers

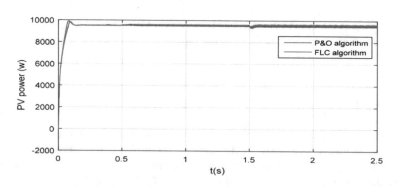

Fig. 8. MPPT tracking response power of the PV system for different controllers

The maximum power point tracking comparison is shown in Fig. 8. The proposed algorithm shows a significant improvement in the time interval corresponding to the period of atmospheric conditions. In addition, the proposed algorithm shows a better response time.

From the results shown in Figs. 9 and 10, we can observe that the output voltage and power tracked respectively the reference voltage and the maximum power point with good performances.

The maximum power point tracking comparison is shown in Figs. 9 and 10. The FLC algorithm shows a significant improvement in the time interval corresponding

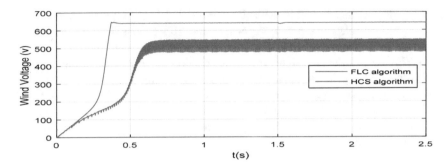

Fig. 9. Voltage of PMSG wind for different controllers

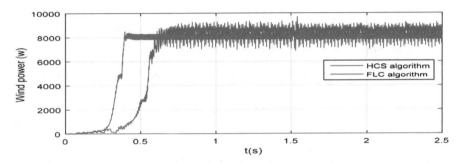

Fig. 10. MPPT tracking response of PMSG wind system for different controllers

to the period of atmospheric conditions. In addition, the proposed algorithm shows a better transient response which can be seen in the time interval 0 s < t < 0.06 s.

It has seen clearly how FLC based MPPT controller reduced the response time of photovoltaic system and has been seen clearly continues oscillation of operation point for the P&O conventional technique. It is a result of the continuous perturbation of the operating voltage in order to reach the MPP. Whereas this phenomenon of oscillation it doesn't observed in FLC based MPPT technique on both signals of power and voltage. This result has reduced power losses.

Second, the performances of the PMSG wind power generation branch are assessed. Figures 9 and 10 show the rectifier output power characteristics under stable speeds 12 m/s. It verifies that the characteristics of the output power versus the output voltage have an obvious maximum power point under speed. Also, it can be observed that both the maximum power point and efficiency increase with the speed. The maximum power point increment is simply due to the increase of input wind power.

The detailed implementation of PMSG based WECS with fuzzy logic controller shown in Fig. 11. To check the tracking ability of MPPT techniques, the wind speed is fixed at 12 m/s. two MPPT approaches in simulated in same varying wind speed conditions. From the analysis it can be stated that the FLC based MPPT technique is most efficient maximum power tracking power when compared with Hill Climb Search (HCS) based technique.

The MPPT point tracking by both algorithms (HCS and FLC) is shown by Figs. 11 and 12. We can clearly see that the race of the MPPT point in most cases is less important for the proposed FLC algorithm compared to the conventional HCS algorithm. This is due to the instability and the oscillation of the HCS algorithm especially around the MPPT point. In addition, the proposed algorithm shows a better transient response which can be seen in the time interval 0 s < t < 0.38 s. The response time for the FLC is equal to 38 ms while it is equal to 78 ms for the HCS conventional algorithm. Figure 12 shows the same power response with important oscillations for the HCS algorithm.

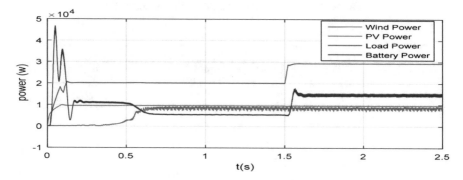

Fig. 11. Power flow in various sections (PW, PPV, PL, PB) with conventional algorithm MPPT controller

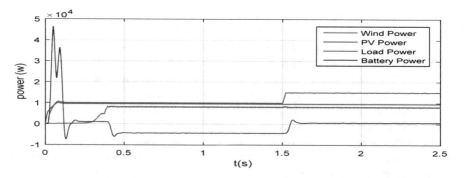

Fig. 12. Power flow in various sections (PW, PPV, PL, PB) FL MPPT controller

Third, the wind–PV hybrid generation characteristics are evaluated. Figure 11 shows that the HCS algorithm's response is less than the FLC controller. The proposed FLC algorithm shows a significant improvement in the time interval corresponding to the period of rapidly changing atmospheric conditions.

To prove the robustness and reliability of our proposed inverter we test its ability to load variation up to 50%. In the first test, we use a classical MPPT controller P&O and HCS to the both PV and wind power generation system respectively.

The THD and power factor measurements of the proposed inverter are measured using power quality analyzer. the THD results, which is measured corresponding to P&O and HCS algorithms to PV system and wind power system respectively, is found that the THD is 27.27% which well very high, The results of THD which is measured corresponding to FLC in the both PV and wind power systems prototype provide better THD level at the inverter output with 5.07%. Therefore, the results show less total harmonic distortion in the FLC case as shows in Fig. 12.

5 Conclusion

In this paper, a new wind–PV hybrid generation system has been proposed and implemented. This stand-alone hybrid generation system can fully utilize the characteristics of the proposed wind generator and the PV panels to extract the maximum power from the wind and solar energy sources.

An FLC based MPPT scheme for PV-PMSG wind hybrid power system has been presented in this paper. A prototype of Solar wind hybrid generation power system is used for implementation of the proposed FLC in MPPT control algorithms. The performance of the proposed controller has been found better than that of the conventional P&O and HCS for the PV and PMSG wind systems respectively. Furthermore, as compared to the conventional algorithms results indicated that the proposed FLC scheme can provide better THD level at the inverter output. Therefore, the proposed FLC based MPPT scheme for PV and wind systems could be a potential candidate for PV-wind inverter applications under variable load conditions.

Acknowledgment. This work was financially supported by the ENERGARID and SGRE laboratory and Tahri Mohammed University Béchar 08000 ALGERIA, under the scientific Programme Project Electro-energetique Industrial Option Renewable Energy, 2015/2016.

References

1. Saidi, A.: Solar-wind hybrid renewable energy systems: evolutionary technique. In: Electrotechnica, Electronica, Automatica (EEA), vol. 64, no. 4, pp. 24–27 (2016). ISSII 1582-5175
2. Liu, C., Chau, K.T., Zhang, X.: An efficient wind–photovoltaic hybrid generation system using doubly excited permanent-magnet brushless machine. IEEE Trans. Industr. Electron. **57**(3), 831–839 (2010)
3. IEA: Technology roadmap, wind energy. Report, Energy Technology Perspectives, 2013 edition, International Energy Agency (2013)
4. IEA-PVPS: Technology roadmap, solar photovolatic energy. Report, Energy Technology Perspectives, 2014 edition, International Energy Agency (2014)
5. Wang, J., Yang, F.: Optimal capacity allocation of standalone wind/solar/battery hybrid power system based on improved particle swarm optimisation algorithm. IET Renew. Power Gener. **7**(5), 443–448 (2013)

6. Kumar, A.V.P., Parimi, A.M., Rao, K.U.: Implementation of MPPT control using fuzzy logic in solar-wind hybrid power system. In: 2015 IEEE International Conference on Signal Processing, Informatics, Communication and Energy Systems (SPICES), Kozhikode, pp. 1–5 (2015)
7. Rajesh, K., Kulkarni, A.D., Ananthapadmanabha, T.: Modeling and simulation of solar PV and DFIG based wind hybrid system. In: SMART GRID Technologies, 6–8 August 2015 (2015)
8. Kim, S.-K., Jeon, J.-H., Cho, C.-H., Kim, E.-S., Ahn, J.-B.: Modeling and simulation of a grid-connected PV generation system for electromagnetic transient analysis. Sol. Energy **83**, 664–678 (2009)
9. Saidi, A., Benachaiba, C.: Comparison of IC and P&O algorithms in MPPT for grid connected PV module. In: 2016 8th International Conference on Modelling, Identification and Control (ICMIC), Algiers, Algeria, pp. 213–218 (2016)
10. Tiwari, R., Babu, N.R.: Fuzzy logic based MPPT for permanent magnet synchronous generator in wind energy conversion system. IFAC Papers OnLine **49**(1), 462–467 (2016)
11. Saib, S., Gherbi, A.: Simulation and control of hybrid renewable energy system connected to the grid. 978-1-4673-7172-8/15/$3 1.00 ©2015. IEEE (2015)
12. Castillo, J.P., Mafiolis, C.D., Escobar, E.C., Barrientos, A.G., Segura, R.V.: Design, construction and implementation of a low cost solar-wind hybrid energy system. IEEE Lat. Am. Trans. **13**(10), 3304–3309 (2015)
13. Abdulkarim, A., Abdelkader, S.M., Morrow, D.J., Falade, A.J., Lawan, A.U., Iswadi, H.R.: Effect of weather and the hybrid energy storage on the availability of standalone microgrid. Int. J. Renew. Energy Res. **6**(1), 189–198 (2016)
14. Villalva, M.G., Gazoli, J.R.: Comprehensive approach to modeling and simulation of photovoltaic arrays. IEEE Trans. Power Electron. **24**, 1198–1208 (2009)
15. Faranda, R., Leva, S.: Energy comparison of MPPT techniques for PV systems. WSEAS Trans. Power Syst. **3**, 447–455 (2008)
16. Mahamudul, H., Saad, M., Henk, M.I.: Photovoltaic system modeling with fuzzy logic based maximum power point tracking algorithm. Int. J. Photoenergy **2013**, 10 (2013). Article ID 762946
17. Bendib, B., Krim, F., Belmili, H., Almi, M.F., Boulouma, S.: Advanced fuzzy MPPT controller for a stand-alone PV system. Energy Procedia **50**, 383–392 (2014). The International Conference on Technologies and Materials for Renewable Energy, Environment and Sustainability, TMREES14

Maximum Power Point Tracking Control of Photovoltaic Generation Based on Fuzzy Logic

Fathia Hamidia[1]([✉]), A. Abbadi[1], and M. S. Boucherit[2]

[1] LREA Laboratory, Yahia Feres University, Medea, Algeria
fe_hamidia@yahoo.fr
[2] LCP Laboratory, Ecole Nationale Polytechnique, ENP, Algiers, Algeria

Abstract. Global warming and environmental pollution in one hand and the depletion of fossil fuel resources in the coming decades in the other hand are now the two main concerns worldwide. Renewable energies are a solution to replace conventional sources of energy that are polluting and tend to disappear. Photovoltaic energy constitutes among renewable energies that it has the broadest potential of development.

This paper proposes a direct torque controlled IM supplied with photovoltaic panel to replace flux oriented control; and to track the maximum power point, this technique (MPPT) is based on fuzzy logic technique to get better performance specially on variation of load and weather condition. Finally, simulation results are given to show the effectiveness and feasibility of the approach.

Keywords: Fuzzy logic · Induction motor · PVG · MPPT

1 Introduction

Since the last century, energy consumption has increased dramatically. However, our resources of oil and gas are not eternal and it is not also best to burn them more to avoid exacerbating pollution. Algeria has the effect of this significant renewable energy resource that can overcome particularly in the context of the production of electrical energy, the main vector of all economic and social development [1].

Nowadays, there are many forms of renewable energy, the most commonly used are: solar, wind and hydraulic. The sun is in the form of heat that is transmitted by infrared radiation, these exhibits under effect of photovoltaic (PV). The speed control of AC drive supplied by PV is one of the most important applications in PV systems, it has become a favorable solution for many applications, because of gaining more acceptance and market share, particularly in rural: areas that have a substantial amount of insolation and have no access to an electric grid. So, photovoltaic systems are used to pump water for livestock, plants or humans. Since the need for water is greatest on hot sunny days, the technology is an obvious choice for this application. Agricultural watering needs are usually greatest during sunnier periods when more water can be pumped with a solar system [2].

© Springer International Publishing AG 2018
M. Hatti (ed.), *Artificial Intelligence in Renewable Energetic Systems*, Lecture Notes in Networks and Systems 35, https://doi.org/10.1007/978-3-319-73192-6_20

New control techniques based on intuition and judgment has emerged. The researchers have named intelligent controllers; these controllers offer the possibility of obtaining reproductive dynamics of a nonlinear complex system only through its input/output, without using a structural model. These intelligent techniques are increasingly used in the design, modeling and control of complex systems, namely neural networks and fuzzy logic.

The induction motor is used more and more for photovoltaic pumping systems. The low cost of the engine, the low maintenance requirements and the increased efficiency for solar pumping systems make it particularly major problem with the use of PV panels is their non-linear nature [3].

In this paper, we present and discuss the application of direct torque control on the induction motor supplied with photovoltaic energy.

2 Induction Model

The transformation of PARK brings back to the equation stator in reference frame related to the rotor.

$$\frac{dI_{sd}}{dt} = \frac{1}{\sigma L_s}\left[-\left(R_s + \frac{M_{sr}^2}{L_r T_r}\right)I_{sd} + \omega_s \sigma L_s I_{sq} + \frac{M_{sr}}{L_r T_r}\varphi_{rd} + \frac{M_{sr}}{L_r}\omega_r \varphi_{rd} + V_{sd}\right]$$

$$\frac{dI_{rq}}{dt} = \frac{1}{\sigma L_s}\left[-\omega_s \sigma L_s I_{sd} - \left(R_s + \frac{M_{sr}^2}{L_r T_r}\right)I_{sq} - \frac{M_{sr}}{L_r}\omega_r \varphi_{rd} + \frac{M_{sr}}{L_r T_r}\varphi_{rq} + V_{sq}\right] \quad \text{with } \left(\sigma = 1 - \frac{M_{sr}^2}{L_s L_r}\right)$$

$$\frac{d\varphi_{rd}}{dt} = \frac{M_{sr}}{T_r}I_{sd} - \frac{1}{T_r}\varphi_{rd} + (\omega_s - \omega_r)\varphi_{rd}$$

$$\frac{d\varphi_{rq}}{dt} = \frac{M_{sr}}{T_r}I_{sq} - \frac{1}{T_r}\varphi_{rq} - (\omega_s - \omega_r)\varphi_{rd}$$

$$\frac{d\omega}{dt} = \frac{P^2 M_{sr}}{L_r J}\left(I_{sq}\varphi_{rd} - I_{sd}\varphi_{rq}\right) - \frac{F}{J}\omega - \frac{P}{J}T_r$$

3 PV Model

A practical PV array consists of a collection of solar cells connected in series and/or parallel. An equivalent circuit model for a solar cell is shown in Fig. 1, it is derived as the key element is the current source generating the photovoltaic current. The model consists of a current source, a diode, shunt resistor R_p and a series resistance R_s.

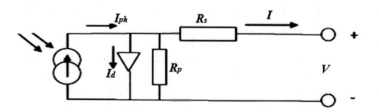

Fig. 1. PV cell model

The topology of boost converter is shown in Fig. 2. For this converter the output voltage is always higher than the input PV voltage. Power flow is controlled by the on/off duty cycle of the switching transistor. This converter topology can be used in conjunction with lower PV voltages. No extra blocking diode is necessary when the boost topology is used.

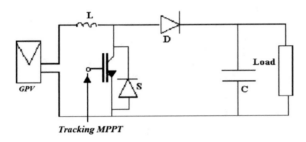

Fig. 2. Boost converter DC/DC

For sizing a photovoltaic system, we have to carry out in the first the motor consumption; in hence to define the energy need of the load (as in our case we need to get $U_{dc} = 514$ V). After, in the second, we must take in account the obtained results and also the meteorological data as the input parameters of the photovoltaic installation of the input program. The sizing of the photovoltaic system is carried out according to the algorithm [4].

The commercialized solar modules are formed generally by a number of cells assembled in parallel N_p or/and in series N_s. In addition, a datasheet is provided and includes the main following information about the product presented in (Table 1).

The panels are assembled in series and parallel to generate a 514 V voltage range in a MPP operation under different load changes. The characteristics of the considered BP MSX-120.

Table 1. Data sheet information of a PV panel BP MSX120.

The nominal open-circuit voltage	42.1 V
The nominal short-circuit current	3.87 A
The voltage at the MPP	33.7 V
The maximum experimental peak output power	120 W
The current at the MPP	3.56 A
The open-circuit voltage/T° coefficient	(80 ± 10) mV%C°
The short circuit current/T° coefficient	(0.065 ± 0.015)%C°
Parallel resistance R_s	0.473 Ω
Serie resistance R_p	1367 Ω

4   MPPT Control Based on Fuzzy Logic

To optimize the power provided by the generator, a static converter, which operates as an adapter, must be added. The Fig. 3 shows the converter, which is a boost chopper. It exploits the MPPT technique; there are many algorithms that are used to control the MPPT. In this work, we propose an intelligent fuzzy logic technique to control duty cycle of the switching transistor, as shown in Fig. 3.

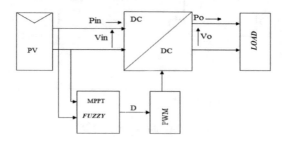

Fig. 3. Proposed fuzzy MPPT

This technique presented in this work is proposed to resolve the problem of the classical controller (P&O), because this type of controller does not require a precise model.

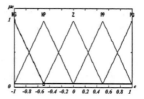

Fig. 4. Membership function of the MPPT Fuzzy logic controller (FLC)

The input linguistic variables change in power error $\Delta e(k)$ and output linguistic variable $D(k)$ member function. The second controller based on fuzzy techniques (FLC, PI-FLC) presented in this work is proposed to control rotor speed of induction motor and to replace the classical PI controller. The input linguistic variables speed error $e(k)$, change in speed error $\Delta e(k)$ and output linguistic variable $u(k)$ member function (Fig. 5).

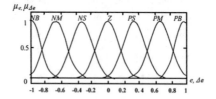

Fig. 5. Membership function of the fuzzy speed controller (FLC)

We propose the PI-FLC controller to delete the static error obtained by using FLC logic controller.

The PI-type FLC control action is shown in Fig. 4, the fuzzy associative memory (FAM) of Mamdani rule base model to develop FLC as fuzzy speed regulator which in term replace the P (Fig. 6).

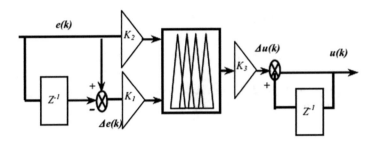

Fig. 6. Basic structure PI-FLC controller

5   Direct Torque Control

In the DTC, the stator flux vector is estimated by taking the integral of difference between the input voltage and the voltage drop across the stator resistance given by [5]:

$$\varphi_s = \int_0^t (V_s - Ri_s)dt \tag{3}$$

Let's us replace the estimate of the stator voltage Vs with the true value and write it as:

$$V_s(S_a, S_b, S_c) = \frac{2}{3} U_o \left(S_a + S_b e^{\frac{j2\pi}{3}} + S_c e^{\frac{j4\pi}{3}}\right) \tag{4}$$

S_a, S_b, S_c, represent the state of the three phase legs 0 meaning that the phase is connected to the negative and 1 meaning that the phase is connected to the positive leg [6]. The stator current space vector is calculated from measured currents i_a, i_b, i_c:

$$i_s = \frac{2}{3}\left(i_a + i_b e^{\frac{j2\pi}{3}} + i_c e^{\frac{j4\pi}{3}}\right) \tag{5}$$

The component α and β of vector φ_s can be obtained:

$$\varphi_{s\alpha} = \int_0^t (V_{s\alpha} - Ri_{s\alpha})dt$$
$$\varphi_{s\beta} = \int_0^t (V_{s\beta} - Ri_{s\beta})dt \tag{6}$$

Stator Flux amplitude and phase angle are calculated in expression (7):

$$\begin{cases} \varphi_s = \sqrt{\varphi_{s\alpha}^2 + \varphi_{s\beta}^2} \\ \angle\varphi_s = arctg\,\dfrac{\varphi_{s\beta}}{\varphi_{s\alpha}} \end{cases} \quad (7)$$

Once the two components of flux are obtained, the electromagnetic torque can be estimated from the relationship cited below:

$$C_{em} = \frac{3}{2}p\left(\varphi_{s\alpha}i_{s\beta} - \varphi_{s\beta}i_{s\alpha}\right) \quad (8)$$

The voltage plane is divided into six sectors so that each voltage vector divides each region into two equal parts (Fig. 7).

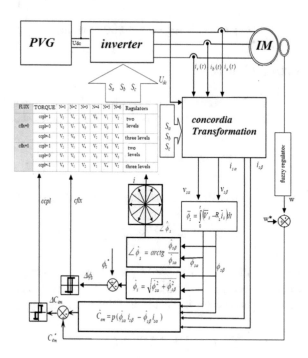

Fig. 7. Schematic diagram of DTC-IM supplied by photovoltaic generator

The stator flux vector can be estimated using the measured current and voltage vectors:

$$\frac{d\varphi_s}{dt} = V_s - R_s I_s \tag{9}$$

Or

$$\varphi_s = \int (V_s - R_s I_s) dt \tag{10}$$

6 Simulation Results

Some tests have been carried out to improve the performances of fuzzy MPPT method. To verify the effectiveness of the proposed technique, simulations are performed in this section by using MATLAB/SIMULINK. In this simulation of asynchronous machine, the nominal power P_n is 1.5 kw, Nominal voltage V_n is 220 V, stator resistances R_s are 4.85 Ω, rotor resistance R_r is 3.805 Ω, stator inductance L_s is 0.274 H, Rotor inductance L_r is 0.274 H, moment of inertia J is 0.031 kg.m^2, and friction coefficient K is 0.008.

As shown Fig. 8, the system is tested under constant load torque (10 N.m). We note that the estimated values of fluxes, torque converge very well to their simulated values. The simulation results show that the proposed structure can provide a very fast torque response. It can be noticed that these results obtained by using P&O algorithm (in open Fig. 8 or closed loop Fig. 9) gives us high torque and flux ripples (Fig. 8) than the results obtained by using fuzzy logic (Figs. 8 and 9). As shown Fig. 10, the system based on fuzzy technique is tested under torque, rotor speed, temperature and irradiation variation. This figure show the effectiveness of our fuzzy approach.

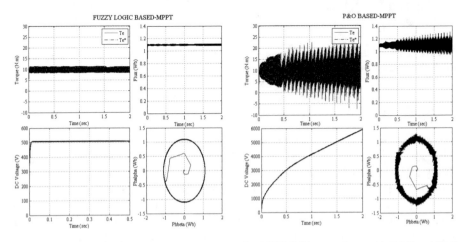

Fig. 8. Torque, flux and DC voltage responses using DTC-IM in open loop with P&O and FLC method (G = 1000 W/m^2, T = 298 K$^\circ$).

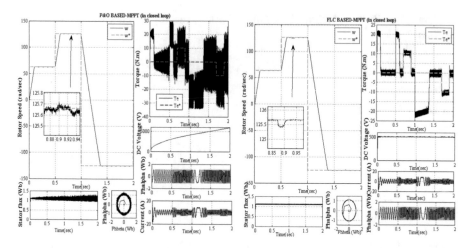

Fig. 9. Torque, flux, rotor speed and current responses using DTC-IM in closed loop by using P&O and FLC (G = 1000 W/m^2, T = 298 K°).

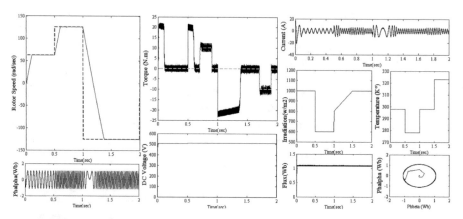

Fig. 10. Torque, flux and rotor speed responses (in closed loop) with load torque, speed, irradiation and temperature variation

7 Conclusion

It has been possible to obtain satisfactory results using a conventional controller. However as this work indicates, that the introduction of these artificial techniques (fuzzy logic) becomes a necessity to achieve high performance invariable speed drive, because these news techniques do not require a precise model. Considering the PVG, many approaches of Maximum Power Point Tracking (MPPT) algorithms have been developed such as Perturb and Observe algorithms (P&O), INcremental Conductance algorithms (INC), and artificial intelligence-based algorithms. In this paper, a direct duty cycle based on fuzzy logic technique is used to replace P&O algorithm and it is employed due to its robust and its adequacy with nonlinear characteristic of the PVG. The simulation results show that proposed techniques give satisfactory results.

References

1. Boukhalafa, S., Bouchafaa, F.: Analysis and control of a further maximum power point (MPPT) of PV array. In: International Conference On Information Processing and Electrical Engineering, Tebessa, Algeria, pp. 188–194 (2012)
2. Bouzeriaa, H., Fethaa, C., Bahib, T., Abadliab, I., Layateb, Z., Lekhchinec, S.: Fuzzy logic space vector direct torque control of PMSM for photovoltaic water pumping system. Energy Procedia **74**, 760–771 (2015)
3. Abouda, S., Nollet, F., Chaari, A., Essounbouli,N., Koubaa, Y.: Direct torque control of induction motor pumping system fed by a photovoltaic generator. In: International Conference on Control, Decision and Information Technologies (CoDit), pp. 404–408 (2013)
4. Barazane, L., Kharzi, S., Malek, A., Larbès, C.: A sliding mode control associated to the field-oriented control of asynchronous motor supplied by photovoltaic solar energy. Revue des Energies Renouvelables **11**(2), 317–327 (2008)
5. Hamidia, F., Abbadi, A., Boucherit, M.S.: Direct torque controlled dual star induction Motors (in open and closed loop). In: The 4th international Conference on Electrical Engineering (ICEE 2015) (2015)
6. Merzoug, M.S., Naceri, F.: Comparison of field-oriented control and direct torque control for permanent magnet synchronous motor (PMSM). World Acad. Sci. Eng. Technol. **45**, 299–304 (2008)

Fuzzy-Logic-Based Solar Power MPPT Algorithm Connected to AC Grid

Amel Abbadi[1($\boxtimes$)], F. Hamidia[1], A. Morsli[1], D. Boukhetala[2], and L. Nezli[2]

[1] Electrical Engineering and Automatic LREA,
Electrical Engineering Department, University of Medea, Medea, Algeria
amel.abbadi@yahoo.fr
[2] Laboratoire de Commande des Processus, Département d'Automatique,
École Nationale Polytechnique, El Harach, Algeria

Abstract. This paper describes a PV system supplied a large scale interconnected grid. A PV array is connected to AC grid via a DC-DC boost converter and a three-phase three-level Voltage Source Converter (VSC). Maximum Power Point Tracking (MPPT) is implemented in the boost converter by means of a Simulink model using fuzzy logic controller by switching the duty cycle of the boost converter.

A three phase VSC converts the 500 V DC link voltage to 260 V AC and keeps unity power factor. The VSC control system uses two control loops: an external control loop which regulates DC link voltage to 260 V AC and an internal control loop which regulates Id and Iq grid currents (active and reactive current components). Id current reference is the output of the DC voltage external controller. Iq current reference is set to zero in order to maintain unity power factor. Vd and Vq voltage outputs of the current controller are converted to three modulating signals used by the PWM Generator.

Keywords: Fuzzy logic controller · Grid-connected photovoltaic system
MPPT · PV array

1 Introduction

The photovoltaic (PV) system is one of the important renewable energy sources because it is available in most area around the world. Also, it is easy to get it as a dc power source. With using inverters, dc power can be converted into ac power which can be used in loads or connect to ac grid system [1]. Photovoltaic power supplied to the utility grid is gaining more and more visibility, while the world's power demand is increasing. When system PV is connected the utility network. Many demands such as power system stability and power quality are primary requests. As a consequence, large research efforts are put into the control of these systems in order to improve their behavior. All photovoltaic systems are interfacing the utility grid through a voltage source inverter [2] and a boost converter. Many control strategies and controller types [3–5] have been investigated. The significant advantage of photovoltaic system is the use of abundant and free energy from the sun [2]. Photovoltaic system has a major

© Springer International Publishing AG 2018
M. Hatti (ed.), *Artificial Intelligence in Renewable Energetic Systems*, Lecture Notes in Networks and Systems 35, https://doi.org/10.1007/978-3-319-73192-6_21

problem: the amount of electric power generated by solar arrays changes continuously according the weather conditions and day time [6]. In order to increase this efficiency, MPPT controllers are used. Such controllers are becoming an essential element in PV systems. A significant number of MPPT control have been elaborated since the seventies, starting with simple techniques such as voltage and current feedback based MPPT to more improved power feedback based MPPT such as the perturbation and observation (P&O) technique or the incremental conductance technique [7, 8]. Recently intelligent based controls MPPT have been introduced.

In this paper, an intelligent control technique using fuzzy logic control is associated to an MPPT controller in order to improve energy conversion efficiency.

This paper is organized as follows. In Sect. 2, the system under study is described. The Fuzzy logic controller design is presented in Sect. 3. In Sect. 4, the proposed control is validated by means of simulation and discussed. Finally, the conclusions are summarized in Sect. 5.

2 Power System Model

Figure 1 shows the detailed model of a 100-kW Grid-connected PV Array. The 100-kW PV array is connected to a 25-kV grid (Fig. 2 depicts the power system utility grid) via a DC-DC boost converter and a three-phase three-level Voltage Source Converter (VSC). Maximum Power Point Tracking (MPPT) is implemented in the boost converter by means of a Simulink model using the fuzzy logic controller.

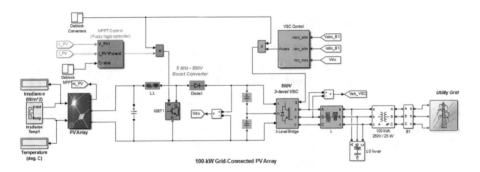

Fig. 1. Power system model under study

The detailed model contains the following components:

- PV array delivering a maximum of 100 kW at 1000 W/m^2 sun irradiance.
- 5-kHz DC-DC boost converter increasing voltage from PV natural voltage 273 V DC at maximum power) to 500 V DC. Switching duty cycle is optimized by a MPPT controller that uses the fuzzy logic controller. This MPPT system automatically varies the duty cycle in order to generate the required voltage to extract maximum power.

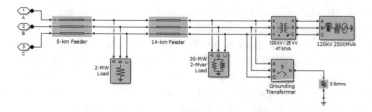

Fig. 2. Power system utility grid

- 1980-Hz 3-level 3-phase VSC. The VSC converts the 500 V DC link voltage to 260 V AC and keeps unity power factor. The VSC control system uses two control loops: an external control loop which regulates DC link voltage to ±250 V and an internal control loop which regulates Id and Iq grid currents (active and reactive current components). Id current reference is the output of the DC voltage external controller. Iq current reference is set to zero in order to maintain unity power factor. Vd and Vq voltage outputs of the current controller are converted to three modulating signals Uabc_ref used by the PWM Generator. The control system uses a sample time of 100 ms for voltage and current controllers as well as for the PLL synchronization unit. Pulse generators of Boost and VSC converters use a fast sample time of 1 ms in order to get an appropriate resolution of PWM waveforms.
- 10-kvar capacitor bank filtering harmonics produced by VSC.
- 100-kVA 260 V/25 kV three-phase coupling transformer.
- Utility grid (25-kV distribution feeder +120 kV equivalent transmission system).

2.1 Photovoltaic Array System

To get the output voltage of 273 V from PV array, the 303 sun Power SPR-305E-WHT-D with 64.2 V open circuit voltage model is used. This array is available in Matlab/Simulink built-in internal libraries. The array consists of 66 strings of 5 series-connected modules connected in parallel (66*5*305.2 W = 100.7 kW). The PV array block has two inputs that allow you varying sun irradiance (input 1 in W/m^2) and temperature (input 2 in deg. C). The irradiance and temperature profiles are defined by a Signal Builder block which is connected to the PV array inputs.

The manufacturer specifications for one module are:

- Number of series-connected cells: 96
- Open-circuit voltage Voc = 64.2 V
- Short-circuit current: Isc = 5.96 A
- Voltage and current at maximum power: Vmp = 54.7 V, Imp = 5.58 A

Solar cell mathematical modeling is an important step in the analysis [9] and design of PV control systems. The electric model of the PV is shown in Fig. 3.

The current of the module is given by the following equation:

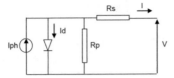

Fig. 3. Equivalent circuit of a real solar cell

$$I = I_{ph} - I_{sat}\left[\exp\left(\frac{V + R_s I}{aV_t} - 1\right)\right] - \frac{V + R_s I}{R_p} \tag{1}$$

Where Isat is the reverse saturation current of the diode, a is non-ideality factor of the p-n junction, $V_t = q/KT$, Rp and Rs are the intrinsic shunt and series resistance of the solar cell respectively, Iph is the generated current under a given irradiation and I and V are the output current and output voltage of the PV cell.

The equation of photo-generated current Iph in Standard Test Conditions (G = 1000 W/m^2 and Ta = 25 °C) is given by:

$$I_{ph} = \left(I_{ph,n} + K_I\Delta(T)\right)\frac{G}{G_n} \tag{2}$$

The conduction current of the diode is given by the following equation

$$I_d = I_{sat}\left[\exp\left(\frac{V + R_s I}{aV_t} - 1\right)\right] \tag{3}$$

The saturation current of the diode is given by the following equation:

$$I_{sat} = \frac{I_{sc,n} + K_I\Delta(T)}{\left[\exp\left(\frac{V_{oc,n} + K_V\Delta(T)}{aV_t}\right) - 1\right]} \tag{4}$$

The curve of Fig. 4 shows the I-V and P-V characteristics for the whole array.

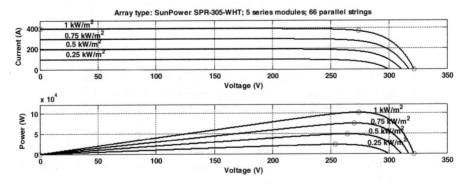

Fig. 4. The I-V and P-V characteristics for the the 303 sun Power SPR-305E-WHT-D

3   Fuzzy Logic Controller Design

MPPT using Fuzzy Logic Control gains several advantages of better performance, robust and simple design. In addition, this technique does not require the knowledge of the exact model of system. The main parts of FLC, fuzzification, rule-base, inference and defuzzification, are shown in Fig. 5a. In the proposed system, the input variables of the FLC are slope of the PV cell's Power-Voltage (P-V) curve ($E(k)$) and variation of slope ($\Delta E(k)$) (Fig. 5b) [10] (Table 1).

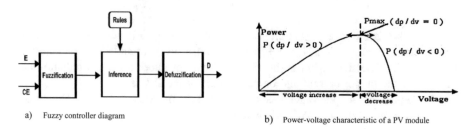

a)   Fuzzy controller diagram

b)   Power-voltage characteristic of a PV module

Fig. 5. Fuzzy logic controller design

Table 1. FLC rules base

Fuzzy rule		NB	NS	ZE	PS	PB
				$E(k)$		
$\Delta E(k$	NB	ZE	PB	PS	ZE	NB
	NS	PB	PS	ZE	ZE	NB
	ZE	PB	PS	ZE	NS	NB
	PS	PB	ZE	ZE	NS	NB
	PB	PB	ZE	NS	NB	ZE

4   Simulation Results

As it has been mentioned before, our application is *100*-kW *array connected to a* *25*-kV grid via a DC-DC boost converter and a three-phase three-level VSC. The simulation is performed under different temperature and sun irradiation conditions (Fig. 6).

- From t = 0 s to t = 0.05 s, pulses to Boost and VSC converters are blocked. PV voltage corresponds to open-circuit voltage (Nser*Voc = 5*64.2 = 321 V). The three-level bridge operates as a diode rectifier and DC link capacitors are charged above 500 V.

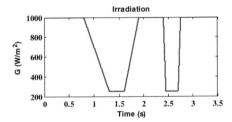
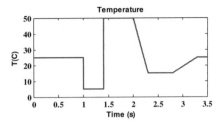

Fig. 6. The different temperature and sun irradiation conditions

- At t = 0.05 s, Boost and VSC converters are de-blocked. DC link voltage is regulated at Vdc = 500 V. Duty cycle of boost converter is fixed (D = 0.5) and sun irradiance is set to 1000 W/m². Steady state is reached at t = 0.25 s. Resulting PV voltage is therefore V_{PV} = (1−D)*Vdc = (1−0.5)*500 = 250 V. The PV array output power is 96 kW whereas maximum power with at 1000 W/m² irradiance is 100.7 kW. The phase A voltage and current at 25 kV bus are in phase (unity power factor).
- At t = 0.4 s MPPT is enabled. The MPPT regulator starts regulating PV voltage by varying duty cycle in order to extract maximum power. Maximum power (100.7 kW) is obtained when duty cycle is D = 0.454.
- At t = 0.43 s, PV array mean voltage = 273.5 V as expected from PV module specifications (Nser*Vmp = 5*54.7 = 273.5 V).
- From t = 0.8 s to t = 1.3 s, sun irradiance is ramped down slowly from 1000 W/m² to 250 W/m². At t = 1 the temperature decreased rapidly from 25 to 5. Fuzzy MPPT controller continues tracking maximum power.
- At t = 1.3 s when irradiance has decreased to 250 W/m², duty cycle is D = 0.49. Corresponding PV voltage and power are Vmean = 258 V and Pmean = 23.3 kW. Note that the fuzzy MPPT controller continues tracking maximum power during this slow irradiance change.
- From t = 1.3 s to t = 1.6 s sun irradiance is 250 W/m². The temperature is increased to 50 C. at t = 1.4 in order to observe impact of temperature increase. Note that when temperature increases from 5 C to 50 C, the array output power decreases from 23.3 kW to 19.3 kW.
- From t = 1.6 s to t = 1.9 s sun irradiance is restored back to 1000 W/m², the array output power increases from 19.3 kW to 91.25 kW.
- From t = 2 s to t = 2.3 s temperature is decreased to 15 C, the array output power increases from 91.3 kW to 100.6 kW.
- From t = 2.4 s to t = 2.45 s, sun irradiance is ramped down fast from 1000 W/m² to 250 W/m² and then restored back to 1000 W/m², from t = 2.7 s to t = 2.75. The array output power decreases from 100 kW to 22.7 kW and increases to 100 kW.
- From t = 2.8 s to t = 3.3 s, the array output power increases from 100 kW to 100.7 kW and the duty cycle is D = 0.45.

Figure 7 shows the PV Voltage and current. Figure 8 shows the duty cycle and the DC link Voltage. Figure 9 depicts the output voltage of VSC (Vab). Figure 10 depicts the grid voltage,grid current and the grid active power.

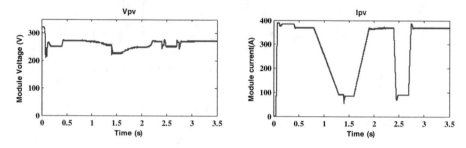

Fig. 7. PV voltage and current

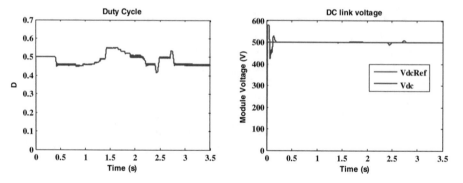

Fig. 8. The duty cycle and the DC link voltage

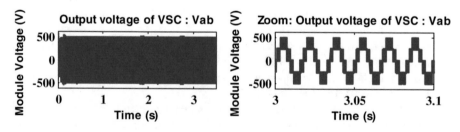

Fig. 9. Vab output voltage of VSC

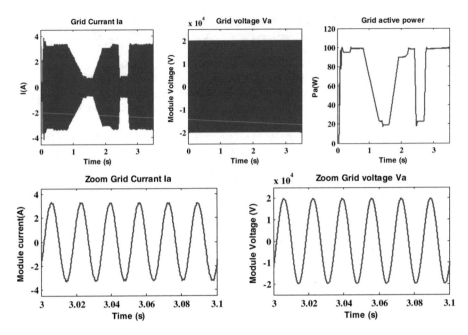

Fig. 10. Grid voltage, grid current and the grid active power

5 Conclusion

This paper proposed a fuzzy approach for controlling MPPT of a photovoltaic system connected to the utility grid. Fuzzy logic based MPPT algorithm has two inputs defined as the P-V Slope and Variation of Slope. The fuzzy MPPT controller is used to extract the optimal photovoltaic power. The obtained results showed that the control system employing fuzzy control is very effective in tracking the MPP when irradiance and temperature levels change. The dc link voltage is well regulated to its reference value using a boost converter. A current and a dc link voltage regulators are used to transfer the photovoltaic power and to synchronise the output inverter with the grid. The simulated model and the results, obtained for differents operating conditions, have shown the performances of the grid connected photovoltaic system.

References

1. Hannan, M.A., Ghani, Z.A., Mohamed, A., Uddin, M.N.: Real-time testing of a fuzzy-logic-controller-based grid-connected photovoltaic inverter system. IEEE Trans. Ind. Appl. **51**(6), 4775–4783 (2015)
2. Jiayi, H., Chuanwen, J., Rong, X.: A review on distributed energy resources and microgrid. Renew. Sustain. Energy Rev. **12**(9), 2472–2483 (2008)
3. Yousef, A.M., El-Saady, G., Abu-Elyouser, F.K.: Fuzzy logic controller for a photovoltaic array system to AC grid connected. In: Smart Grid (SASG), Saudi Arabia, pp. 1–8 (2016)

4. Liang, J., Green, T., Weiss, G., Zhong, Q.C.: Evaluations of repetitive control for power quality improvements of distributed generation. In: Proceedings of Pesc 2002, vol. 4, Cairns, Qld, pp. 1803–1808 (2002)
5. Mattavelli, P., Marfao, F.: Repetitive-base control for selective harmonic compensation in active power filters. IEEE Trans. Indus. Electron. **51**(5), 1018–1024 (2004)
6. Kataria, A.K., Panday, C.K.: Study of ground-reflected component and it is contribution in diffuse solar radiation incident on inclined surfaces over India. Int. J. Energy Environ. **1**, 547–554 (2010)
7. Hohm, D.P., Ropp, M.E.: Comparative study of maximum power point tracking algorithms. Progr. Photovolt. Res. Appl. **11**, 47–62 (2003)
8. Hussein, K.H., Muta, I., Hoshino, T., Osakada, M.: Maximum photovoltaic power tracking: an algorithm for rapidly changing atmospheric conditions. In: IEE Proceedings–Generation, Transmission and Distribution, vol. 142, pp. 59–64 (1995)
9. Quaschning, V.: Renewable Energy and Climate Change. Wiley, New York (2010)
10. Shiau, J.-K., Wei, Y.-C., Chen, B.-C.: A study on the fuzzy-logic-based solar power MPPT algorithms using different fuzzy input variables. Algorithms **8**, 100–127 (2015)

Higher Performance of the Type 2 Fuzzy Logic Controller for Direct Power Control of Wind Generator Based on a Doubly Fed Induction Generator in Dynamic Regime

Belkacem Belabbas[1,2]([⊠]), Tayeb Allaoui[1], Mohamed Tadjine[2], and Mouloud Denaï[3]

[1] Laboratoire de Génie Energétique et Génie Informatique L2GEGI, University of Ibn Khaldoun Tiaret, Tiaret, Algeria
belabbas_1986@yahoo.fr
[2] Laboratoire de Commandes des Processus, Ecole Nationale Polytechnique Algiers, El Harrach, Algeria
[3] School of Engineering and Technology, University of Hertfordshire, Hatfield, UK

Abstract. This paper presents an application of a new control based on Type 2 Fuzzy Logic Controller (T2FLC) for controlling the both powers (active, reactive) of the Doubly Fed Induction Generator (DFIG) in Wind Energy Conversion System (WECS). In addition, a comparison applied between T2FLC and a classical Type 1 Fuzzy Logic Controller (T1FLC) to the purpose of showing the performance of the T2FLC controller as compared to T1FLC controller. The proposed controllers implemented and tested using MATLAB/Simulink. The simulation results show that the controller based on T2FLC characterized by good performance with better convergence and it is fast than those obtained with the T1FLC method.

Keywords: WECS · DFIG · T2FLC · T1FLC

1 Introduction

Doubly Fed Induction Generators (DFIGs) have been widely used for large-scale Wind Energy Conversion System (WECS) applications and are currently the dominant technology in the production of Renewable Energies due to their advantages such as variable speed operation, reduced converter cost, improved system efficiency and reduced electrical power fluctuations [1, 2]. The control of powers (Active, reactive) of the DFIG has been the subject of extensive research [2–4]. Conventional Proportional-Integral (PI) controller are the most commonly used to control the powers of DFIG due to their implementation simple [5]. However, conventional strategies based on PI controller usually based on linearized models, which are difficult to obtain and may not give satisfactory performance under challenging operating conditions such as Parameter Variations (PV) [6]. More robust control methods have proposed in the literature to improve the operation performance of the DFIG. T1FLC has successfully

© Springer International Publishing AG 2018
M. Hatti (ed.), *Artificial Intelligence in Renewable Energetic Systems*, Lecture Notes in Networks and Systems 35, https://doi.org/10.1007/978-3-319-73192-6_22

applied to control the DFIG [7–11]. This controller presents a better performance compared with the traditional PI controller. However, the main disadvantage of T1FLC is the limitation in the face of the significant uncertainties in the system parameters. This limitation due mainly to the use of simple Membership Function (MF) characterized by the T1FLC controller [12]. To manage the uncertainties, we used the T2FLC controller to control the powers of DFIG. This controller characterized by using the other MF of three-dimensional to the purpose of ensuring an additional degree of freedom to manage the uncertainties [13, 14]. For the WECS based on a DFIG, it is necessary to produce a maximum or optimal power according to the Wind Speed (WS). So, several Maximum Power Point Tracking (MPPT) control strategies have proposed in the literature. The most popular based on the Tip Speed Ratio (TSR) algorithm with the control of the Mechanical Speed (MS) of the DFIG that be used in this paper. This method requires an accurate knowledge of the wind turbine parameters and measurement of the WS in order to determine the required generator's speed to extract the maximum power [15–17].

2 Modeling of the Wind Energy Conversion System

The configuration of wind system based of DFIG is consists to a variable speed wind turbine with MPPT controller, a power control of the DFIG is ensured by a T(2&1) FLC controllers, and the Rotor Side Converter (RSC) connected to the rotor of DFIG is controlled by a SVM (Space Vector Modulation) control.

2.1 Wind Turbine Model and Characteristics

The wind turbine is a system that transforms the kinetic power of wind into aerodynamic power P_a. It expressed by the following equation:

$$P_a = \frac{1}{2} C_p(\lambda, \beta) \rho \pi R^2 V^3 \tag{1}$$

$$\lambda = \frac{R \Omega_t}{V} \tag{2}$$

2.2 Modelling of the DFIG

The classical DFIG model described in the Park reference frame given by the following set of electrical equations and electromagnetic torque of DFIG given by [18]:

$$\begin{cases} V_{ds} = R_s i_{ds} + \frac{d\psi_{ds}}{dt} - \omega_s \psi_{qs} \\ V_{qs} = R_s i_{qs} + \frac{d\psi_{qs}}{dt} + \omega_s \psi_{ds} \\ V_{dr} = R_r i_{dr} + \frac{d\psi_{dr}}{dt} - \omega_r \psi_{qr} \\ V_{qr} = R_r i_{qr} + \frac{d\psi_{qr}}{dt} - \omega_r \psi_{dr} \end{cases} \tag{3}$$

$$\begin{cases} \psi_{ds} = L_s i_{ds} + L_m i_{dr} \\ \psi_{qs} = L_s i_{qs} + L_m i_{qr} \\ \psi_{dr} = L_r i_{dr} + L_m i_{ds} \\ \psi_{qr} = L_r i_{qr} + L_m i_{qs} \end{cases} \tag{4}$$

$$T_{em} = \frac{3}{2} \frac{P.L_m}{L_s} \left(\psi_{qs} i_{dr} - \psi_{ds} i_{qr} \right) \tag{5}$$

$$\begin{cases} P_s = V_{ds}.i_{ds} + V_{qs}.i_{qs} \\ Q_s = V_{qs}.i_{ds} - V_{ds}.i_{qs} \end{cases} \tag{6}$$

2.3   DFIG With Stator Field Orientation Strategy

To make the control of the DFIG equivalent that of the DC machine, for which there exist a natural decoupling between flux and torque. An orientation strategy of stator flux following the d-axis is applied. One can write:

$$\begin{cases} \psi_{ds} = \psi_s \\ \psi_{qs} = 0 \end{cases} \tag{7}$$

$$\begin{cases} V_{ds} = 0 \\ V_{qs} = V_s \approx \omega_s \psi_s \end{cases} \tag{8}$$

$$\begin{cases} \psi_s = L_s i_{ds} + L_m i_{dr} \\ 0 = L_s i_{qs} + L_m i_{qr} \end{cases} \tag{9}$$

$$T_{em} = -\frac{3}{2} \frac{P.L_m}{L_s} \psi_{ds} i_{qr} \tag{10}$$

$$\begin{cases} i_{ds} = \frac{V_s}{\omega_s L_s} - \frac{L_m}{L_s} i_{dr} \\ i_{ds} = -\frac{L_m}{L_s} i_{qr} \end{cases} \tag{11}$$

$$\begin{cases} V_{dr} = R_r i_{dr} + L_r \sigma \frac{di_{dr}}{dt} - F_{emd} \\ V_{qr} = R_r i_{qr} + L_r \sigma \frac{di_{qr}}{dt} + F_{emq} \end{cases} \tag{12}$$

With:

$$\begin{cases} F_{emd} = g \omega_s L_r \sigma i_{qr} \\ F_{emq} = g \omega_s L_r \sigma i_{dr} + g \frac{L_m V_s}{L_s} \end{cases}$$

Finally, the Eq. 6 become:

$$\begin{cases} P_s = -\frac{V_s L_m}{L_s} i_{qr} \\ Q_s = -\frac{V_s L_m}{L_s} i_{dr} + \frac{V_s^2}{L_s \omega_s} \end{cases} \tag{13}$$

3 Control Strategies

3.1 MPPT with Speed Control

The goal of the MPPT strategy is to extract the maximum power from the wind. The TSR is the method most used for MPPT. The optimal turbine speed achieved when TSR remains at optimal value.

$$\Omega_t^* = \frac{\lambda_{opt} \cdot V}{R} \tag{14}$$

3.2 DFIG Control Strategy

This section presents the proposed control strategy based on T2FLC. We have propose to use the direct control of power illustrated in Fig. 1. Which contains a single loop, this loop allows controlling the powers of the DFIG. Therefore, the T2FLC be first tested and will be the reference compared to the T1FLC controller. Figures 2 and 3 show the basic structure of the two controllers T2FLC and T1FLC respectively. These controllers consist of four main elements are: Fuzzification, Defuzzification, Inference and rule base. But the only difference between the two controllers is in the type of MF. For the concept of T2FLCnt coroller is well detailed in [19, 20].

The limitation of T1FLC in the face of the significant uncertainties in the system parameters requires the introduction of new tools in the T1FLC controller by using the three-dimensional fuzzy sets to solve the uncertainty problem. These MF are characterized by an additional degree of freedom to manage uncertainties. To control the powers of the DFIG we have used the two controllers (T2FLS, T1FLC) with MF which are shown in Figs. 4 and 5 respectively. These MF are triangular and trapezoidal. As long as we have chosen three MF, we obtain nine basic rules which are shown in Table 1. For both controllers (T2FLC, T1FLC) we used the same methods for fuzzy inference is the method of Mamdani (max-min) and that of defuzzification is the method of center of gravity.

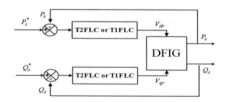

Fig. 1. Proposed DFIG control structure.

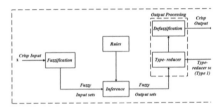

Fig. 2. Structure of a T2FLC system.

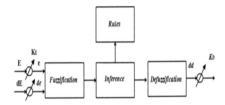

Fig. 3. Structure of a T1FLC system.

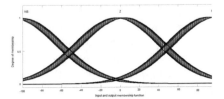

Fig. 4. MF of input and output of T2FLC.

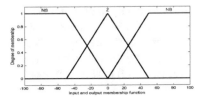

Fig. 5. MF of input and output of T1FLC.

Table 1. Fuzzy role of T1FLC.

Output		dε		
		NS	Z	NB
ε	NS	NS	NS	Z
	Z	NS	Z	NB
	NB	Z	NB	NB

4   Simulation Results and Discussion

To test the performance of the new T2FLC-based controller used to control the two powers of the DFIG. A simulation study for the DFIG-based of WECS has done under the Matlab/Simulink environment. To better present the advantages and performances of T2FLC, a study of four tests be described in the following subsections:

4.1   Control the Power by T2FLC

In the first section of these simulation results, we have simulate the system with the controller T2FLC in order to control the two powers of the DFIG. For this purpose we have chosen a variable wind profile which is presented in Fig. 6. it is characterized by an average speed of 6.5 m/s. This WS will drive the shaft of the DFIG by a MS via the gearbox shown in Fig. 7. After this figure, it can be seen that the MS of the DFIG follows its reference successfully, and with a static error negligible in the permanent regime and with a fast response time in the dynamic regime.

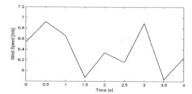

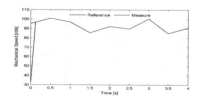

Fig. 6. Random of the wind.   **Fig. 7.** MS: reference (black) and actual (blue). (Color figure online)

With the MPPT algorithm that will extract an Optimal Power from the wind turbine according to the WS. So, we chose this optimum power as an active reference power. Thus, in order to ensure a unit power factor, it is necessary to choose a reference of zero for the reactive power. After Figs. 8(a) and (b) show the reponse of two powers (Active, Reactive) of the DFIG respectively. We find that the two powers follow these references successfully, and with a quick response time. These results show the efficiency of the T2FLC controller to control the two powers of the DFIG.

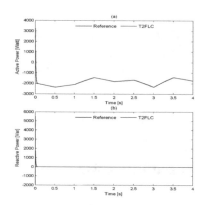

Fig. 8. Powers of DFIG controlled by T2FLC: (a) Active and (b) Reactive.

4.2   Comparative Study Between T2FLC and T1FLC

In this second section of these simulation results, which was based on a comparative study between the two controllers (T2FLC, T1FLC), used to control the two powers of the DFIG. For the purpose of shows the benefits and performance of the new control based on T2FLC compared with the classical controller based on T1FLC. Figures 9 and 10 present the responses of the two controllers T2FLC and T1FLC used to control the two powers of the DFIG respectively with these zones in a permanent and dynamic regime. After these figures, it has observed that the two controllers (T2FLC and T1FLC) are assure the control of the two powers of the DFIG. In addition, they are characterized by good performances in the permanent regime. However, in the dynamic regime, it is clear that the T2FLC controller is more efficient compared to the

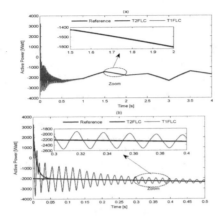

Fig. 9. (a) Response of the active power with T (1&) FLC controllers, (b) Dynamic regime of (a).

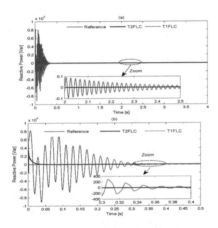

Fig. 10. (a) Response of the reactive power with T (1&) FLC controllers, (b) Dynamic regime of (a).

conventional T1FLC controller, because with the T2FLC the two powers converge rapidly towards the power references and do not possess any overshoot or oscillation.

5   Conclusion

This present work of this paper based on a study of the performance of the T2FLC controller used to control the two powers of the DFIG. A qualitative analysis carried out by a simulation test series to the purpose of showing the advantages of the T2FLC controller. After a quantitative result of simulation, the T2FLC controller shows a good performance by its precision and its speed compared with T1FLC.

References

1. Zin, A.A.B.M., Pesaran, H.A.M., Khairuddin, A.B., Jahanshaloo, L., Shariati, O.: An overview on doubly fed induction generators' controls and contributions to wind based electricity generation. Renew. Sustain. Energy Rev. **27**, 692–708 (2013)
2. Janssens, N., Lambin, G., Bragard, N.: Active power control strategies of DFIG wind turbines. In: Power Technology, IEEE Lausanne, pp. 516–521 (2007)
3. Jerbi, L., Krichen, L., Ouali, A.: A fuzzy logic supervisor for active and reactive power control of a variable speed wind energy conversion system associated to a flywheel storage system. Electr. Power Syst. Res. **79**(6), 919–925 (2009)
4. Hu, J., Nian, H., Hu, B., He, Y., Zhu, Z.Q.: Direct active and reactive power regulation of DFIG using sliding-mode control approach. IEEE Trans. Energy Convers. **25**(4), 1028–1039 (2010)
5. Tamaarat, A., Benakcha, A.: Performance of PI controller for control of active and reactive power in DFIG operating in a grid-connected variable speed wind energy conversion system. Front. Energy **8**(4), 371–378 (2014)
6. Jeong, H.-G., Kim, W.S., Lee, K.-B., Jeong, B.C., Song, S.-H.: A sliding-mode approach to control the active and reactive powers for a DFIG in wind turbines. In: Power Electronics Specialists Conference. IEEE, pp. 120–125 (2008)
7. Dida, A., Benattous, D.: A complete modeling and simulation of DFIG based wind turbine system using fuzzy logic control. Front. Energy **10**, 1–12 (2016)
8. Lekhchine, S., Bahi, T., Soufi, Y.: Indirect rotor field oriented control based on fuzzy logic controlled double star induction machine. Int. J. Electr. Power Energy Syst. **57**, 206–211 (2014)
9. Dida, D.A., Attous, B.: Doubly-fed induction generator drive based WECS using fuzzy logic controller. Front. Energy **9**(3), 272–281 (2015)
10. Hamane, B., Doumbia, M.L., Bouhamida, M., Draou, A., Chaoui, H., Benghanem, M.: Comparative study of PI, RST, sliding mode and fuzzy supervisory controllers for DFIG based wind energy conversion system. Int. J. Renew. Energy Res. IJRER **5**(4), 1174–1185 (2015)
11. Tir, Z., Malik, O.P., Eltamaly, A.M.: Fuzzy logic based speed control of indirect field oriented controlled double star induction motors connected in parallel to a single six-phase inverter supply. Electr. Power Syst. Res. **134**, 126–133 (2016)
12. Castillo, O., Amador-Angulo, L., Castro, J.R., Garcia-Valdez, M.: A comparative study of type-1 fuzzy logic systems, interval type-2 fuzzy logic systems and generalized type-2 fuzzy logic systems in control problems. Inf. Sci. **354**, 257–274 (2016)
13. Kayacan, M.E., Khanesar, A.: Fundamentals of type-2 fuzzy logic theory. In: Fuzzy Neural Networks for Real Time Control Applications, pp. 25 35. Butterworth-Heinemann (2016). Chap. 3
14. Villanueva, I., Ponce, P., Molina, A.: Interval type 2 fuzzy logic controller for rotor voltage of a doubly-fed induction generator and pitch angle of wind turbine blades. Int. Fed. Autom. Control **48**(3), 2195–2202 (2015)
15. Eltamaly, H.A.M., Farh, H.M.: Maximum power extraction from wind energy system based on fuzzy logic control. Electr. Power Syst. Res. **97**, 144–150 (2013)
16. Abo-Khalil, A.G., Lee, D.-C.: MPPT control of wind generation systems based on estimated wind speed using SVR. IEEE Trans. Ind. Electron. **55**(3), 1489–1490 (2008)
17. Narayana, M., Putrus, G.A., Jovanovic, M., Leung, P.S., McDonald, S.: Generic maximum power point tracking controller for small-scale wind turbines. Renew. Energy **44**, 72–79 (2012)

18. Merahi, E.F., Berkouk, E.M.: Back-to-back five-level converters for wind energy conversion system with DC-bus imbalance minimization. Renew. Energy **60**, 137–149 (2013)

19. Sabahi, K., Ghaemi, S., Pezeshki, S.: Application of type-2 fuzzy logic system for load frequency control using feedback error learning approaches. Appl. Soft Comput. **21**, 1–11 (2014)

20. Sabahi, K., Ghaemi, S., Badamchizadeh, M.: Designing an adaptive type-2 fuzzy logic system load frequency control for a nonlinear time-delay power system. Appl. Soft Comput. **43**, 97–106 (2016)

Fuzzy Logic Controller for Improving DC Side of PV Connected Shunt Active Filter Based on MPPT Sliding Mode Control

Abdelbasset Krama[1]([⊠]), Laid Zellouma[1], Boualaga Rabhi[2], and Abdelbaset Laib[3]

[1] LEVRES Laboratory, Electrical Engineering Department, El-Oued University, El-Oued, Algeria
Krama.ab@gmail.com
[2] LMSE Laboratory, Electrical Engineering Department, Biskra University, Biskra, Algeria
[3] LEPCI Laboratory, Electronics Department, Setif University, Setif, Algeria

Abstract. This paper presents the simulation study of three phase shunt active filter connected to photovoltaic array that based on DPC approach. The proposed system has the ability to inject active power into the grid, however it improves the quality of energy by correcting the power factor and reducing harmonic currents in distribution network. This work interests in improving the control of DC side where it is proposed to use fuzzy logic controller in DC bus voltage regulation loop, this later have been developed to provide better performance compared to conventional PI controller. Otherwise, to extract the maximum power of PV system it is proposed to associate the PV panels by a DC–DC boost converter. Therefore, the control of the boost is based on sliding mode that has good accuracy on following the maximum power point. The results of simulation obviously demonstrate that the designed fuzzy logic controller can be used for improving DC bus voltage under external parameter variation.

Keywords: Direct power control · Fuzzy logic · Shunt active power filter
Photovoltaic system · Harmonic

1 Introduction

The increasing use of power electronic devices in manufacturing activities and by customers leads to deterioration of power quality; it consumes non sinusoidal currents that causes harmonic and poor power factor [1]. These later cause numerous drawbacks such as: increase of line losses, saturation of distribution transformers and affect the reliability of power electronic devices [2], however, the demand of good power quality is increasing due to the growth of using sensitive devices that require smooth sinusoidal waveforms [3]. To fit these requirements passive filters have been used to suppress current harmonics and to mitigate the power pollution in grid. Although, this kind of filters offer numerous shortcomings such as huge size, higher cost, resonance problems, and sensitivity to parameters variation that lead resonance problems [3, 4]. The

© Springer International Publishing AG 2018
M. Hatti (ed.), *Artificial Intelligence in Renewable Energetic Systems*, Lecture Notes in Networks and Systems 35, https://doi.org/10.1007/978-3-319-73192-6_23

abovementioned problems have been effectually relieved by using a active filters that was established in around 1970s [5, 6] by Akagy. Shunt active power filter is a flexible solution for harmonic currents mitigation as well as power factor correction. Moreover, It has a perfect ability to adapt with different type of loads [3].

Currently, the use of photovoltaic is increasing quickly and becomes ever more vital issue for energetic and economic development. Photovoltaic (PV) energy seems to be the most favorable source, because it is accessible almost everywhere [7], in addition to its simplicity in implementation with low maintenance requirement. Otherwise, the photovoltaic technology still faces efficiency limits due to its high cost, low performance and nonlinear behavior of PV panel [8]. To overcome these difficulties technique of Maximum Power Point Tracking (MPPT) based on sliding mode control (SMC) have been recommended, this algorithm provide high robustness and good accuaracy in extracting the maximum power available from PV array.

Recently, efforts have been conducted to associate the active filters with Renewable energy sources to benefit from advantages of both a renewable energy source and an improvement of power quality [9].

There have been several techniques suggested for controlling SAPF connected PV system and among them are; P-Q theory as in [10] and The second-order generalized integrator (SOGI) as in [11]. In this work, it is suggested to use technique of direct power control that has become more interesting technique since the last few years due to its advantages. It shows good performance and high proficiency in harmonic and reactive power compensation. Furthmore, it is simple and fast to implement because it relies on direct control of power, so it does not need complex harmonic currents identification and current control loops as in previous mentioned techniques.

The control of DC capacitor voltage in this system is very important because it permits to obtain the reference value of active power that allows to give better performance in time response and system steadiness. For this reason this research is concentrated on DC bus voltage regulation, for which it is proposed to use technique of soft computing that gives better results than conventional controller. A new fuzzy logic controller is constructed to regulate DC bus voltge that is capable to reduce voltage variation especially in external changing condition. Results of this simulation are compared to those obtained by conventional PI controller.

2 General Description of Proposed System

The main task of the studied system is to transfer the power from PV module into the grid, and also to assure the compensation of the harmonic currents and reactive power. The structure in Fig. 1 describes the general configuration of global system based on a three-phase inverter, this later is used to connect PV system into the grid via DC-DC boost converter. Hence, it ensures the injection of the maximum of PV power, in addition to the enhancement of power quality that was deteriorated by nonlinear load. The load is represented by three phase diode rectifier with resistive load.

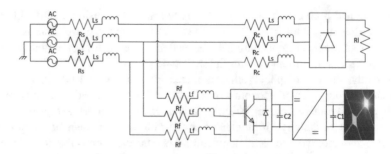

Fig. 1. General configuration of proposed system.

3   Shunt Active Power Filter

3.1   The Theoretical Principle of DPC

The idea of the DPC was proposed by Ohnishi [12], for the first time, he used the values of active and reactive power as control variables instead of instantaneous three-phase currents.

The control strategy of DPC applied to the SAPF is illustrated in Fig. 2.

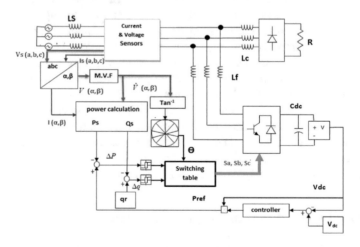

Fig. 2. General scheme of DPC for SAPF

3.2   Calculation of Instantaneous Powers

The calculation of the instantaneous active and reactive power is based on the measuring the voltage and the current of the source, then they can be calculated by the following expressions (1), (2):

$$P_s(t) = V_{sa}.i_{sa} + V_{sb}.i_{sb} + V_{sc}.i_{sc} \tag{1}$$

$$q_s(t) = -\frac{1}{\sqrt{3}}[(V_{sa} - V_{sb}).i_{sa} + (V_{sb} - V_{sc}).i_{sb} + (V_{sc} - V_{sa}).i_{sc} \tag{2}$$

3.3   Choice of Sector

The influence of each output vector resulting from SAPF on active and reactive power is very dependent on the real position of the source voltage vector. To increase accuracy and also to avoid the problems encountered at the borders of each vector of control, the plan of the vector space is divided into 12 sectors of 30° for each one Fig. 3, where the first sector is defined between (3):

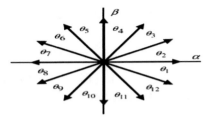

Fig. 3. Sectors on stationary coordinates.

$$-\frac{\pi}{3} < \theta < 0 \tag{3}$$

The sectors can be numerically expressed as (4):

$$(n - 2)\frac{\pi}{6} < \theta_n \le (n - 1)\frac{\pi}{6} \quad n = 1, 2. \ldots 12 \tag{4}$$

The angle is calculated using the inverse trigonometric function, based on the components of the voltage vector in the (α, β) reference frame, indicated by the following Eq. (5):

$$\theta = \arctan \frac{V_\alpha}{V_\beta} \tag{5}$$

Where $V\alpha$ and $V\beta$ are the source voltage on (α, β) frame which are calculated according to the following Eq. (6):

$$\begin{bmatrix} V_\alpha \\ V_\beta \end{bmatrix} = \sqrt{\frac{2}{3}} \begin{bmatrix} 1 & -\frac{1}{2} & -\frac{1}{2} \\ 0 & \frac{\sqrt{3}}{2} & -\frac{\sqrt{3}}{2} \end{bmatrix} \begin{bmatrix} V_a \\ V_b \\ V_c \end{bmatrix} \tag{6}$$

3.4   Switching Table

Two-level inverter generates seven voltage vectors for eight different combinations, the calculation of each voltage based on a combination of the respective switches state and the DC bus voltage. By employing Table 1, the optimum state of the inverter switches can be chosen in each state of commutation according to the combination of the digital signals Sp, *and* Sq *and* sector number.

Table 1. Switching table.

S_p	S_q	θ1	θ2	θ3	θ4	θ5	θ6	θ7	θ8	θ9	θ10	θ11	θ12
1	0	101	111	100	000	110	111	010	000	011	111	001	000
	1	110	111	010	000	011	111	001	000	101	111	100	000
0	0	101	100	100	110	110	010	010	011	011	001	001	101
	1	100	110	110	010	010	011	011	001	001	101	101	100

3.5   Self Tunning Filter (STF)

Abdusalam et al. [13] have presented in their work the principle, frequency and dynamic response of the STF under distorted conditions. Voltage signal, before and after filtering can be expressed by (7), (8):

$$\hat{V}_\alpha(s) = \frac{k}{s}\left[V_\alpha(s) - \hat{V}_\alpha(s)\right] - \frac{w_c}{s}\hat{V}_\alpha(s) \tag{7}$$

$$\hat{V}_\beta(s) = \frac{k}{s}\left[V_\beta(s) - \hat{V}_\beta(s)\right] - \frac{w_c}{s}\hat{V}_\beta(s) \tag{8}$$

Where, wc is the pulsation of STF.

4   Photovoltaic System

4.1   Modeling of Solar PV Module

The combination of several photovoltaic cells in series and in parallel provides a photovoltaic generator. The behavior of photovoltaic cell has been extensively studied for more than 25 years. Many articles on modeling modules exist in the literature. The model for a single diode is the most commonly used due to good results. The equivalent model of a PV cell can be seen in Fig. 4 [14].

The characteristic (I-V) for a PV Cell is described by the nonlinear Eq. (9), [14, 15].

$$I_{pv} = I_{sc} - I_0 \cdot [\exp(\frac{V + R_s.I_{pv}}{V_t.\alpha}) - 1] - \frac{V + R_s.I_{pv}}{R_p} \tag{9}$$

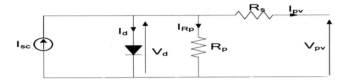

Fig. 4. Single-diode model of the theoretical PV cell.

Where:

Isc: the current generated by the incident light. k: the Boltzmann constant $(1.3806503 \times 10^{-23}$ J/K).

I0: the reverse saturation or leakage current of the diode. α: the diode ideality constant.

$V_t = N_s.k.T/q$ is the thermal voltage of the array with. Rs: the equivalent series resistance.

N_s: number of cells connected in series. R_p: the equivalent parallel resistance.

q: the electron charge $(1.60217646 \times 10^{-19}$ C). T: the temperature of the p–n junction in Kelvin

Each module used in the simulations provides 55 W at maximum power point MPP, in Standard Test Conditions mode. Table 2 shows the main information of this module.

Table 2. The main information of proposed PV module.

Parameters of the PV module SOLARA SM 220 S/M 55	
Maximum power Pmax	55 W ($\pm$ 10%)
Maximum power point Voltage (Vmpp)	17.8 V
Open circuit voltage (Uoc)	21.7 V
Maximum power point Current (Impp)	3.1 A
Short circuit current (Ioc)	3.2 A

4.2 MPPT Based on Slinding Mode Control

Recently, the sliding mode becomes the most widely used in the control of boost converter in order to extract MPP of PV generator.

The advantages of sliding mode control are various: high precision, good stability, simplicity, invariance, robustness etc. [16, 17].

It is found, in literature, Many different switching surfaces that provide both stability and adequate dynamics for a given application. We choose a a non-linear surface, as in (10), due to the ease of implementation [18].

$$S = \frac{dI_{pv}}{dV_{pv}} V_{pv} V_{pv} + I_{pv} \tag{10}$$

The simple configuration of the first order sliding mode control bases on two main parts as can be seen in (1):

$$u = u_{eq} + K.sign(S) \tag{11}$$

The first one is called the equivalent control quantity, it is derived from the condition (12) that verifys Lyapounov stability theorem.

$$d\dot{S} = 0 \tag{12}$$

The equivalent control part is defined as (13).

$$du_{eq} = 1 - \frac{V_{pv}}{V_{dc}} \tag{13}$$

K: is a constant that represents the maximum controller output required to ensure the stabilization in steady state.

The real control signal is suggested as (14):

$$u = \begin{vmatrix} 1 & if & u_{eq} + K.sign(S) > 1 \\ u_{eq} + k.sign(S) & if & 0 < u_{eq} + K.sign(S) < 1 \\ 0 & if & u_{eq} + K.sign(S) < 0 \end{vmatrix} \tag{14}$$

5   DC-Bus Regulation Using Fuzzy Logic Controller

Fuzzy logic control, which was introduced firstly by Zadeh in 1965 [19], is inspired from fuzzy set theory that is besed on transformation in membership function. A scope of application of fuzzy logic which becomes frequent is that of the adjustment and controlling industrial regulations. This method makes it possible to obtain an effective manner of control, without having to call important theoretical developments. It presents the interest to take into account the experiences acquired by users and operators of the process to be controlled [20].

The Construction fuzzy logic controller (FLC) is based on three main steps: fuzzification, inference system and defuzzification [21], as shown in Fig. 5.

Fig. 5. Main structure of fuzzy logic controller.

In fuzzification phase, the digital inputs are converted to fuzzy variables, as in this study there are two inputs represented by the error and its variation, where the fuzzy variables are depicted by seven triangular membership functions for both inputs and output, as represent Figs. 6 and 7.

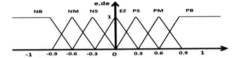   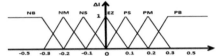

Fig. 6. Membership function of inputs (e, de).   **Fig. 7.** Membership function of output.

Thus, the inference system phase determines the fuzzy output according to the kind of inputs based on rule table represented by Table 3, Technique of inference chosen is Sum – product, that is articulated on Mamdani approach [21].

Table 3. Rules table

Error	Error variation						
	NB	NM	NS	EZ	PS	PM	PB
NB	NB	NB	NB	NB	NM	NS	NB
NM	NB	NB	NB	NM	NS	EZ	NM
NS	NB	NB	NM	NS	EZ	PS	NS
EZ	NB	NM	NS	EZ	PS	PM	EZ
PS	NM	NS	EZ	PS	PM	PB	PS
PM	NS	EZ	PS	PM	PB	PB	PM
PB	EZ	PS	PM	PB	PB	PB	PL

The last phase is the deffuzification where the fuzzy variable after the inference system are transformed to digital value by using center of gravity method.

Ge, Gde et Gc: Represent the gains of adaptation. These gains are determined by adjusting to having the desired response. Generally, They play an extremely important role. Indeed, it is the last stage that fix the performance of the control. The quantities e and de are normalized In a universe of discourse [−1; +1], these quantities must be converted into linguistic variables [21, 22].

6 Simulation Results and Discussion

The proposed PV model is simulated using Matlab/Simulink, at standard test conditions, simulated I-V and P-V characteristics are shown in Fig. 8.

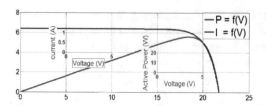

Fig. 8. I-V and P-V characterestics of PV model.

To compare and analyze the proposed fuzzy logic controller against conventional PI controller, dc- bus voltage for each controller are represented in the same figure (Fig. 9).

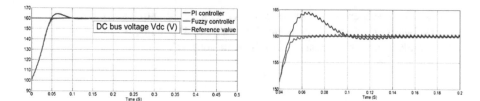

Fig. 9. Obtained DC bus voltage after using Fuzzy logic and PI controller.

Correspond to figure it is observed that the fuzzy logic controller has a better result than conventional controller, it is clear that fuzzy logic reduces time response and voltage overshoot is omitted, furthermore, the steady state error is near to zero (Fig. 10).

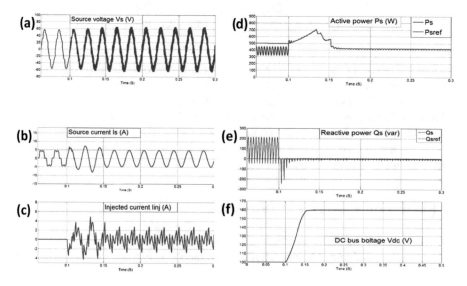

Fig. 10. Simulation results at filter switched on: (a, b, c, d, e, f).

To validate the proposed control technique applicability presented in this work, simulations were carried out by means of MATLAB software under different operation modes: before filtering and after filtering, then after introducing PV system when solar power is injected to the grid.

It is observed that before inserting shunt active filter with fuzzy logic controller, the source current has not a sinusoidal waveform due to presence of non-linear load that is represented by diode bridge rectifier, thereby the grid provides active and reactive power to the load. Once the SAPF is switched on at t = 0.4 S, the source current becomes sinusoidal, moreover the reactive power is compensated at the grid side whereas the active power follows its nominal value.

Dc bus voltage is stabilized to its reference value after short transient time (0.08 S) that proves the proficiency of the proposed controller.

To evaluate the performance of the present system in transferring solar power into the grid, simulation results have been carried out under variable illumination, in fig the solar irradiation is changed respectively (0 W/m2, 1000 W/m2, 400 W/m2), otherwise the temperature value is saved constant "25 °C" all over the experiment.

According to Fig. 11, it can be seen that the source active power is decreased once the PV system is inserted at t = 0.8 S, after t = 1S (when the solar irradiation will be the optimum G = 1000 W/m2) the PV panel provides its maximum power P_{max} = 220 W which proves the capability of MPPT to track the optimum power. However the PV power is decreased when the solar irradiation is changed (1000 w/m2 to 400 w/m2).

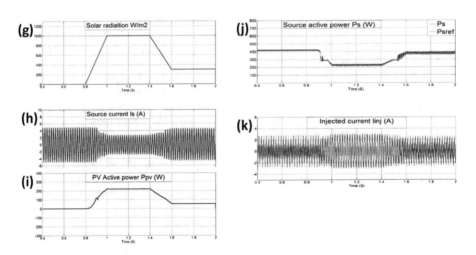

Fig. 11. Simulation results under variable illumination: (g, h, i, j, k).

The current of utility grid after inserting PV system has been analyzed to obtain its THD and compared to those obtained in the case of conventional PI controller. Figure 12 demonstrates the capability of proposed controller in reducing harmonic currents and providing lower THD.

(l)   **(m)**

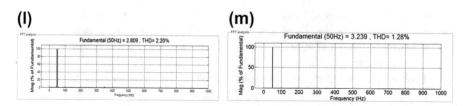

Fig. 12. Spectral analysis of source current: PI (l), fuzzy logic (m).

7   Conclusion

In this paper theoretical and simulation study of a fuzzy logic controller for PV connected shunt active power filter is presented. The proposed system has perfect ability in power quality enhancement in addition to solar power injection. The obtained results show that fuzzy logic controller gives better result especially in transient state where overshoot is significantly reduced compared to conventional PI controller, and it gives fast response, moreover the THD of supply current is reduced to 1.28% that confirms IEEE 519-1992, the simulation results under changing climate condition demonstrate that the studied system adapts easily to other more severe constraints.

References

1. Arulkumar, K., Palanisamy, K., Vijayakumar, D.: Recent advances and control techniques in grid connected PV system–a review. Int. J. Renew. Energy Res. **6**(3), 1037–1049 (2016)
2. Chaoui, A., Gaubert, J.-P., Krim, F.: Power quality improvement using DPC controlled three-phase shunt active filter. Electr. Power Syst. Res. **80**(6), 657–666 (2010)
3. Uyyuru, K.R., Mishra, M.K., Ghosh, A.: An optimization-based algorithm for shunt active filter under distorted supply voltages. IEEE Trans. Power Electron. **24**(5), 1223–1232 (2009)
4. Yi, H., et al.: A source-current-detected shunt active power filter control scheme based on vector resonant controller. IEEE Trans. Ind. Appl. **50**(3), 1953–1965 (2014)
5. Mishra, M.K., Karthikeyan, K.: A study on design and dynamics of voltage source inverter in current control mode to compensate unbalanced and non-linear loads. In: International Conference on Power Electronics, Drives and Energy Systems, PEDES 2006, New Delhi, pp. 1–8 (2006)
6. Akagi, H., Nabae, A.: Control strategy of active power filters using multiple voltage source PWM converters. IEEE Trans. Ind. Appl. **IA-22**, 460–465 (1986)
7. Harrag, A., Messalti, S.: Variable step size modified P&O MPPT algorithm using GA-based hybrid offline/online PID controller. Renew. Sustain. Energy Rev. **49**, 1247–1260 (2015)
8. Rizzo, S.A., Scelba, G.: ANN based MPPT method for rapidly variable shading conditions. Appl. Energy **145**, 124–132 (2015)
9. Reisi, A.R., Moradi, M.H., Showkati, H.: Combined photovoltaic and unified power quality controller to improve power quality. Sol. Energy **88**, 154–162 (2013)
10. Bouzelata, Y., Kurt, E., Chenni, R., Altın, N.: Design and simulation of a unified power quality conditioner fed by solar energy. Int. J. Hydrogen Energy **40**(44), 15267–15277 (2015). (in press)

11. Singh, B., Kumar, S., Jain, C.: Damped-SOGI-based control algorithm for solar PV power generating system. IEEE Trans. Ind. Appl. **53**(3), 1780–1788 (2017)
12. Ohnishi, T.: Three phase PWM converter/inverter by means of instantaneous active and reactive power control. In: International Conference of the IEEE Industrial Electronics Society (IECON), Kobe, Japan, November 1991, vol. 1, pp. 819–824 (1991)
13. Abdusalam, M., Poure, P., Karimi, S., Saadate, S.: New digital reference current generation for shunt active power filter under distorted voltage conditions. Electr. Power Syst. Res. **79**(5), 759–765 (2009)
14. Villalva, M.G., Gazoli, J.R., Filho, E.R.: Comprehensive approach to modeling and simulation of photovoltaic arrays. IEEE Trans. Power Electron. **24**(5), 1198–1208 (2009)
15. Rauschenbach, H.S.: Solar Cell Array Design Handbook. Van Nostrand Reinhold, NewYork (1980)
16. Rekioua, D., Achour, A.Y., Rekioua, T.: Tracking power photovoltaic system with sliding mode control strategy. Energy Procedia **36**, 219–230 (2013)
17. Abderrezek, H., Harmas, M.N.: Comparison study between the terminal sliding mode control and the terminal synergetic control using PSO for DC-DC converter. In: 2015 4th International Conference on Electrical Engineering (ICEE), Boumerdes, pp. 1–5 (2015)
18. Slotine, J.J.E.: Applied Nonlinear Control. Addison Wesley, Reading (2005)
19. Zadeh, L.A.: Fuzzy sets. Inf. Control **8**, 338–353 (1965)
20. Saad, L., Zellouma, L.: Fuzzy logic controller for three-level shunt active filter compensating harmonics and reactive power. Electr. Power Syst. Res. **79**(10), 1337–1341 (2009)
21. Ünsal, S., Alişkan, İ.: Performance analysis of fuzzy logic controllers having Mamdani and Takagi-Sugeno inference methods by using unique software and toolbox. In: 2016 National Conference on Electrical, Electronics and Biomedical Engineering (ELECO), Bursa, pp. 237–241 (2016)
22. Zellouma, L., Rabhi, B., Saad, S., Benaissa, A., Benkhoris, M.F.: Fuzzy logic controller of five levels active power filter. Energy Procedia **74**, 1015–1025 (2015)

Improving PV Performances Using Fuzzy-Based MPPT

Abdelghani Harrag[1,2(✉)] and S. Messalti[3]

[1] Physics Department, Faculty of Sciences, Ferhat Abbas University Setif 1,
El-Bez, 19000 Setif, Algeria
a.b.harrag@gmail.com
[2] CCNS Laboratory, Electronics Department, Faculty of Technology,
Ferhat Abbas University Setif 1, El-Maabouda, Setif, Algeria
[3] Electrical Engineering Department, Mohamed Boudiaf University,
Route BBA, 28000 Msila, Algeria
Messalti.sabir@yahoo.fr

Abstract. In this paper, a new MPPT controller based on variable step size fuzzy controller have been proposed and investigated. The performance of the proposed algorithm are analyzed under different atmospheric conditions using Matlab/Simulink environment using boost DC-DC converter connected to Solarex MSX-60 panel. The results have demonstrated the high performances of the proposed technique showing good improvements of the proposed algorithm to track effectively the maximum power point with low oscillation, low ripple, low overshoot and good rapidity in slow or fast changing atmospheric conditions.

Keywords: PV systems · MPPT · Fuzzy logic · FLC

1 Introduction

The fossil fuels are quickly being drained pushing the development of green renewable energy sources like sunlight and wind that are regularly recharged. Environmental change, coupled with worries about high oil and energy costs, is driving a worldwide pattern towards the expanded utilization of renewable energy. Among renewable energy sources, solar energy is playing an increasingly important role in the decarbonisation of most industrialized countries [1]. A photovoltaic system is made up of several photo-voltaic solar cells, capable of generating between 1 and 2 W depending on the type of solar cell material. For higher power output, PV cells can be connected in series and parallel to form higher power modules. Although the modules are from the same manu-factures or from the same materials, the module characteristics varies and can affect the entire system performance. In addition, the nonlinear characteristics presenting only one maximum power point called (MPP) requires the use of an intermediate conversion stage interfacing the PV array and the power system that uses the electrical power produced. This stage known as Maximum Power Point Tracking controller (MPPT) must be capable of maximum of power from the PV generator to the load and able to perform this adaptation in the presence of time-varying operating conditions affecting the PV generator [2].

© Springer International Publishing AG 2018
M. Hatti (ed.), *Artificial Intelligence in Renewable Energetic Systems*, Lecture Notes in Networks and Systems 35, https://doi.org/10.1007/978-3-319-73192-6_24

Over the last two decades, a range of MPPT algorithms has been proposed using conventional techniques like perturbation and observation (P&O) algorithm [4], Hill Climbing (HC) algorithm [5], Incremental Conductance (IC) [6], or artificial intelligence algorithms like neural networks [7], genetic algorithm (GA) [8], particle swarm optimization algorithm [9] (Fig. 1).

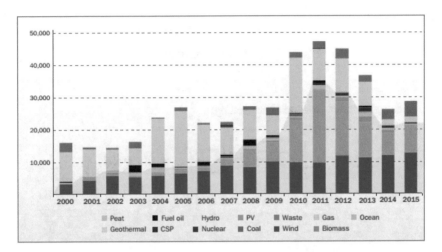

Fig. 1. Installed renewable power capacity [3].

In this paper, a fuzzy variable step-size MPPT controller for PV systems has been proposed in order to improve the static and dynamic MPPT performances by the adjustment of the boost DC-DC duty cycle. The proposed fuzzy MPPT was developed and simulated using a boost converter connected to Solarex MSX-60 module. The results have demonstrated the high performances of the proposed technique showing good improvements of the proposed algorithm to track effectively the maximum power point with low oscillation, low ripple, low overshoot and good rapidity in slow or fast changing atmospheric conditions. The rest of this paper is organized as follows: Sect. 2 describes the PV modeling. The proposed variable step size fuzzy MPPT controller is detailed in Sect. 3. The simulation results and discussion are presented in Sect. 4. Finally, Sect. 5 concludes the paper giving some comments and future directions.

2   PV Modeling

The well-known and widely used model based on the well-known Shockley diode equation is presented below (Fig. 2).

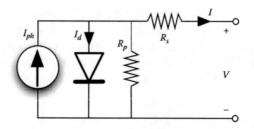

Fig. 2. Solar cell single-diode model [10].

The output current I can be expressed by:

$$I = N_p I_{ph} - N_p I_{rs} \left[e^{\left(\dfrac{q(v + R_S.I)}{A.k.T.N_S} \right)} - 1 \right] - N_p \left(\dfrac{q(v + R_S I)}{N_S.R_p} \right) \tag{1}$$

where

V is the cell output voltage;

q is the electron charge ($1.60217646 \times 10^{-19}$C);

k is the Boltzmann's constant ($1.3806503 \times 10^{-23}$J/K);

T is the temperature in Kelvin;

I_{rs} is the cell reverse saturation current;

A is the diode ideality constant;

N_p is the number of PV cells connected parallel;

N_s is the number of PV cells connected in series.

The generated photocurrent I_{ph} is related to the solar irradiation by the following equation:

$$I_{ph} = \left[I_{sc} + k_i (T - T_r) \right] \dfrac{S}{1000} \tag{2}$$

where

k_i is the short-circuit current temperature coefficient;

S is the solar irradiation in W/m²;

I_{sc} is the cell short-circuit current at reference temperature;

T_r is the cell reference temperature.

The cell's saturation current is varies with temperature according to the following equation:

$$I_{rs} = I_{rr} \left[\dfrac{T}{T_r} \right]^3 \exp \left(\dfrac{q.E_G}{k.A} \left[\dfrac{1}{T_r} - \dfrac{1}{T} \right] \right) \tag{3}$$

where
    E_G is the band-gap energy of the semiconductor used in the cell;
    I_{rr} is the reverse saturation at T_r.

3   Fuzzy MPPT Controller Implementation

Fuzzy logic have the benefits of operating with general inputs instead of a correct mathematical model and handling nonlinearities mathematical logic management typically consists of three stages: fuzzification, rule base, inference method and defuzzification [11, 12].

In this study, the proposed fuzzy MPPT requires as inputs the error E and the change in error ΔE defined by:

$$E_k = \frac{P(k) - P(k-1)}{V(k) - V(k-1)} \tag{4}$$

$$\Delta E_k = E_k - E_{k-1} \tag{5}$$

where
    P(k) and P(k-1) are the power at instant k and k-1;
    V(k) and V(k-1) are the voltage at instant k and k-1;
    The output is the PWM duty cycle step variation Δd.

Fuzzification: The universe of discourse for input variables (E_k and ΔE_k) as well as for the output variable (Δd) is divided into seven fuzzy set s: PL (Positive Large), PM (Positive Medium), PS (Positive Small), Z (Zero), NS (Negative Small), NM (Negative Medium) and NL (Negative Large).

Rule base: The Fuzzy algorithm tracks the MPP based on the rule-base consisting of 49 rules as shown below (Fig. 3):

E_k / ΔE_k	NL	NM	NS	Z	PS	PM	PL
NL	PL	PL	PL	PL	NM	Z	Z
NM	PL	PL	PL	PM	PS	Z	Z
NS	PL	PM	PS	PS	PS	Z	Z
Z	PL	PM	PS	Z	NS	NM	NL
PS	Z	Z	NM	NS	NS	NM	NL
PM	Z	Z	NS	NM	NL	NL	NL
PL	Z	Z	NM	NL	NL	NL	NL

Fig. 3. Rule base.

Inference method: in this study, we use the Mamdani's inference system is utilized with the max-min creation strategy [13].

Defuzzification: we use the centroid method for defuzzification considered as one of the normally utilized defuzzification routines.

4   Results and Discussions

To illustrate the effectiveness of proposed fuzzy MPPT controller, the model of the entire system designed using Matlab/Simulink software is considered. It is composed of PV panel operating at variable temperature and irradiation conditions and fuzzy controller driving the variable step size needed by the MPPT algorithm that drives the duty cycle of DC-DC converter.

Figure 4 shows the array output power corresponding to the input irradiation scheme defined. While Figs. 5, 6, 7 and 8 shows the zoom of specific points A, B, C and D.

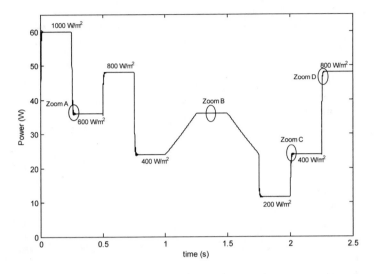

Fig. 4. PV output power corresponding to the input irradiation scheme defined

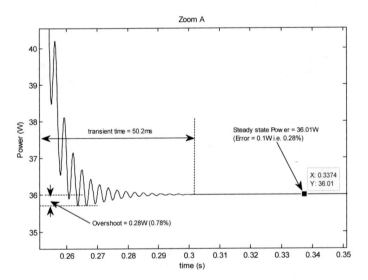

Fig. 5. Point A.

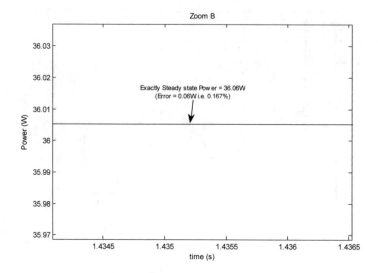

Fig. 6. Point B.

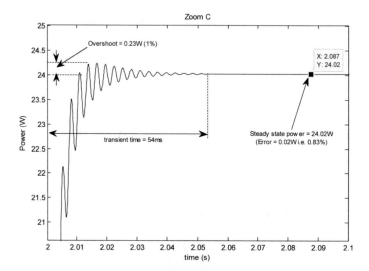

Fig. 7. Point C.

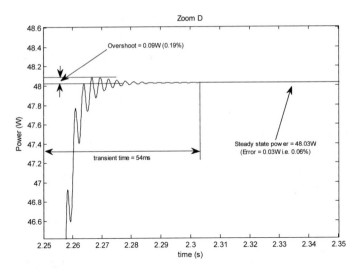

Fig. 8. Point D.

From Figs. 4, 5, 6, 7 and 8, the proposed fuzzy MPPT controller shows a good performances with low overshoot, low response time and no ripple in steady state. The results prove the effectiveness of the proposed algorithm in transient response as well in steady state regime improving the MPPT tracking reducing as consequence the power losses. Table 1. summarizes the important points.

Table 1. Summarized results.

Point	Overshoot (W)	Resp. time (ms)	Steady state Error (W)
Point A	0.28	50.2	0.1
Point B	–	–	0.06
Point C	0.23	54	0.02
Point D	0.09	54	0.03

5 Conclusion

In this paper, a fuzzy maximum power point controller for photovoltaic systems under fast changing atmospheric conditions has been presented and investigated. The proposed fuzzy variable step-size MPPT algorithm is simulated using Matlab/Simulink model, where the entire system has been implemented including a boost DC-DC converter connected to Solarex MSX-60 panel. The results have demonstrated the high performances of the proposed technique showing good improvements of the proposed algorithm to track effectively the maximum power point with low oscillation, low ripple, low overshoot and good rapidity in slow or fast changing atmospheric conditions. As future works, we plan to compare the proposed fuzzy MPPT to other types of MPPT algorithms in order to confirm and validate the effectiveness of the proposed philosophy.

References

1. IEA: Technology Roadmaps Bioenergy for Heat and Power (2014). http://www.iea.org/publication/
2. Rekioua, D., Matagne, E.: Optimization of Photovoltaic Power Systems: Modelization, Simulation and Control. Springer-Verlag, London (2012)
3. IRENA_Roadmap.   http://www.irena.org/DocumentDownloads/Publications/IRENA_RE map_2016_edition_report.pdf
4. Harrag, A., Messalti, S.: Variable step size modified P&O MPPT algorithm using GA-based hybrid offline/online PID controller. Renew. Sustain. Energy Rev. **49**, 1247–1260 (2015)
5. Eltawil, M.A., Zhao, Z.: MPPT techniques for photovoltaic applications. Renew. Sustain. Energy Rev. **25**, 793–813 (2013)
6. Tey, K.S., Mekhilef, S.: Modified incremental conductance MPPT algorithm to mitigate inaccurate responses under fast-changing solar irradiation level. Sol. Energy **101**, 333–342 (2014)
7. Messalti, S., Harrag, A., Loukriz, A.: A new neural networks MPPT controller for PV systems. In: 2015 6th International Renewable Energy Congress (IREC), Sousse Tunisia, pp. 1–6, 24–26 March 2015
8. Shaiek, Y., et al.: Comparison between conventional methods and GA approach for maximum power point tracking of shaded solar PV generators. Sol. Energy **90**, 107–122 (2013)
9. Letting, L.K., Munda, J.L., Hamam, Y.: Optimization of a fuzzy logic controller for PV grid inverter control using s-function based PSO. Sol. Energy **86**, 1689–1700 (2012)
10. Femia, N., Petrone, G., Spagnuolo, G., Vitelli, M.: Power Electronics and Control Techniques for Maximum Energy Harvesting in Photovoltaic Systems. Taylor & Francis, London (2003)
11. Pedrycz, W.: Fuzzy Control and Fuzzy Systems, 2nd edn. Research Studies Press Ltd, Taunton (1993)
12. Gounden, N.A., Peter, S.A., Nallandula, H., Krithiga, S.: Fuzzy logic controller with MPPT using line-communicated inverter for three-phase grid-connected photovoltaic systems. Renew Energy **34**, 909–915 (2009)
13. Mamdani, Ebrahim H.: Application of fuzzy algorithms for control of simple dynamic plant. Proc. Inst. Electr. Eng. **121**(12), 1585–1588 (1974)

FPGA-Based Implementation of an Intelligent Fault Diagnosis Method for Photovoltaic Arrays

Wafya Chine[1(✉)], Adel Mellit[1(✉)], and Rabah Bouhedir[1,2(✉)]

[1] Electronics Department, Jijel University, Jijel, Algeria
wafia.chine@gmail.com, adelmellit2013@gmail.com,
bouhedirrabah@gmail.com
[2] Electronics Department, Blida1 University, Blida, Algeria

Abstract. Fault diagnosis in photovoltaic (PV) installations is a fundamental task to protect the components of PV systems (modules, strings and inverters), from damage and to eliminate possible fire risks. In this paper, an intelligent fault detection and diagnosis method has been presented and implemented into a reconfigurable Field Programmed Gate Array (FPGA). Only faults that can be appeared in PV arrays are examined in this work. The designed method consists of two parts: the first one is based on signal threshold approach, and the second one is based on an artificial neural network (ANN). The whole parts of the system have been implemented into FPGA board (named ZYNQ XC7Z010-1CLG400C). Xilinx System Generator (XSG) and VIVADO tools have been used to simulate and implement the algorithm. Results demonstrate with success the possibility implementation of the designed diagnosis method.

Keywords: Photovoltaic systems · Fault diagnosis · Neural networks · FPGA

1 Introduction

Photovoltaic energy has become the most available source all over the world since the photovoltaic (PV) modules have acceptable performance, easy to install, do not caused the pollution. As reported in [1], at the end of 2016 the total installed PV capacity was around 300GWp. The energy produced by a PV system depends on a number of factors such as; the nominal characteristics of the components of the PV system, geometrical configurations, weather conditions, the local horizon and the near-field shading and faults that may occur during its operation, …etc. Most faults occurred in a PV system can be related to the PV array, inverter, grid, and the connection between the different elements of the PV system. These faults reduce significantly the produced power, availability, reliability and effectively the "security" of the whole system. For this reason, fault diagnosis techniques are become today more and more proposed in the literature. Generally, fault detection and diagnosis methods can be grouped mainly into two groups [2]: electrical and non-electrical methods. Electrical methods can classified into four categoris: (1) methods that do not require climate data (solar radiation and module temperature) [3], (2) methods based on the I-V characteristics [4], (3) methods used the maximum power point tracking (MPPT) and power losses [5], (4) and statistical and artificial intelligence approaches [6–8].

© Springer International Publishing AG 2018
M. Hatti (ed.), *Artificial Intelligence in Renewable Energetic Systems*, Lecture Notes in Networks and Systems 35, https://doi.org/10.1007/978-3-319-73192-6_25

Most fault diagnosis methods are developed and simulated using software, such as Labview, Matlab etc. for real time applications. Integrating methods into electronics hardware devices, using Microcontroller, DSP or Field Programmed Gate Arrays (FPGAs), for real time applications play a very important role in this field. Thus, FPGAs based hardware solutions, using the device's inherent parallelism, have been received increased attention, as they allow engineers/designers to develop efficient hardware architectures based on flexible software. Additional FPGA advantages include the fact that their hardware logic is extremely fast, much faster than software-based logic. They are easier to interface to the outside world, either through custom peripherals or via glue logic to custom co-processors. They are also better suited for bit-level operations than a microprocessor [9].

The main objective of this paper is to present the different steps for designing and implementing the fault detection and diagnosis method proposed in [10], into FPGA board. In our previous work [9], both algorithms have been implemented separately. However in the present work we will implement all parts of the designed method including modeling and parameters extraction, fault detection and fault diagnosis. A ZYNQ XC7Z010-1CLG400C board has been chosen due to its good performances.

2   Methodology

The fault detection and diagnosis technique developed in [9] is able to identify one normal and eight faulty modes. It consists of two parts: fault detection algorithm and fault diagnosis.

2.1   Fault Detection

Figure 1 shows the diagram block of the fault detection method [9], firstly the difference between the measured and the simulated PV array output power is compared with a threshold (Th) in order to detect the possible presence of a fault.

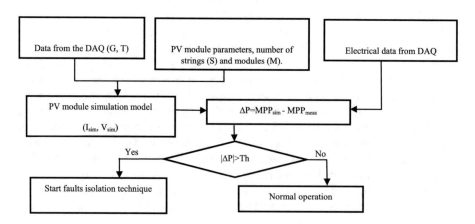

Fig. 1. Block diagram of the fault detection technique [9]

2.2 Fault Diagnosis

The fault diagnosis method is divided into two algorithms: model based algorithm and ANN based algorithm.

Method-based algorithm (Algorithm-I). The first algorithm aims to analysis five symptoms which are: reduction in the short circuit current (C1), reduction in the open circuit voltage (V1), reduction in the output current (C2), reduction or an increase in the output voltage (V2), an increased number of MPPs in the I-V characteristic (N). First, the measured and simulated PV strings attributes are calculated and the relative differences are compared with some thresholds. The thresholds are calculated on the base of the maximum relative error introduced by the model uncertainty and the measurement noise. Figure 2 shows the flowchart of the algorithm-I which allows the isolation of the following six different situations:

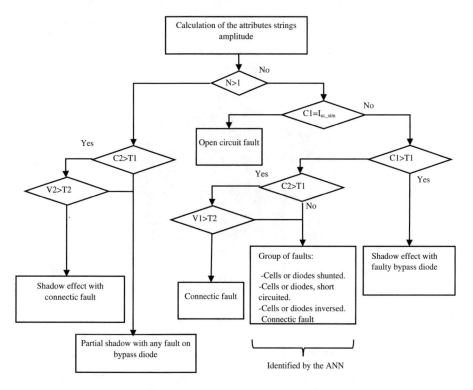

Fig. 2. The fault isolation technique [9]

- Module open circuited: F4;
- Connectic fault: F5;
- Partial shadow (bypass diodes work correctly): F6;
- Partial shadow (when a bypass diode is faulted): F7;

- Partial shadow with connectic fault: F8;
- A group of faults including: F1, F2, F3, and F5.

Where the thresholds T1 and T2 are calculated based of the maximum error introduced by the model uncertainty and the measurement noise.

ANN based algorithm (Algorithm-II). With reference to Fig. 2, the first algorithm cannot distinguish the faults: F1, F2, F3, and F4, thus in other to isolate these faults, a classification technique was needed; thus, an ANN technique has been employed. A Multi-Layer Perceptron (MLP) architecture is used, it consists of three neurons in the input layer corresponding to the ratio between the measured and the simulated values of the open circuit voltage (R_{voc}), the maximum power point current (R_{impp}) and the maximum power point voltage (R_{vmpp}). Four neurons in the output layer corresponding to the fault class. The network with 13×13 hidden nodes was trained with Levenberg-Marquardt (LM) algorithm. The minimum Mean Square Error (MSE) achieved during the training and test processes are 0.008 and 0.009 respectively. The classification confusion matrix reveals that the correct and the false classification rates obtained during the test process are 90.3% and 9.7% respectively (Fig. 3). The employed dataset are prepared and measured at Jijel University, Renewable Energy Laboratory (REL) [9].

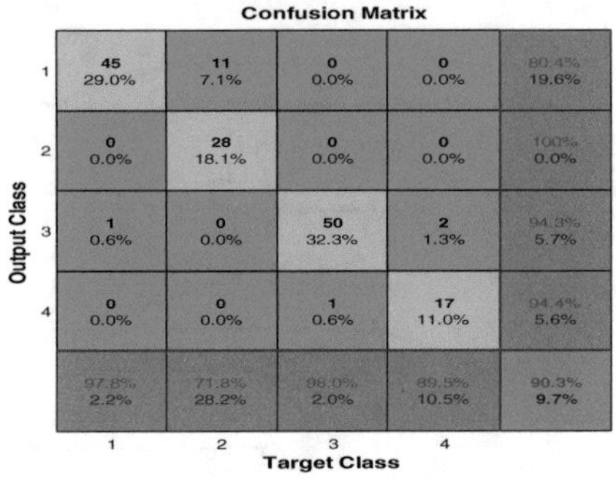

Fig. 3. Confusion matrix for ANN network

3   FPGA-Based Implementation and Results

3.1   Implementation Steps

The fault detection and diagnosis method is implemented into ZYBO (zynq xc7z010-1clg400c) board, using Xilinx system generator. Figure 4 presents the architecture of the whole system; it consists mainly of five blocks:

- Parameters extraction of the measured I-V characteristic based on the comparison of the power point in three successive points of the I-V characteristic.
- Modeling the I-V characteristic using translation equations, because electrical models (one diode and two diode models), although they are characterized by a good precision, they use exponential blocks and require the resolution of nonlinear

equation system, so their implementation with Xillinx blocks was complicated. Furthermore, the performances of the employed model and the single diode model are very closer.

- Fault detection algorithm
- Model based algorithm for fault diagnosis (Algorithm I)
- ANN based algorithm for fault diagnosis (Algorithm II). The sigmoid function is implemented by dividing the sigmoid in multiple segments. Each segment is presented by a simple function as given in Eq. (1).

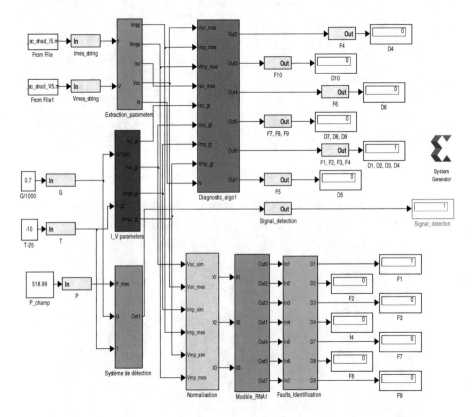

Fig. 4. Architecture of the fault detection and diagnosis method based on XSG tool

3.2 Results

Synthesis. Synthesis involves "compiling" the VHDL code, with tools (e.g. Xilinx Foundation VIVADO 2015.2) which is a commercially available tool, into Basic logic gates and flip flops. The obtained RTL schematic of the fault detection and diagnosis method is shown in Fig. 5.

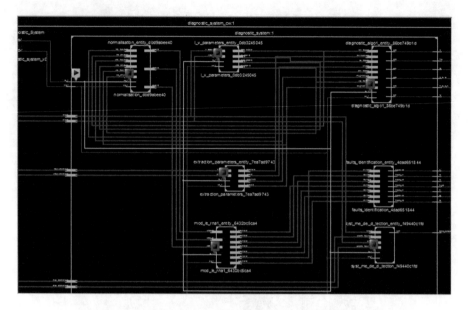

Fig. 5. Synthetized « RTL » schematic of fault detection and diagnosis method

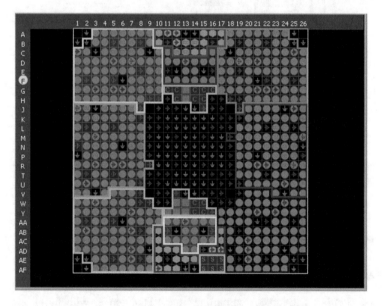

Fig. 6. Overview of the tool pin assignment of inputs/outputs of fault detection and diagnosis method

Placement and Routing. The placement consists to specify points on the circuit corresponding to the blocks used and choose the legs of the circuit corresponding to the input and output ports, and the routing consists to establish electrical connections between the blocks used. Figure 6 shows an overview of the tool pin assignment of inputs/outputs of our project.

Generating the Programming File and Test. During this step, we generated the bit stream file and loaded it into a Zynq XC7Z010-1CLG400C through the option "Configure Target Device" using a JTAG cable. Once the program has been loaded on the circuit, the project has been tested successfully on the board as shown in Fig. 7.

Fig. 7. Generating the Bitstream and programming the FPGA Board

4   Conclusions and Perspectives

In this paper we have demonstrated the possibility of implementing an intelligent fault detection and diagnosis method for a photovoltaic array into a FPGA board. The FPGA device with the advantageous features of flexibility, parallelism of computations robustness, and concurrent operation, presents the most suitable choice for rapid prototyping of digital control system than conventional DSP and MC devices. Generalization of the methodology for large-scale PV plants applications remain a big challenge. The future work will focus on the online validation of the implemented method for real time applications

References

1. IEA: A Snapshot of Global Markets, Report IEA PVPS T1-29 (2017)
2. Tina, G.M., Cosentino, F., Ventura, C.: Monitoring and diagnostics of photovoltaic power plants. In: Sayigh, A. (eds.) Renewable Energy in the Service of Mankind, vol. II, pp 505–516 (2016)

3. Takashima, T., Yamaguchi, J., Otani, K., Oozeki, T., Kato, K., Ishida, M.: Experimental studies of fault location in PV module strings. Sol. Energy Mater. Sol. Cells **93**, 1079–1082 (2009)
4. Chine, W., Mellit, A., Pavan, A.M., Lughi, V.: Fault diagnosis in photovoltaic arrays. In: 2015 International Conference on Clean Electrical Power (ICCEP). IEEE, pp. 67–72 (2015)
5. Chouder, A., Silvestre, S.: Automatic supervision and fault detection of PV systems based on power losses analysis. Energy Convers. Manage. **51**, 1929–1937 (2010)
6. Bonsignore, L., Davarifar, M., Rabhi, A., Tina, G.M., El-hajjaji, A.: Neuro-fuzzy fault detection method for photovoltaic systems. Energy Procedia **62**, 431–441 (2014)
7. Davarifar, M., Rabhi, A., El-Hajjaji, A., Dahmane, M.: Real-time model base fault diagnosis of PV panels using statistical signal processing. In: International Conference on Renewable Energy Research and Applications (ICRERA). IEEE, pp. 599–604 (2013)
8. Chine, W., Mellit, A.: ANN-based fault diagnosis technique for photovoltaic strings. In: 4th IEEE International Conference on Electrical Engineering (ICEE), pp. 1–6 (2015)
9. Mellit, A., Kalogirou, S.: MPPT-based artificial intelligence techniques for photovoltaic systems and its implementation into FPGA chips: review of current status and future perspectives. Energy **70**, 1–21 (2014)
10. Chine, W., Mellit, A., Lughi, V., Malek, A., Massi, Pavan A.: A novel fault diagnosis technique for photovoltaic systems based on artificial neural networks. Renew. Energy **90**, 501–512 (2016)

Modeling Hourly Solar Radiation at Ground Level by Semi Empirical Models in Algeria

Kaoula Talbi[(⊠)] and Samia Harrouni[(⊠)]

Laboratoire d'Instrumentation (LINS), Faculté d'Electronique et d'Informatique,
USTHB, BP 32 EL Alia, Bab Ezzouar, 16111 Algiers, Algeria
Talbi.kha@gmail.com, sharrouni@yahoo.fr

Abstract. Global solar radiation is very important variable in agricultural meteorology and many applied sciences, is measured at very limited number of meteorological Algeria stations. The aim of our study is the determination of global solar radiation for short time scale by some geographical and meteorological parameters, for this purpose we developed tree semi empirical models: Capderou, Atwater&Ball and Lieu&Jordan models. MATLAB was the simulation tool where the program was developed to calculate the hourly global radiation. The results obtained were confirmed by comparing them with the measured results by two statistical indicators: coefficient of determination R^2 and the percentage root mean square error RMSE% for two Algerian sites: Tamanrasset and Algiers.

Keywords: Global solar radiation · Empirical models · Capderou
Liu&Jordan · Atwatter&Ball · Coefficient of determination
Percentage root mean square error

1 Introduction

The need for solar radiation data became more and more important mainly as a result of the increasing number of solar energy applications. A large number of solar radiation computation models were developed like semi empirical models, few of them based on the simulation of global radiation for short time scale. These models require some transmittance and absorptance coefficients witch need the availability of meteorological and geographical parameters. To check the suitability of these models under Algeria conditions we were compared the estimated global solar radiation values to measured values in two Algerian locations Tamanrasset and Algiers using three models: Capderou [1, 2], Lieu&Jordan [2, 3] Atwater&Ball [4, 5], models.

2 Geographical Coordinates

Any point of the terrestrial sphere can be spotted by geographical coordinates as follows (Table 1):

Tahifet region situated in south of Algeria exactly in Tamanrasset the data was during the year 1992 and stored every 10 min on a tilted surface of 10°. For Bab

M. Hatti (ed.), *Artificial Intelligence in Renewable Energetic Systems*, Lecture Notes
in Networks and Systems 35, https://doi.org/10.1007/978-3-319-73192-6_26

Table 1. Terrestrial coordinates of regions

Place	Latitude	Longitude	Altitude (m)
Tahifet	22°53'N	06°00'E	1400
Bab Ezzouar	36°43'N	03°11'E	12

Ezzouar's, it located in Algiers north of Algeria, the data are collected during the year 2014 on a 30°-tilted surface and stored every 15 min. These data were obtained from the acquisition system of an experimental photovoltaic installation.

3   Estimation Models

3.1   Capderou Model

The total radiation incident on an unspecified level of orientation is indicated by the sum of two terms as indicated by the relation 1:

$$G = I + D \tag{1}$$

I is the direct radiation, it is expressed by [1, 2]:
where

$$I = I_{sc} exp \left| -T_l \left(0.9 + \frac{9.4}{(0.89)^z} \sin(h) \right) \right| \cos(i) \tag{2}$$

I_{sc} is extraterrestrial constant solar calculated from the Eq. (3):

$$I_{sc}(nj) = I_0 [1 + 0.03 \cos(0.98(nj - 3))] \tag{3}$$

Such as $I_0 = 1376 \, w/m^2$ and nj is the number of day.

T_l and h are respectively the Link turbidity factor and the sun height angle which is calculated by using the latitude of the location φ, the time angle ω and the declination of the sun δ as follow:

$$\sin h = \cos \omega \cos \varphi \cos \delta + \sin \varphi \sin \delta \tag{4}$$

Knowing that for a horizontal plane we have $\cos(i) = \sin(h)$, consequently, Eq. 2 becomes:

$$I = I_h = I_{sc} exp \left| -T_l \left(0.9 + \frac{9.4}{(0.89)^z} \sin(h) \right)^{-1} \right| \sin(h) \tag{5}$$

D is The diffuse radiation composed of three parts [1, 2]:

$$D = D_1 + D_2 + D_3 \tag{6}$$

D_1 is the diffuse radiation on behalf of the sky expressed by [1]:

$$D_1 = \delta_d \cos(i) + \delta_i \frac{1 + \sin \gamma}{2} + \delta_h \cos \gamma \tag{7}$$

D_2 is the diffuse radiation on behalf of the ground defined by:

$$D_2 = \delta_a \frac{1 - \sin \gamma}{2} \tag{8}$$

For a horizontal plane, the diffuse radiation on behalf of the ground is null. D_3 is the retro-diffused diffuse radiation expressed by [2]:

$$D_3 = \delta_R \frac{1 - \sin \gamma}{2} \tag{9}$$

3.2   Lieu&Jordan Model

The general formula suggested by Lieu&Jordan for the calculation of the total solar radiation is given by the relation [2, 3]:

$$G = I_h \cdot R_b + D_h \cdot \left(\frac{1 + \cos\beta}{2}\right) + \left(\frac{1 - \cos\beta}{2}\right) \cdot \rho \tag{10}$$

For horizontal plane [2]:

$$G = G_h = I_h + D_h \tag{11}$$

Such as I_h is the direct radiation on a horizontal level, given by the equation:

$$I_h = A \sin(h) \exp\left(\frac{-1}{c \sin(h + 2)}\right) = \frac{I}{R_b} \tag{12}$$

I is the direct radiation on a tilted level of an angle β

$$I = R_b I_h \tag{13}$$

Where R_b is the factor of orientation [2] expressed by:

$$R_b = \frac{\cos(\varphi - \beta) \cos \delta \cos \omega + \sin(\varphi - \beta)\sin\delta}{\cos \varphi \cos \delta \cos \omega + \sin\varphi \sin\delta} \tag{14}$$

The diffuse radiation D_h is obtained by the following relation:

$$D_h = B(\sin(h))^{0.4} \tag{15}$$

The general expression of the diffuse radiation D on a tilted level of an angle β is obtained as follows:

$$D = D_h \left(\frac{1 + \cos \beta'}{2} \right) \qquad (16)$$

In (12) and (15) A, B and C are constants which take account the nature of the sky, they are given according to Table 2:

Table 2. Coefficients A, B and C for Lieu and Jordan model [2]

Nature of sky	A	B	C
Very clear sky	1300	87	6
Average sky	1230	125	4
Polluted sky	1200	187	5

The reflected radiation is given by:

$$R = (I_h + D_h)(\frac{1 - \cos \beta'}{2})\rho \qquad (17)$$

Where ρ is the soil Albedo, for horizontal plan, the value of the reflected radiation will be null [2].

3.3   Atwater&Ball Model

Ball and Atwater suggested a general formula for the calculation of the total solar radiation, this formula given by [4]:

$$G = I_{sc.} \cos \theta_{z.} [(T_M - a_w) \cdot \tau_A / (1 - 0.0685 * p)] \qquad (18)$$

Such as I_{sc} is the extraterrestrial solar constant calculated by (3), θ_z is the zenith angle,

T_M the transmittance for all molecular effects except water vapor absorption [4, 5] has a relationship with the air mass of ozone m_0 and the surface pressure (p), is calculated as follow:

$$T_M = 1.021 - 0.0824[m_0(949 * 10^{-6} * p + 0.051)]^{0.5} \qquad (19)$$

The absorption coefficient of direct radiation by the vapor of water (a_w) has the following relation [5]:

$$a_w = 0.077 * (U_w * m_0) \qquad (20)$$

Where U_w is the amount of water vapor in a vertical path through the atmosphere. τ_A the transmittance of aerosols is obtained as follow [2, 5]:

$$\tau_A = \exp(-K_a * m_a) \tag{21}$$

Such as

$$K_a = 0.2758 * k_{a\lambda/\lambda=0.38\,\mu m} + 0.35 * k_{a\lambda/\lambda=0.5\,\mu m} \tag{22}$$

$k_{a\lambda/\lambda=0.38\,\mu m}$ and $k_{a\lambda/\lambda=0.5\,\mu m}$ are the attenuation coefficients defined by experimental measures [2].

m_a is the corrected air mass calculated by Kasten&Young as the equation:

$$m_a = \frac{1 - 0.1 * z}{\sin(h) + 0.50572(h + 6.07995)^{-1.6364}} \tag{23}$$

4   Results and Discussion

The evolution of hourly solar irradiance measured and estimated by the three models versus time for clear and cloudy sky for the two studied sites presented by Figs. 1 and 2:

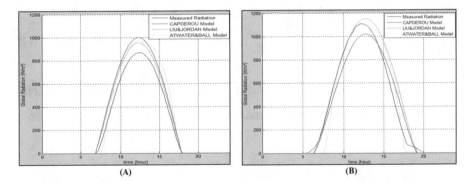

Fig. 1. Comparison between measured and computed global solar irradiance using three models for clear days. (A) day of November 14 for Tahifet. (B) day of June 04 for Bab Ezzouar.

For the two regions, Figures show us that the three models have obtained different global radiation, Liu&Jordan and Atwater&Ball estimate better in days with clear sky conditions. For cloudy sky days we note that Capderou model have good estimation results.

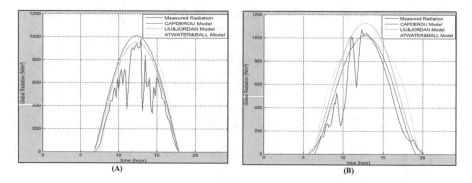

Fig. 2. Comparison between measured and computed global solar irradiance using three models for cloudy sky days. (A) day of November 19 for Tahifet. (B) day of June 21 for Bab Ezzouar.

To validate these results, we give the measurement values as a function of each model for clear and cloudy sky respectively, which is represented in the Figs. 3, 4, 5 and 6:

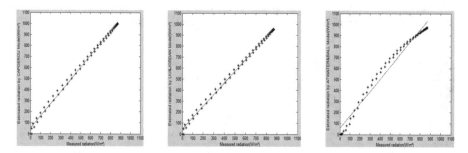

Fig. 3. Estimated irradiance by the three models versus the measured values for clear sky day at Tamanrasset

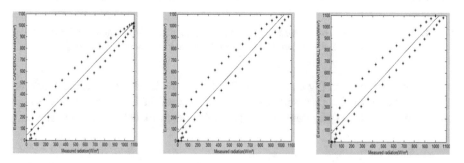

Fig. 4. Estimated irradiance by the three models versus the measured values for clear sky day at Algiers

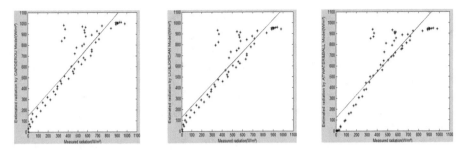

Fig. 5. Estimated irradiance by the three models versus the measured values for cloudy sky day at Tamanrasset

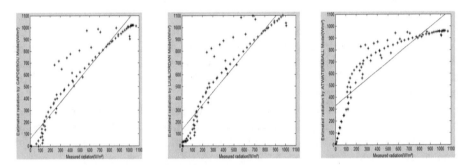

Fig. 6. Estimated irradiance by the three models versus the measured values for cloudy sky day at Algiers

For more conformation, we have used two statistical indicators which are the coefficient of determination R^2 and the percentage root mean square error $RMSE\%$.

A comparison of the results obtained from the three semi empirical models and our measured data for clear and cloudy sky conditions is given in Tables 3 and 4:

Table 3. Error statistical indicators of the three models for clear sky days in two Algerian regions

	Tamanrasset (Tahifet)			Algiers (Bab Ezzouar)		
	Capderou	Liu&Jordan	Atwater&Ball	Capderou	Liu&Jordan	Atwater&Ball
R^2	0.9950	0.9958	0.9766	0.9505	0.9494	0.9770
$RMSE\%$	21.29	14.91	25.79	17.22	44.87	21.51

Table 4. Error statistical indicators of the three models for cloudy sky days in two Algerian regions

	Tamanrasset (Tahifet)			Algiers (Bab Ezzouar)		
	Capderou	Liu&Jordan	Atwater&Ball	Capderou	Liu&Jordan	Atwater&Ball
R^2	0.8160	0.8158	0.8044	0.8926	0.8875	0.8096
$RMSE\%$	39.47	44.45	46.50	30.87	45.88	60.65

Model is more efficient when R^2 closer to 1. For $RMSE\%$, ideally a 20 value should be obtained.

The regression analysis shows that the coefficient of determination is high for Liu&Jordan model (0.9494–0.9958) this is in agreement with the $RMSE\%$ values which vary in the intervals [14.91,44.87] this prove that this model performed better than the other models in days with clear sky conditions.

For Atwater&Ball we note that the coefficient of determination is greater than 0.9770 for Algiers site.

For days with covered sky conditions, Capderou model estimated better for the two Algerian sites where we have (0.8160–0.8926) and $RMSE\%$ vary from 30.87 to 30.87.

5 Conclusion

The estimation of solar irradiance is a crucial step in solar energy projects, but the distribution of the solar radiation is not well known for regions where no measured values are available, it is practice to estimate the solar irradiance using several semi empirical models like Capderou, Atwater&Ball and Lieu&Jordan models for short time scale like hourly estimation.

In this work we find that Liu&Jordan and Atwater&Ball model have good estimation in days with clear sky conditions, while Capderou model estimate better in days with cloudy sky conditions. These three models could be applicable for sites with similar climatic conditions where the solar irradiance data is unavailable.

References

1. Capderou, M.: Atlas solaire de l'Algérie. In: TOM 1, vol. 1. Modèles Théoriques et Expérimentaux. Office des Publication Universitaires. Alger, Aout (1987)
2. Mokhtaria, C.A., Ilyes, M.R.: Introduction au gisement solaire algérien Théorie et applications. Université Amar Telidji – Laghouat–Faculté de Technologie Département d'Electronique
3. Liu, B.Y.H., Jordan, R.C.: The long-term average performance of flat-plate solar-energy collectors. Sol. Energy 7(2), 53–74 (1963)
4. Mesri-Merad, M., Rougab, I., Cheknane, A., Bachari, N.I.: Estimation du rayonnement solaire au sol par des modèles semi-empiriques. Département de Génie Electrique, Faculté des Sciences et des Sciences de l'Ingénieur. Université Amar Tilidji, Route de Ghardaïa, Laghouat, Algérie
5. Bird, R.E., Hulstrom, R.L.: Simplified clear sky model for direct and diffuse insolation on horizontal surfaces. Technical report N°SERI/TR-642-761, Golden, Colorado: Solar Energy Research Institute (1981)
6. Bird, R.E., Hulstrom, R.L.: Direct insolation models. Solar Energy Research Institute, U.S. Department of Energy, January 1980
7. Koussa, D.S., Koussa, M., Belhamel, M.: Reconstitution du rayonnement solaire par ciel clair. Revue des Energies Renouvelables 9(2), 91–97 (2006)

8. Koussa, M., Malek, A., Haddadi, M.: Validation de Quelques Modèles de Reconstitution des Eclairements dus au Rayonnement Solaire Direct, Diffus et Global par Ciel Clair. Revue des Energies Renouvelables **9**(4), 307–332 (2006)

9. Souza, A.P., Escobedo, J.F.: Estimates of hourly diffuse radiation on tilted surfaces in Southeast of Brazil. Institute of Agricultural and Environmental Sciences, Federal University of MatoGrosso, Department of Natural Sciences, Faculty of Agronomic Sciences, State University of São Paulo (2013). Int. J. Renew. Energ. Res.

10. Talbi, K., Harrouni, S.: Modeling of solar radiation received at ground level using semi empirical models for short time scales. In: 8th International Conference on Modeling, Identification and Control, ICMIC-2016, Algiers, Algeria, 15–17 November 2016

11. http://www.solar-med-atlas.org/solarmed-atlas/socioeconomic_policy.htm#s=DZ

Modular Platform of e-Maintenance with Intelligent Diagnosis: Application on Solar Platform

M. Chalouli[1]([⊠]), N. Berrached[1], and M. Denaï[2]

[1] LARESI Laboratory, Electronics Department,
University of Sciences and Technologies Oran -MB, Oran, Algeria
medchalouli@live.fr, laresi.usto.2015@gmail.com
[2] University of Hertfordshire, Hatfield, Hertfordshire, UK
m.denai@herts.ac.uk

Abstract. The "e-maintenance" is the modern definition of the maintenance, to face the in-creasing complexity of the industrial troubleshooting by adding a stronger dimension of cooperation at information's level. Besides, it permits to remote control the maintenance tasks which will be a gain in time and costs. In this paper, we present "e-MIED for e-Maintenance Intelligent & Diagnostics" as an e-maintenance platform presenting a clear guideline for researcher in this field; we centered the conception of our platform on a completely modular framework in order to allow easily adjusting the global environment behavior in accordance with the requirements of client resources. Moreover, the software platform will be applied on a novel hardware photovoltaic solar platform made by the research team. Fault diagnosis module using Support Vector Machine is applied to distinguish the different health states of a machine or component and it's considered as the main part of e-MIED platform.

Keywords: e-Maintenance · Photovoltaic · Diagnostics · CBM
Solar power platform · e-MIED

1 Introduction

The photovoltaic generator is considered as a machine in the maintenance process and will be diagnostic according to its features which are Current, Voltage and Temperature in this case of study. And, there isn't a machine without maintenance; thus, the intelligent maintenance is applied to minimize the downtime and the fault occurrence [1]. The maintenance mainly can be divided on two categories, reactive or proactive where the first one represented as the response to work requests or identified need. It focuses on restoring the failed machine or system to its functional operating condition. The second approach responds primarily to equipment assessment, root cause analysis, and predictive procedures. In both cases, there is no data entry from machines before the failure occurs. These two categories are considered wasteful [2]. The fail of machine isn't sudden as it seems. The science and the new technologies proved that the failure happened after certain degradation related to causes often unseen by humans [3].

© Springer International Publishing AG 2018
M. Hatti (ed.), *Artificial Intelligence in Renewable Energetic Systems*, Lecture Notes in Networks and Systems 35, https://doi.org/10.1007/978-3-319-73192-6_27

The e-maintenance is an extension of the maintenance in order to move it from cost to gain task due to the relatively high costs from one side. And from the other side, the costs of damage related to the failure. The industry todays require a high productivity with near zero downtime of maintenance which request to master the machines diagnosis as well as the store management with the ability to replace the failed component and the Human resources management which help to appoint the adequate operator for maintenance tasks. Maintaining a machine is keeping it working in good condition according to norms or repairing a machine after its failure, and from that we distinguish two type of maintenance preventive and corrective. The Preventive maintenance relies on monitoring the sensors of a machine detecting any abnormal behavior and/or any sign of faults before even reaching the failure point where the machine stops working. The other type is the corrective maintenance. It occurs once the failure point is reached, where it aims to restore the health state of a machine to its good conditions.

In easy words, the e-maintenance is bench of maintenance services related to the production process such as human resources and inventory management with the advantage to bring the maintenance out the production system, where the diagnostics can be applied on the machines without any interruption of the production line providing all the necessary information.

This paper is organized as follows: after the introduction, the most agreed definition is presented in Sect. 2, followed in the next section by a state of art. Where, the fourth section presents the architecture of our e-maintenance platform "e-MIED". Section 5 illustrates the result and implementation of the proposed architecture with a Support Vector Machine classifier for the diagnostics task. Then this paper includes the advantages and drawbacks of the proposed e-MIED platform. Finally, the conclusion is given at the sixth section.

2 Definitions

The maintenance is applied to retain or restore to functional state an operational unit during its lifetime using technical and managerial services according to the French national organization for standardization AFNOR -EVS-EN 13306:2010 EN 13306:2010-. It can be represented by applications or services of diagnostics according to a defined operational unit. Applying the maintenance tasks from the control room on a distant operational unit over network support (Internet often) is known as the remote maintenance. The e-maintenance is considered as the evolution of remote maintenance. Sharing information and knowledge is what make the difference between the e-maintenance and the remote maintenance [4]. In this section, we will take the most used definition of e-maintenance and remote maintenance as well. Full review about the existed e-maintenance platform can be checked here [5–8]. In this paper and from current related work we show the interest of e-maintenance platform and also the lake of guidelines in this field. Research literature allows us to propose an architecture covering the different layers of a full e-maintenance platform to better lead the upcoming researches.

3   Architecture of a e-Maintenance Intelligent and Diagnostics Platform (e-MIED)

The conception of any e-maintenance platform is based on multiple tasks or layers aiming to perform various functions including and not exclusively to: Data Acquisition, Health Assessment, Store Management and Aide-Decision. The proposed architecture follows in many ways the OSA-CBM model [9, 10]. Inspired mostly from the Model-View-Controller pattern and the related works cited previously. The platform is conceived for real-time diagnostic, supporting knowledge sharing between different services and automated management system for inventory and human resource. The Fig. 1 represents the main services classed by the type of data they treated whether metric or management data. The metric data services are grouped in the Tele-Maintenance section and the Managerial data are grouped in the Enterprise Resource Planning section.

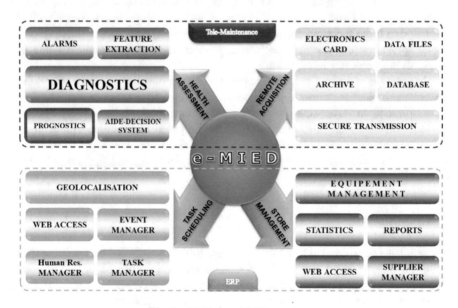

Fig. 1. Multiple e-MIED services.

The most important services provided by the e-MIED platform are cited:

3.1   Photovoltaic Solar Power Platform

The novel platform presented here consists on a test rig for diagnosis and maintenance support. The photovoltaic generator is built using 04 rows of 05 columns with the possibility to disconnect any row at any time just by a press of button. Multiple combinations of BUCK and BOOST are connected to the output of the platform.

Also to mention a dynamic load of 35 W is used for load variation and feature extraction of the solar power platform.

The Fig. 2 demonstrates the solar power platform and the Remote Acquisition Card.

Fig. 2. Solar power platform.

3.2 Flow Chart of e-MIED Services

This structure is made in such a way as to centralize the knowledge base, while connecting each customer to its own dedicated server and also accesses simultaneously the knowledge database from many services. Then, the client will have a monitoring module (View) to visualize the data processed by the multiple services (Server), previously acquired and introduced by the acquisition layer (Controller). This structure is intended specifically for the industrial field and technical domains in general.

The platform will be fully implemented inside the factory which means that the platform can't be accessed outside the factory providing less access but more security and high speed communication with a controlled connection. For external access, the View layer can be allowed or connected to the internet (Fig. 3).

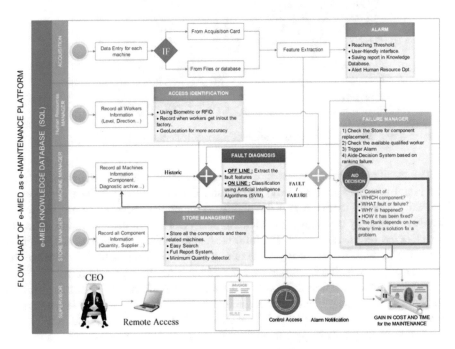

Fig. 3. Flow chart of e-MIED.

4   Implementation and Discussion

After a brief description of the various modules in the (e-MIED) platform, the suggested architecture is simulated through the implementation of different modules. It should be noted that in this article, the components of this architecture are defined with respect to their function, not their technical specifications.

Once the diagnosis done, a description of needs is performed and the system interacts with the maintenance operator for a detailed description of the failure. The system analyzes the data and suggests to the operator a different solution to that failure if it exists or establishes a video connection with the expert selected by the platform. An automatic memorization step of suggested solutions after validation of selection results takes place in order to build up a more complete knowledge base.

Diagnosis, particularly useful in the context of the e-maintenance concept. It allows to go further than triggering a simple alarm, and goes to characterize the failure by determining the exact nature of default met, severity and urgency of the intervention [11]. The Fig. 4 represents the diagnostic task done by an artificial intelligence tools, the Support Vector Machine (SVM) in this case.

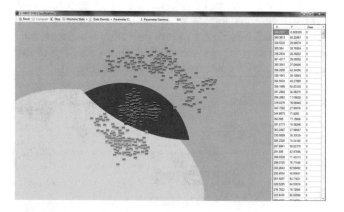

Fig. 4. SVM classification of the e-MIED diagnostic service.

The machines support vectors exploit concepts of the theory of statistical learning theory and terminals first used by Vapnik and Chervonenkis [12]. The compromise between the ability to learn and the ability to generalize these machines is accomplished respectively by minimizing the empirical error and at the same time, trying to maximize the geometric margin. The intuitive justification of this method of learning is this: if the training set is linearly separable, it seems perfectly natural to separate the elements of the two classes so that they are as far from the selected border.

SVM is a classification method that shows good performance in solving various problems. In-depth researches about SVM in this field are done by [13, 14]. The results shows that SVM outperforms Artificial Neural Networks (ANN) and has proved effective in many application areas such as diagnostics and prognostics [15] and even on very large data sets. For huge datasets data mining algorithms perform better as illustrated in [16].

The Fig. 5 represents the acquisition services with the plotting of the characteristics of I-V and P-V and assuring real monitoring of Current, Voltage and Temperature.

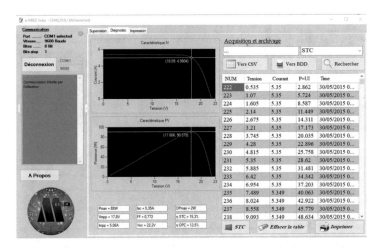

Fig. 5. Acquisition module with IV and PV curves

The Fig. 6 represents some services from the platform, citing from head to bottom: Inventory Manager, Human Resource Manager, Machine Manager, Event Manager and diagnostic

Fig. 6. Screenshot of multiple services menu of the e-MIED platform

The following figure (Fig. 7.) represents a very simplified sequence diagram and class diagram of the e-MIED platform:

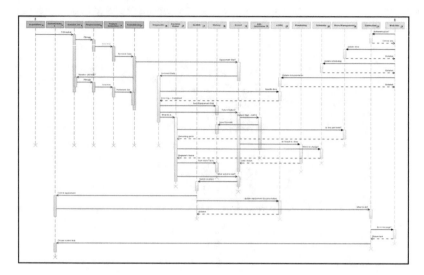

Fig. 7. Sequence diagram of very simplified diagrams of e-MIED platform

5   Conclusion

Improving maintenance to e-maintenance is a new research area that requires bibliographic research in different domains such as industrial maintenance and knowledge management. After analyzing the existing maintenance systems, it is noted that most of them are established according to the characteristics of remote maintenance. There is almost no clear guideline platform for new researcher in this field establishing all e-maintenance modules in details. For that purpose, the architecture of e-Maintenance Intelligent & Diagnostics (e-MIED) platform meeting the above requirements has been presented in this study. Integrating new services in the platform and a better coherence management of the system make up the subject of our future work.

References

1. Ribeiro, L., Barata, J.: A high level e-maintenance architecture to support on-site teams, pp. 129–138 (2008)
2. Lee, J., Ni, J., Djurdjanovic, D., Qiu, H., Liao, H.: Intelligent prognostics tools and e-maintenance. Comput. Ind. **57**(6), 476–489 (2006)
3. Chalouli, M., Berrached, N., Denai, M.: Intelligent health monitoring of machine bearings based on feature extraction. J. Fail. Anal. Prev. **17**(5), 1053–1066 (2017)
4. Mahiddine, L., Hafed, A., Achour, A., Khouri, S., Boudiba, B.: Plate-forme logicielle de e-maintenance et télédiagnostic, pp. 2–7, November 2007

5. Muller, A., Marquez, A.C., Iung, B.: On the concept of e-maintenance: review and current research. Reliab. Eng. Syst. Saf. **93**(8), 1165–1187 (2008)
6. Chang, Y.H., Chen, Y.T., Hung, M.H., Chang, A.Y.: Development of an e-operation framework for SoPC-based reconfigurable applications. Int. J. Innov. Comput. Inf. Control **8** (5B), 3639–3660 (2012)
7. Levrat, E., Iung, B.B.: TELMA: a full e-maintenance platform. In: Second World Congress on Engineering Asset Management and the Fourth International Conference on Condition Monitoring, WCEAM/CM 2007, CDROM (2007)
8. López-Campos, M., Cannella, S., Bruccoleri, M.: e-maintenance platform: a business process modelling approach. DYNA **81**(183), 31–39 (2014)
9. Mouzoune, A., Taibi, S.: Introducing E-Maintenance 20. Int. J. Comput. Sci. Bus. Inf. **9**(1), 80–90 (2014)
10. Kajko-Mattsson, M.: Fundamentals of the eMaintenance concept. In: 1st International Workshop, pp. 22–24 (2010)
11. Boulenger, A.: Maintenance conditionnelle par analyse des vibrations. Tech. l'Ingénieur, vol. 33 (2006)
12. Batista, L., Badri, B., Sabourin, R., Thomas, M.: A classifier fusion system for bearing fault diagnosis. Expert Syst. Appl. **40**(17), 6788–6797 (2013)
13. Yang, W.X., Wang, P.: Application study of EMD-AR and SVM in the fault diagnosis. In: Proceedings of 2014 Prognostics and Health Management Society, PHM 2014, pp. 93–96 (2014)
14. Kankar, P.K., Sharma, S.C., Harsha, S.P.: Fault diagnosis of ball bearings using machine learning methods. Expert Syst. Appl. **38**(3), 1876–1886 (2011)
15. Galar, D., Kumar, U., Fuqing, Y.: RUL prediction using moving trajectories between SVM hyper planes. In: 2012 Proceedings of Annual Reliability and Maintainability Symposium, pp. 1–6 (2012)
16. Kamsu-Foguem, B., Rigal, F., Mauget, F.: Mining association rules for the quality improvement of the production process. Expert Syst. Appl. **40**(4), 1034–1045 (2013)

Artificial Neural Network in Renewable Energy

Weather Forecasting Using Genetic Algorithm Based Artificial Neural Network in South West of Algeria (Béchar)

Youssef Elmir[✉]

Department of Mathematics and Computer Science,
University Tahri Mohammed of Béchar, UTMB, Béchar, Algeria
elmir.youssef@yahoo.fr

Abstract. As one of artificial intelligence technologies, Artificial Neural Networks (ANNs) are widely used as an alternative solution for ill-defined problems. Training using examples make it possible to deal with noisy and replace missed data. Nonlinear problems can be solved using artificial neural networks and, once trained, forecasting.

This paper investigates the application of Genetic Algorithm (GA) based ANN in weather forecasting and compares the whole performance with the one of an ordinary ANN where both of them have been used for the prediction of air temperature, atmospheric pressure, relative humidity and mean wind speed. The proposed system uses ANN with GA based generated weights.

The obtained results using this model are good enough to prove that the proposed system can be used for modelling in other fields of renewable energy problems.

Keywords: Artificial neural network · Genetic algorithms
Weather forecasting · Artificial intelligence · Machine learning

1 Introduction

Today, ANNs have multiple applications in diverse fields; image processing, signal processing, automatic language processing, control, optimisation, simulation, classification, learning modelling, teaching methods enhancement and the approximation of unknown function or modelling of known complex function.

Fonte et al. [1] presented the problem with the introduction of a large quantity of wind generators on the electric grid. A method based in artificial neural networks (ANN) is used to predict the average hourly wind speed. The study starts by choosing the patterns set length to predict de wind speed. The ANN structure and the learning method are chosen as well as the dimensions of the sets of data, training, validation and test. The ANN is tested to archive an acceptable ANN based model. This model is afterwards used to predict the wind speed. The results archived are discussed and the future work perspectives are present.

Jursa and Rohrig [2] introduced a new short-term prediction method based on the application of evolutionary optimization algorithms for the automated specification of

© Springer International Publishing AG 2018
M. Hatti (ed.), *Artificial Intelligence in Renewable Energetic Systems*, Lecture Notes in Networks and Systems 35, https://doi.org/10.1007/978-3-319-73192-6_28

two well-known time series prediction models, i.e., neural networks and the nearest neighbor search. Two optimization algorithms are applied and compared, namely particle swarm optimization and differential evolution. To predict the power output of a certain wind farm, this method uses predicted weather data and historic power data of that wind farm, as well as historic power data of other wind farms far from the location of the wind farm considered. Using these optimization algorithms, we get a reduction of the prediction error compared to the model based on neural networks with standard manually selected variables. An additional reduction in error can be obtained by using the mean model output of the neural network model and of the nearest neighbor search based prediction approach.

Kalogirou et al. [3] trained a suitable ANN to predict the mean monthly wind speed in regions of Cyprus where data are not available. Data for the period 1986–1996 have been used to train the network whereas data for the year 1997 were used for validation. Both learning and prediction were performed with an acceptable accuracy. Two multilayered ANN architectures of the same type have been tried, one with five neurons in the input layer (month, wind speed at 2 m and 7 m for two stations) and one with 11. The additional input data for the 11-inputs are the x and y coordinates of the meteorological stations. The 5-input network proved to be more successful in the prediction of the mean wind speed. The network using only five input parameters is more successful, giving a maximum percentage difference of only 1.8%. The proposed network can be used to fill missing data from a database to predict weather in other nearby locations.

The objective of this paper is to study the possibility of weather forecasting using artificial intelligence techniques such as ANN and GA in the area of south west of Algeria (Béchar city). This is also important for renewable energy applications due the high importance of weather impact on the whole performance of renewable energetic systems.

Based on the work of Kalogirou et al., the proposed system is an artificial neural network with optimized weights using genetic algorithm before training. This system is compared with an ordinary artificial neural network using the same architecture and initial weights. The rest of the paper is structure as follow; in the second section, the architecture of the proposed system is presented; the third section presents the data used for system evaluation and discusses the obtained results of prediction and the comparison with the real measured data and the last section gives the conclusion and some future works.

2 Architecture of the Proposed System

The proposed system is based on a Multi-Layer Perceptron (MLP) as described in Fig. 1.

This ANN contains three layers of neurons, four neurons in the first layer for inputs (day, month, year and time) and four neurons in the third layer for outputs (air temperature, atmospheric pressure, relative humidity and mean wind speed) and the number in the hidden layer is set empirically. As usual, weights are initiated using random values.

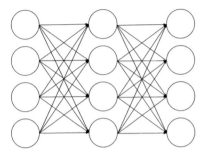

Fig. 1. The proposed architecture of the artificial neural network.

2.1 Artificial Neural Network

Inspired by biological neural networks, ANNs are massively parallel computing systems consisting of an extremely large number of simple processors with many interconnections. ANN models attempt to use some "organizational" principles believed to be used in the human. One type of network sees the nodes as 'artificial neurons'. These are called artificial neural networks (ANNs). An artificial neuron is a computational model inspired in the natural neurons. Since the function of ANNs is to process information, they are used mainly in fields related with it. There are a wide variety of ANNs that are used to model real neural networks, and study behavior and control in animals and machines, but also there are ANNs that are used for engineering purposes, such as pattern recognition, forecasting, and data compression. These basically consist of inputs (like synapses), which are multiplied by weights. Weights assigned with each arrow represent information flow. These weights are then computed by a mathematical function which determines the activation of the neuron. Another function (which may be the identity) computes the output of the artificial neuron (sometimes in dependence of a certain threshold). The neurons of this network just sum their inputs. Since the input neurons have only one input, their output will be the input they received multiplied by a weight [4].

Both ANNs are trained using parameters in Table 1.

2.2 Weights Optimization Using Genetic Algorithm

The training phase often takes long time to find best weights that match best to the problem, according to the parameters defined by the user because of the initialization of weights using random values. This time can be reduced if the initialized weight are appropriates to the problem and that can be done using weights generated by GA.

This goal is investigated by the proposed system that uses the architecture described in Fig. 1 with the same initialized values of weights. These weights are used as initial population of the GA and they are replaced with GA outputs which are the optimized weights.

A genetic algorithm (GA) is a metaheuristic inspired by the process of natural selection that belongs to the larger class of evolutionary algorithms (EA). Genetic algorithms are commonly used to generate high-quality solutions to optimization and

Table 1. Training parameters.

Parameter	Value
Maximum epochs	1000
Performance goal	0
Maximum validation checks	1000
Mu decrease ratio	0.1
Maximum mu	10000000000
Maximum training time	Inf
Minimum gradient	1e-07
Mu	0.001
Mu increase ratio	10

search problems by relying on bio-inspired operators such as mutation, crossover and selection [5].

The process of weights optimization starts by launching the GA with the initial population and generation of new vector of weights by applying genetic operators (mutation, crossover and selection) [5]. In every iteration, the performance of the ANN is evaluated using the new generated weights and this process is repeated until the achievement of the max number of generations or the value of tolerance in fitness function as defined in Table 2.

Table 2. GA parameters.

Parameter	Value
Initial population	Initialized randomly by ANN
Generations number	10000
Tolerance on fitness function	0

Furthermore, optimized weights based ANN is trained using the same parameters used for the first ANN and described in Table 1. The purpose of this step is enhancing the overall performance of the new ANN.

3   Dataset and Experiment Results

3.1   Dataset

The original data are obtained from Reliable Prognosis website; it has been designed and supported by Raspisaniye Pogodi Ltd., St. Petersburg, Russia, since 2004. The company has the license for activity in hydrometeorology and in adjacent fields [6].

An hydrometeorology archive and data of béchar city that has been used in this work was selected from the period between January, 1st, 2010 and December, 31th, 2016 as samples are listed in Table 3. The data of the first five years (2010–2014) were used for training purpose where the two last years (2015–2016) were used for tests and evaluation purpose.

Table 3. Data samples

Local time	Air temperature (Degrees Celsius)	Atmospheric pressure (Millimeters of mercury)	Relative humidity (%)	Mean wind speed (Meters/Second)
31.12.2016 22:00	4.5	700.5	77	0
31.12.2016 19:00	10.0	700.2	53	2
31.12.2016 16:00	13.8	700.6	37	1
31.12.2016 13:00	11.0	701.8	49	2
31.12.2016 10:00	3.0	701.8	90	0
31.12.2016 07:00	0.2	701.4	89	1
31.12.2016 04:00	1.6	701.5	83	2
31.12.2016 01:00	3.3	702.2	76	0
30.12.2016 22:00	5.8	701.9	62	1
30.12.2016 19:00	12.0	701.4	27	1
30.12.2016 16:00	14.1	701.5	31	2
30.12.2016 13:00	11.3	702.7	52	2
30.12.2016 10:00	4.0	702.9	97	2
30.12.2016 07:00	1.4	702.4	100	2
30.12.2016 04:00	3.4	702.6	92	0
30.12.2016 01:00	4.2	702.6	90	0
...				
20.07.2013 22:00	38.6	689.4	16	5
20.07.2013 19:00	42.0	688.6	10	13
20.07.2013 16:00	42.6	689.0	15	8
20.07.2013 13:00	42.2	690.6	17	9

(*continued*)

Table 3. (*continued*)

Local time	Air temperature (Degrees Celsius)	Atmospheric pressure (Millimeters of mercury)	Relative humidity (%)	Mean wind speed (Meters/Second)
20.07.2013 10:00	37.5	690.6	22	2
20.07.2013 07:00	28.3	689.9	50	0
20.07.2013 04:00	31.6	689.6	25	0
20.07.2013 01:00	37.0	690.0	19	4
...				
01.01.2010 19:00	16.5	693.8	40	0
01.01.2010 13:00	18.8	694.0	23	3
01.01.2010 07:00	6.1	692.2	67	2
01.01.2010 01:00	9.0	691.9	58	0

3.2   Experiment Result

It was not easy to set the parameters of ANN and GA in order to obtain the best results but after several tests, some good results were obtained as they are shown in Table 4.

These results show that the mean errors made by both of systems are acceptable, from 2.79 to 3.86° in air temperature, 3.06 to 3.04% in relative humidity, 10.32 to

Table 4. Forecasting performance of the proposed systems.

		Air temerature (Degrees celsius)	Relative humidity (%)	Atmospheric pressure (Millimeters of mercury)	Mean wind speed (Meters/second)	Training time (seconds)
GA-ANN	Mean error	2.7918	3.0454	10.3271	2.3599	41.02
	Maximum error	17.2979	14.3760	67.7513	17.8737	
	Minimum error	0.0006	0.0001	0.0008	0.0018	
ANN	Mean error	3.8632	3.0006	11.0101	2.4065	176.58
	Maximum error	21.5202	14.5708	73.4389	17.7164	
	Minimum error	0.0020	0.0004	0.0001	0.0000	

11.01 in atmospheric pressure and 2.35 to 2.40 meters per second in mean wind speed. The same thing can be noted regarding the min and the max errors.

It is also clear that there is a decreasing of errors made by ANN when using GA where it does not need as time as the ordinary ANN for training.

Figure 2 shows sub-plots of the predicted data obtained by the GA-ANN and ANN on the real measured data.

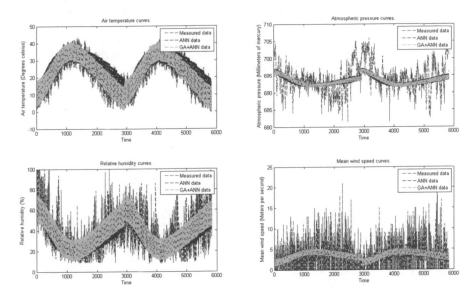

Fig. 2. Comparison of the predicted data using GA-ANN and ANN with the real measured data.

4 Conclusion

The present work concluded that the artificial neural networks are useful for the forecast of weather. This method can be enhanced in by reducing training time and enhancing the overall performance by using genetic algorithm, but it needs in the other side optimization phase before doing training. The weather forecast has great impact on the performance of the operators of the renewable energy power systems because the production and the consumption of the electric energy happens just-in-time. The weather variation throughout the many periods in a day, the proposed system gives the possibility to predict and optimize the production of energy. In future works it is possible to investigate the use of some other different neural networks as Radial Basis Function Networks that some authors refer as best suitable for forecasting problems.

References

1. Fonte, P.M., Silva, G.X., Quadrado, J.C.: Wind speed prediction using artificial neural networks (2005)
2. Jursa, R., Rohrig, K.: Short-term wind power forecasting using evolutionary algorithms. Int. J. Forecast. **24**, 694–709 (2008)
3. Kalogirou, S.A., Neocleous, C., Paschiardis, S., Schizas, C.: Wind speed prediction using artificial neural networks. In: European Sympsium on Intelligent Techniques, Crete, Greece (1999)
4. Neha, G.: Artificial neural network. Netw. Complex Syst. **3**(1), 24–28 (2013)
5. Mitchell, M.: An Introduction to Genetic Algorithms. MIT Press, Cambridge (1996)
6. Raspisaniye Pogodi Ltd, "Weather for 243 countries of the world". https://rp5.ru/Weather_in_the_world. Accessed 4 June 2017

Intra-hour Forecasting of Direct Normal Solar Irradiance Using Variable Selection with Artificial Neural Networks

Hanane Atmani[1], Hassen Bouzgou[2(✉)], and Christian A. Gueymard[3]

[1] Department of Electronics, University of Batna 2, Batna, Algeria
atmani_h@yahoo.com
[2] Department of Industrial Engineering, University of Batna 2, Batna, Algeria
bouzgou@gmail.com, h.bouzgou@univ-batna2.dz
[3] Solar Consulting Services, Colebrook, NH, USA
Chris@SolarConsultingServices.com

Abstract. Renewable Energy Sources (RES) are one of the key solutions to handle the world's future energy needs, while decreasing carbon emissions. To produce electricity, large concentrating solar power plants depend on Direct Normal Irradiance (DNI), which is rapidly variable under broken clouds conditions. To work at optimum capacity while maintaining stable grid conditions, such plants require accurate DNI forecasts for various time horizons. The main goal of this study is the forecasting of DNI over two short-term horizons: 15-min and 1-h. The proposed system is purely based on historical local data and Artificial Neural Networks (ANN). For this aim, 1-min solar irradiance measurements have been obtained from two sites in different climates. According to the forecast results, the coefficient of determination (R^2) ranges between 0.500 and 0.851, the Mean Absolute Percentage Error (MAPE) between 0.500 and 0.851, the Normalized Mean Squared Error (NMSE) between 0.500 and 0.851, and the Root Mean Square Error (RMSE) between 0.065 kWh/m^2 and 0.105 kWh/m^2. The proposed forecasting models show a reasonably good forecasting capability, which is decisive for a good management of solar energy systems.

Keywords: Direct Normal Irradiance · Forecasting · Time series analysis
Variable selection · Artificial neural networks

1 Introduction

Due to environmental and sustainability issues, energy supplies from renewable (such as solar, thermal, photovoltaic, wind, hydro, bio fuels or ocean waves…) are indispensable for every nation's energy strategy, while curbing carbon emissions. Renewable energy resources, and most particularly the solar resource, are abundant over wide geographical areas, contrary to other energy sources. The increasing development of renewable energy sources and energy efficiency strategies improves energy security, brings economic benefits, and mitigates climate change impacts (Kaur et al. 2016). Solar energy has established itself as a competitive provider of electricity to both small-scale and large-scale grid systems. Electricity can be produced from solar irradiance by either photovoltaic (PV)

© Springer International Publishing AG 2018
M. Hatti (ed.), *Artificial Intelligence in Renewable Energetic Systems*, Lecture Notes in Networks and Systems 35, https://doi.org/10.1007/978-3-319-73192-6_29

modules or by Concentrated Solar Thermal (CST) systems. The latter concentrate the direct normal irradiance (DNI) to reach high temperatures (Law et al. 2014), but DNI can vary rapidly under broken-cloud conditions, which complicates its prediction.

Several models have been proposed to forecast DNI using various meteorological variables and methods (Law et al. 2014). Chu (2013) proposed a novel smart forecasting model for DNI, combining an optimized Artificial Neural Network (ANN) with sky image processing. In parallel, Peruchena et al. (2017) proposed an advanced locally-adapted procedure for high-frequency DNI forecasting method, which connects numerical weather prediction (NWP) models and local statistical information derived from site measurements.

The present study introduces a new machine-learning methodology to forecast DNI time series for short-term horizons (15-min and 1-h ahead). The proposed system is composed of two separate blocs: the first one seeks to select the input variables of the historical time series by proposing four different scenarios; the second one consists of an ANN of the Extreme Learning Machine (ELM) kind to forecast the future DNI time series. The following sections present general information about DNI forecasting and input variable selection, as well as a discussion about numerically forecasting DNI time series from real measurements. Then, the performance of the best DNI forecasts for particular forecast horizons is discussed and summarized.

2   Forecasting Direct Normal Irradiance

DNI is related to global horizontal irradiance (GHI) and diffuse horizontal irradiance (DIF) through a fundamental closure equation:

$$GHI = DIF + DNI\,cos\theta_z \qquad (1)$$

where θ_z is the solar zenith angle.

GHI forecasts are often used to predict the performance of non-concentrating solar applications such as photovoltaic (PV) systems and flat-plate solar thermal collectors. In contrast, DNI forecasts are required to estimate the future performance of concentrated solar thermal (CST) systems. Once in service, such systems include at least one weather station that monitors DNI and other meteorological variables. Hence, it is always appropriate to consider that, in practice, a historical time series of DNI is available at any site where DNI forecasts are necessary.

Clouds are the main cause of DNI intensity reduction, and aerosols become the main cause of variation when clouds are absent. The aerosol concentration is usually quantified in terms of aerosol optical depth (AOD). The reduction in DNI intensity due to aerosols is a direct function of AOD (Gueymard 2012), whereas their impact on GHI is much less. Similarly, the overall impact of clouds is much larger on DNI than on GHI. This difference in temporal behavior between GHI and DNI is one key reason why GHI forecasting techniques and results should not be directly applied to DNI forecasting (Twidell and Weir 2015; Kleissl 2013).

Depending on application, various forecasting time horizons are appropriate. For electricity production and grid stability, short-term forecasts typically include the 15-min and 1-h horizons. These two specific horizons are exclusively considered in what follows.

3 Variable Selection

During the data pre-processing step, variable selection (VS) methods are useful to effectively reduce the amount of significant input data. This leads to the development of precise, yet simple data models, since an exhaustive search for the optimal variable subset is infeasible in most cases. The usual applications of VS are in classification, clustering, and regression tasks. Dataset size reductions can be performed in one of two ways: variable set reduction or sample set reduction. VS methods can also be classified in a number of ways. The most common one is the classification into filters, wrappers, embedded, and hybrid methods (Jović et al. 2015). The present study uses and evaluates the performance of four competitive VS strategies:

- **Full Window (FW):** Latest 50 variables (i.e., 50 most recent DNI observations).
- **Small Window (SW):** Latest 5 variables.
- **Forward Search (FS):** In this case, the selection process starts from the oldest variable of the corresponding full window and goes forward until the newest variable. The optimal subset is chosen based on the performance of the ELM Model applied to the validation set.
- **Backward Search (BS):** In this case, conversely, the selection procedure starts from the newest variable and goes back until the last (oldest) variable of the corresponding historical window is reached.

4 Extreme Learning Machine

Feedforward neural networks (FNN) are among the most popular architectures in forecasting. An FNN has three distinctive characteristics: (i) an input layer that gets excitations from the external environment; (ii) one or more hidden layers; and (iii) an output layer to transfer the output of the network to the external environment.

The connections between the layers are associated with weights, which ensure the optimum circulation of information. The determination of the NN weights can be performed through a learning process, after establishing desirable input-output relationships. One of these NN methods, called Extreme Learning Machine (ELM), is selected here. It is an FNN with a single hidden layer used for predicting the outcome of the forecasting process. An ELM is constructed with K hidden neurons and an activation function $g(x)$ to train N different samples (X_i, t_i), where $X_i = [x_{i1}, x_{i2}, \ldots, x_{in}] \in R^n$ and $t_i = [t_{i1}, t_{i2}, \ldots, t_{im}] \in R^m$. The input weights and hidden biases of the ELM are created randomly, and are not tuned, allowing the conversion from a nonlinear system to a linear system:

$$H\beta = T, \tag{2}$$

where $H = \{h_{ij}\}$ $(i = 1,..., N$ and $j = 1,..., K)$ is the hidden-layer output matrix; $h_{ij} = q(w_j x_I + b_j)$ represents the output of the jth hidden neuron related to x_i; $w_j = [w_{j1}, w_{j2}, ..., w_{jn}]^T$ is the weight vector connecting the jth hidden neuron to the input neurons; b_j represents the bias of the jth hidden neuron; $w_j x_I$ denotes the inner product between the weights and the input vector (i.e., historical variables); $\beta = [\beta_1, \beta_2,..., \beta_K]^T$ is a transposed matrix in which $\beta_j = [\beta_{j1}, \beta_{j2},..., \beta_{jm}]^T$ $(j = 1,..., K)$ represents the weight vector connecting the jth hidden neuron and the output neurons; and finally $T = [t_1, t_2,..., t_N]^T$ is the matrix of targets (desired output). Based on Eq. (2), the determination of the output weights (linking the hidden layer to the output layer) is the least-squares solution of the given linear system. The minimum-norm least-squares (LS) solution to the linear system in Eq. (2) is

$$\hat{\beta} = H^+ T, \tag{3}$$

where H^+ is the Moore–Penrose generalized inverse of matrix H. Note that the minimum norm LS solution is unique and has the smallest norm among all the LS solutions (Huang et al. 2006).

5   Results and Discussion

5.1   Time Series Analysis

Time series analysis (TSA) methods produce forecasts by analyzing statistical trends in historical time series, named exogenous variables, which influence the forecast variable. Solar irradiance data time series normally exist for a long period relatively to their temporal resolution (e.g., several days at 15-min resolution or several months at 1-h resolution). Hence, they may be used to produce forecasts using TSA methods. The simplest TSA model is the persistence model, which assumes that conditions in the future will be exactly the same as those in the previous time step.

$$\hat{Y}_p(t + \Delta t) = Y(t). \tag{4}$$

5.2   Evaluation Criteria

The accuracy of forecasting methods is typically reported using the Root Mean Square Error (*RMSE*), Mean Absolute Percentage Error (*MAPE*), Normalized Mean Squared Error (*NMSE*) and Coefficient of Determination (R^2), which are defined as follows:

$$RMSE = \sqrt{\frac{1}{M} \sum_{t=1}^{M} \left(y_t - \hat{y}_t\right)^2} \tag{5}$$

$$MAPE = \frac{1}{M} \sum_{t=1}^{M} \frac{|y_t - \hat{y}_t|}{|y_t|} \times 100 \tag{6}$$

$$NMSE = \frac{\sqrt{\frac{1}{M} \sum_{t=1}^{M} (y_t - \hat{y}_t)^2}}{var(y)} \tag{7}$$

$$R^2 = 1 - \frac{var(y_t - \hat{y}_t)}{var(y_t)}. \tag{8}$$

5.3 Solar Radiation Data

Two datasets containing observations from sites in very different climates are used here. The irradiance dataset recorded at Tucson, Arizona, USA, is representative of an arid climate and covers a period of 21 months. In contrast, Brasilia, Brazil experiences a tropical savanna climate with a dry and a rainy season. A 22-month observation period is available there. The two datasets are averaged to obtain 15-min and 1-h DNI time series, corresponding to the two horizons selected here. Table 1 provides some information about the two radiometric stations just mentioned.

Table 1. Information on the experimental stations.

Site	Latitude (°N)	Longitude (°E)	Elevation (m)	Period	Climate
Brasilia (Brazil)	−15.6010	−47.7130	1023	2006–2007	Tropical
Tucson (USA)	32.2297	−110.9553	786	2011–2012	Arid

As discussed elsewhere (Gueymard and Ruiz-Arias 2016), there is no definitive or widely accepted approach for the necessary quality control of irradiance data. For the present study, the data points that do not pass the tests proposed in that publication were rejected. The night periods were also eliminated, since no forecast is then necessary. Figure 1 provides a comparison between actual DNI observations and the corresponding estimates of clear-sky DNI (Ineichen and Perez 2002) for four successive days at each site.

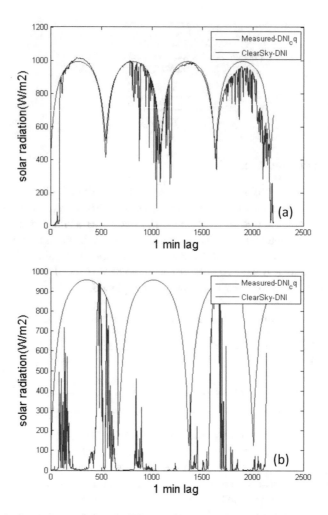

Fig. 1. Sample of quality-controlled 1-min DNI observations vs. clear-sky DNI time series at (a) Tucson and (b) Brasilia.

5.4   Result

Before starting the experiments, each data series is subdivided into two complementary sets, for training and testing purposes. The learning phase is conducted with the training dataset, whereas the model's performance is separately evaluated on the testing dataset. In the present case, approximately two-thirds of the data are assigned to the training set, and the remaining third to the testing set.

The four distinct variable selection scenarios (FW, SW, FS and BS) that were discussed in Sect. 3 are tested here to select the best method. For the sake of comparison, Tables 2 and 3 show the results obtained by the four models, respectively for the two experimental datasets. The BS method provides the best results in nearly all cases.

Table 2. Statistical results for Tucson and Brasilia (15-min time step). Best results are in boldface.

VS method	Tucson				Brasilia			
	FW	SW	FS	BS	FW	SW	FS	BS
R^2	0.7648	0.8355	0.7193	**0.8432**	0.6596	0.7295	0.6228	**0.7561**
MAPE	0.1721	0.1299	0.1968	**0.1254**	0.6562	0.4743	0.7112	**0.4381**
NMSE	0.3131	0.2169	0.3718	**0.2068**	0.2509	0.1971	0.2792	**0.1783**
RMSE	162.8295	135.5459	177.4576	**132.3272**	166.7914	147.8335	175.9446	**140.5784**

Table 3. Statistical results for Tucson and Brasilia (1-h time step). Best results are in boldface.

VS method	Tucson				Brasilia			
	FW	SW	FS	BS	FW	SW	FS	BS
R^2	0.3945	0.5838	0.4789	**0.6564**	0.3931	0.4372	0.2647	**0.5107**
MAPE	0.3515	**0.2479**	0.3213	0.2513	0.8913	0.7677	0.9742	**0.6735**
NMSE	0.7799	0.5312	0.6658	**0.4425**	0.4267	0.3956	0.5158	**0.3434**
RMSE	265.8310	219.3822	245.6131	**200.2282**	205.0965	197.4955	225.5065	**183.9853**

For the two experimental sites, Figs. 2 and 3, display sample results using the 15-min and 1-h horizons, respectively. As could be expected, clear-sky periods are much easier to forecast than periods of rapidly changing cloudiness. The optimal set of variables is selected based on the performance of the ELM on the training set. For both horizons and both sites, the backward selection (BS) performs best compared to the three other scenarios. This can be explained by the fact that the BS variable selection process starts from the observations closest to the predicted variable. These most recent observations are those with the highest persistence degree with respect to the variable to be predicted, and thus are well correlated with it, most generally. Furthermore, the comparison of the two figures, as well as the results in Tables 2 and 3, clearly indicate that the 15-min horizon is much easier to forecast than the 1-h horizon. This can be tentatively explained by the fact that a hourly period may consist of 60 min of clear sky, 60 min of stable cloudiness, or many short periods of variable cloudiness. All these conditions result in very different hourly DNIs. In comparison, a 15-min period is more likely to be stable.

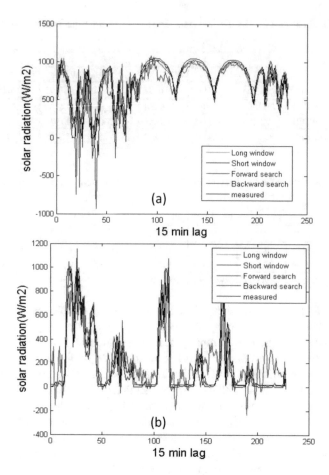

Fig. 2. Measured and predicted DNI values for 15-min ahead horizon using the LW, SW, FS and BS methods for (a) TUCSON and (b) BRASILIA.

In summary, the BS technique produces an optimal selection of inputs, which in turn results in superior model performance. This is important because it confirms the fact that advanced variable selection algorithms can significantly improve the forecasting accuracy.

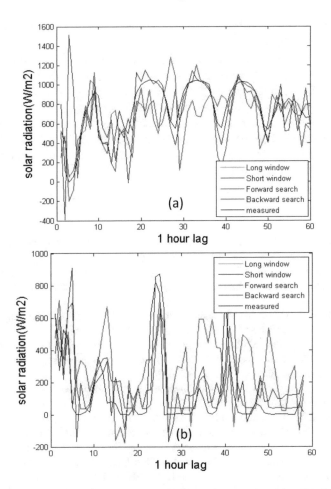

Fig. 3. Measured and predicted DNI values for 1-h ahead horizon using the LW, SW, FS and BS methods for (a) TUCSON and (b) BRASILIA.

6   Conclusion

This contribution introduces a new method for intra-hour forecasting of DNI time series. The method is applied to measurements obtained at two stations in different climates and for two short-term horizons, namely 15-min and 1-h time steps. The proposed methodology consists of two blocks: a variable selection step and a regression method based on an artificial intelligent technique using ELMs. The same forecasting horizon of one-step ahead is used in both cases.

Four different methods are tested to optimize the variable selection step. The statistical results show that the Backward Search (BS) method outperforms the three other variable selection scenarios. The major advantage of BS, and the reason behind its better results, is that the variables it selects are more linked or correlated (from a persistence standpoint) to the forecasted variable. This in turn positively impacts the forecasting performance of the model.

The present results indicate that the proposed method is very successful in case of relatively clear-sky periods, and degrades during periods of rapidly variable cloudiness, during which DNI changes abruptly. This could be expected, considering the known difficulty of forecasting DNI (compared to GHI forecasting). The forecasts are also significantly more accurate at the 15-min horizon. This is a significant finding considering the importance of this horizon for grid stability and for the electricity market (price trading). Future research will compare the present results to those based on other forecasting techniques of the literature.

References

Kaur, A., Nonnenmacher, L., Pedro, H.T., Coimbra, C.F.: Benefits of solar forecasting for energy imbalance markets. Renew. Energy **86**, 819–830 (2016)

Law, E.W., Prasad, A.A., Kay, M., Taylor, R.A.: Direct normal irradiance forecasting and its application to concentrated solar thermal output forecasting–a review. Sol. Energy **108**, 287–307 (2014)

Chu, Y., Pedro, H.T., Coimbra, C.F.: Hybrid intra-hour DNI forecasts with sky image processing enhanced by stochastic learning. Sol. Energy **98**, 592–603 (2013)

Peruchena, C.M.F., Gastón, M., Schroedter-Homscheidt, M., Kosmale, M., Marco, I.M., García-Moya, J.A., Casado-Rubio, J.L.: Dynamic paths: towards high frequency direct normal irradiance forecasts. Energy **132**, 315–323 (2017)

Gueymard, C.A.: Temporal variability in direct and global irradiance at various time scales as affected by aerosols. Sol. Energy **86**(12), 3544–3553 (2012)

Twidell, J., Weir, T.: Renewable energy resources. Routledge, London (2015)

Kleissl, J.: Solar energy forecasting and resource assessment. Academic Press, Boston (2013)

Jović, A., Brkić, K., Bogunović, N.: A review of feature selection methods with applications. In: Information and Communication Technology, 38th International Convention on Electronics and Microelectronics (MIPRO), pp. 1200–1205. IEEE (2015)

Huang, G.B., Zhu, Q.Y., Siew, C.K.: Extreme learning machine: theory and applications. Neurocomputing **70**(1), 489–501 (2006)

Ineichen, P., Perez, R.: A new airmass independent formulation for the Linke turbidity coefficient. Sol. Energy **73**(3), 151–157 (2002)

Gueymard, C.A., Ruiz-Arias, J.A.: Extensive worldwide validation and climate sensitivity analysis of direct irradiance predictions from 1-min global irradiance. Sol. Energy **128**, 1–30 (2016)

A GRNN Based Algorithm for Output Power Prediction of a PV Panel

Kamal Kerbouche[1](✉), S. Haddad[2], A. Rabhi[3], A. Mellit[2],
M. Hassan[3], and A. El Hajjaji[3]

[1] Energetic Systems Modeling Laboratory, University of Biskra, Biskra, Algeria
kemel.kerbouche@univ-biskra.dz
[2] Renewable Energy Laboratory, Department of Electronics, University of Jijel,
Jijel, Algeria
[3] Modeling, Information and Systems Laboratory, UPJV, Amiens, France

Abstract. In this paper we investigated the reliability of a GRNN algorithm for the power prediction of a PV panel in order to minimize the effect of fast changing of the meteorological conditions. An experimental database of meteorological data (irradiance and module temperature) as input and electrical measure (power delivered by PV Panel) as output has been used. A database composed of two sets 97 values each one is used for training and validating the proposed GRNN model. The data used to develop the proposed algorithm are attained during two separated days from a PV panel within the MIS-Lab of UPJV, France. According to the gained results the algorithm can help to predict real instantaneous power even during temporary change in meteorological data.

Keywords: Neural networks · GRNN · Power supervision · Small PV

1 Introduction

The increasing energy demands combined with rising conventional fuel costs and environmental awareness have contributed to the emergence of renewable energy sources during the last decade. Photovoltaic (PV) systems have sustained a remarkable annual growth rate, driven by several factors including technological innovation, improved cost effectiveness and government incentives. From 2000 to 2011, the International Energy Agency (IEA) reports that global PV installed capacity increased from 1 GW to 67 GW. Nevertheless, significant constraints still hinder the large-scale integration of PV in the electricity mix. In particular, the unpredictability and variability of the solar energy cause major problems to the reliability and stability of existing grid-connected power systems.

The intermittent nature of solar energy poses many challenges to renewable energy system operators in terms of operational planning and scheduling. PV output forecasting is therefore essential for utility companies to plan the operations of power plants properly so as to ensure the stability, reliability and cost effectiveness of the system. Predicting the output of photovoltaic systems is therefore essential for managing the operation and assessing the economic performance of power systems [1].

© Springer International Publishing AG 2018
M. Hatti (ed.), *Artificial Intelligence in Renewable Energetic Systems*, Lecture Notes in Networks and Systems 35, https://doi.org/10.1007/978-3-319-73192-6_30

Artificial intelligence techniques are becoming useful as alternate approaches to conventional techniques or as components of integrated systems. They have been used to solve complicated practical problems in various areas and are becoming more popular nowadays. They can learn from examples, are fault tolerant in the sense that they are able to handle noisy and incomplete data, are able to deal with nonlinear problems and once trained can perform prediction and generalization at high speed. AI-based systems are being developed and deployed worldwide in a wide variety of applications, mainly because of their symbolic reasoning, flexibility and explanation capabilities. Artificial intelligence has been used in different sectors, such as engineering, economics, medicine, military, marine, etc. They have also been applied for modeling, identification, optimization, prediction, forecasting and control of complex systems [2].

We aim in this work to develop an algorithm that can predict the output power of a solar panel using meteorological data only. Thus, we installed the system with sensors, collected data and used them to train our algorithm so it can predict the power once we give it the module temperature and the solar irradiance.

2   Generalized Regression Neural Network Model

A GRNN is a variation of the radial basis neural networks (RBFNs), which is based on kernel regression networks. A GRNN does not require an iterative training procedure as back propagation networks. It approximates any arbitrary function between input and output vectors, drawing the function estimate directly from the training data. In addition, it is consequent that as the training set size becomes large, the estimation error approaches zero, with only mild restrictions on the function [3].

Typically, the purpose of training is to make estimations for future cases in which only the inputs to the network are known. The result of conventional network training is a single set of weights that can be used to make such estimations.

The GRNN is one of the simplest neural networks in term of network architecture and learning algorithm. The advantage is that the learning is instantaneous. GRNN is based on one-pass learning algorithm; it is a highly parallel network of radial basis. GRNN is composed of input layer, radial basis hidden layer and linear output layer; its architecture is shown in Fig. 1 [4]. It is similar to the radial basis network, but has a slightly different second layer.

In Fig. 1, the number of the units in the hidden layer is equal to the training sample size M, and the weight function of this level is the Euclidean distance measuring

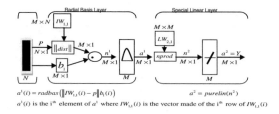

$$a^1(i) = radbas\left(\left\|IW_{1,1}(i) - p\right\|b_1(i)\right) \qquad a^2 = purelin(n^2)$$

$a^1(i)$ is the ith element of a^1 where $IW_{1,1}(i)$ is the vector made of the ith row of $IW_{1,1}(i)$

Fig. 1. GRNN architecture (MATLAB Neural Networks toolbox)

function $\|dist\|$, its function is to calculate the distance between the input of the network and the hidden layer weight matrix IW1, 1, b1 is the threshold of the hidden layer. Symbol (.*) in Fig. 1, indicates the product element per element of the output of IW1,1. The result of the product n1 is the net input of the transfer function. Hidden layer transfer function is the Radial Basis Function (RBF), Gaussian function is often used.

The weight function of the output layer is the standardization of the right point multiplication function (with nprod denoted). n2 indicates the vector of the network; each element of it can be valued first by making dot-product between a1 and the element in each line of the matrix LW2,1 then doing the division between the result and the sum of every element of a1. Finally, the results of n2 is offered to the linear transfer function a2 = purlin(n2) to calculate the network output. Figure 2 shows the linear transfer function, purelin(n) and radial basis transfer function, radbas(n).

The estimation model takes the following form:

$$P = f([T,G],W) + \varepsilon \qquad (1)$$

Where P is the vector of model outputs (PV power output); [T,G] is the vector of model inputs (module temperature and irradiance); W is the vector of model parameters (connection weights), f($\bullet$) are the functional relationship between model outputs, inputs and parameters, and ε is the vector of model errors (Fig. 3).

The ANN was simulated in MATLAB using 'newgrnn' function, which creates a new GRNN (two-layer network).

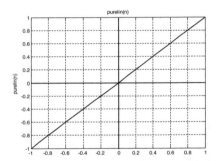

Fig. 2. Linear transfer function, purelin(n)

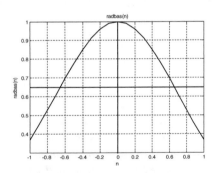

Fig. 3. Radial basis transfer function, radbas(n)

As part of the ANN model development process, the available data are generally divided into training, testing and validation subsets. The training set is used to estimate the unknown connection weights, the testing set is used to decide when to stop the training process in order to avoid over-fitting and/or which network structure is optimal, and the validation set is used to assess the generalization ability of the trained network.

3   System Description

In order to conduct a comprehensive study of the relationship between PV output power and the meteorological conditions (irradiance and temperature) a system has been installed in the MIS-Lab of UPJV, France. The PV system selected for the present study consists of five main components namely a Sanyo 215 W solar panel, meteorological measurement devices (thermocouple and pyrometer), Electrical current and voltage sensors, a communication board, a programmable electric load and a PC equipped with a Labview and MATLAB interfaces as shown in Fig. 4 and described in [5].

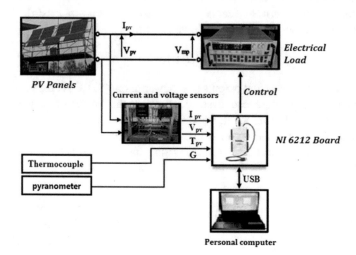

Fig. 4. Experimental set [5]

The data acquisition process was established during a day from 9:15 until 17:15 with a measurement interval of 5 min which effect a database of 97 values for each parameter (Temperature and irradiance) then a similar database of maximum power was calculated from the current and voltage measurements data.

The experiment was established again in the most typical irradiation day (no clouds) in order to compare between the two sets of data results.

4   System Description

In this section we will describe the steps and data used to train and validate the developed algorithm.

4.1   The Training Process

In this work we choose the General Regression Neural Networks (GRNN) because it does not require iterative training, so it is simple, stable with fast training speed and good description for the characteristics of dynamic. Moreover, the parameters need to be adjusted has only one which determines the network can maximally avoid the influence to the results caused by man-made subjective assumptions [6].

First the Algorithm has been trained with the typical day database as following:

– input parameters are the module temperature and the solar irradiance (97 values for each victor).
– output parameter is the maximum extracted power calculated using current and voltage databases (97 values).

Then, the second database of another day is used to test the performances of the algorithm.

4.2   The Inputs

The Figs. 5 and 6 show the input parameters for the two days. The first day is a typical day in terms of meteorological conditions (irradiation and temperature).

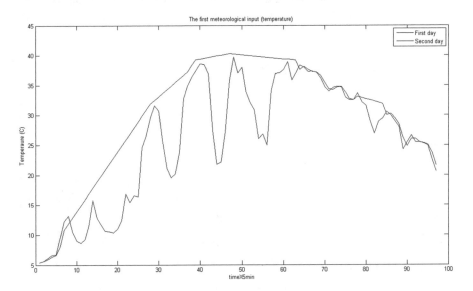

Fig. 5. First meteorological input (temperature)

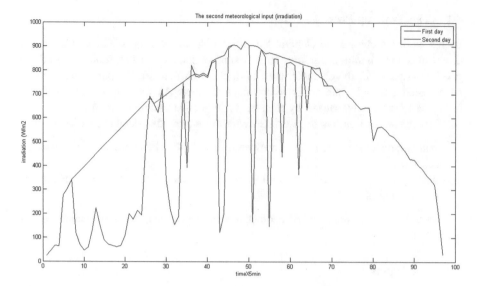

Fig. 6. Second meteorological input (irradiation)

The Fig. 7 shows the maximum produced power obtained from the calculated power based on the measured currents and voltages.

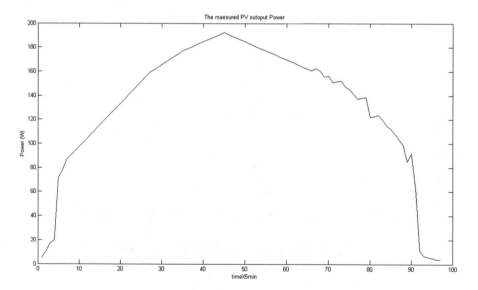

Fig. 7. Measured PV panel output power

5   Results and Discussion

After training the developed algorithm using the first day database. We used the second day's database to test its efficiency then the MSE error between the two graphs is calculated. The Fig. 8 shows both measured power and predicted power using the developed algorithm.

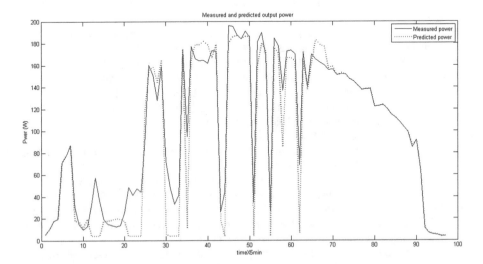

Fig. 8.   Measured and predicted output power

The RMS calculated between the is: 435.4299.

The above results show the efficiency of this Algorithm in predicting the output power of PV panel based on meteorological data only.

6   Conclusion

In this work, A GRNN algorithm have been successfully developed for predicting the output power of a 215 W PV panel installed on the MIS-Lab of UPJV, France. It has been demonstrated that the algorithm is able to predict the output power of the Panel based on available solar irradiance and temperature with good accurate results.

The advantage of the proposed GRNN algorithm is that it does not require much calculations and use only meteorological data.

Further works can use the advantages to use this algorithm to implement it for real time power supervision for diagnosis aims.

Acknowledgment. The authors would like to thank MIS-Lab of UPJV, France, for providing the facilities and databases to conduct this research.

References

1. Ramsami, P., Oree, V.: A hybrid method for forecasting the energy output of photovoltaic systems. Energy Convers. Manage. J. **95**, 574–632 (2015)
2. Mellit, A., Kalogirou, S.A.: Artificial intelligence techniques for photovoltaic applications: a review. Progress Energy Combust. Sci. J. **34**, 406–413 (2008)
3. Firat, M., Gungor, M.: Generalized regression neural networks and feed forward neural networks. Adv. Eng. Softw. J. **40**, 731–737 (2009)
4. Mathworks, Inc.: Matlab Documentation Center, Neural Network Toolbox, User's Guide (2013)
5. Boutana, N., Mellit, A., Haddad, S., Rabhi, A., Massi Pavan, A.: An explicit I-V model for photovoltaique module technologies. Energy Convers. Manage. J. **138**, 400–412 (2017)
6. Kuang, X., Xu, L., Huang, Y., Liu, F.: Real-time forecasting for short-term traffic flow based on general regression neural network. In: Proceedings of the 8th World Congress on Intelligent Control and Automation, July 6–9 2010, Jinan, China (2010)
7. Chine, W., Mellit, A., Lughi, V., Malek, A., Sulligoi, G., Massi Pavan, A.: A novel fault diagnosis technique for photovoltaic systems based on artificial neural networks. Renew. Energy J. **90**, 501–512 (2016)
8. Zhou, W., Yang, H., Fang, Z.: A novel model for photovoltaic array performance prediction. Appl. Energy J. **84**, 1187–1198 (2007)
9. Yadav, A.K., Chandel, S.S.: Solar radiation prediction using artificial neural network techniques: a review. Renew. Sustain. Energy Rev. J. **33**, 772–781 (2014)
10. Qazi, A., Fayaz, H., Wadi, A., Raj, R.G., Rahim N.A.: The artificial neural network for solar radiation prediction and designing solar systems: a systematic literature review. Cleaner Prod. J. https://doi.org/10.1016/j.jclepro.2015.04.041
11. Mellit, A., Massi Pavan, A.: Performance prediction of 20 kW p grid-connected photovoltaic plant at Trieste (Italy) using artificial neural network. Energy Convers. Manage. J. **51**, 2431–2441 (2010)
12. Karabacak, K., Cetin, N.: Artificial neural networks for controlling wind–PV power systems: a review. Renew. Sustain. Energy Rev. J. **29**, 804–827 (2014)
13. Kalogirou, S.A.: Artificial neural networks in renewable energy systems applications: a review. Renew. Sustain. Energy Rev. J. **05**, 373–401 (2001)
14. Samarasinghe, S.: Neural Networks for Applied Sciences and Engineering, 2nd edn. Auerbach Publication (2009). ISBN 978-0-8493-3375-X
15. Palm III, W.J.: Introduction to Matlab 7 for Engineers. MC Craw Hill, New York (2000)

Modeling and Operation of PV/Fuel Cell Standalone Hybrid System with Battery Resource

Saidi Ahmed[✉], Cherif Benoudjafer, and Chellali Benachaiba

Department of Electrical Engineering, Faculty of Technology, Tahri Mohammed University,
BP 417 Route de Kenadsa, 08000 Béchar, Algeria
ahmedsaidi@outlook.com, benoudjafer.cherif@mail.univ-bechar.dz,
chellali99@yahoo.fr

Abstract. The main challenge in replacing legacy systems with the newer alternatives is to capture maximum energy and deliver maximum power at minimum cost for the given load. Solar energy which is free and abundant in most parts of the world has proven to be an economical source of energy in many applications. Among the renewable energy technologies, the solar energy coupling with fuel cell technology will be the promising possibilities for the future green energy solutions. The new efficient photovoltaic array (PVA) has emerged as an alternative measure of renewable green power, energy conservation and demand-side management. However in photovoltaic power generation system the control problems arise due to large variances of output under different insulation. This problem can be overcome by hybrid photovoltaic generation system i.e. use of photovoltaic arrays with fuel cells and power storage such as battery bank. The stand-alone hybrid system aims to provide power efficiency supply to the consumers with a constant voltage and frequency along with proper power management using simple control techniques. The modeling and control strategies of the hybrid system are realized in MATLAB/Simulink.

Keywords: PV · SOFC · MPPT · Battery resource · Hybrid system

1 Introduction

From the first AC power system - Niagra Falls, 1895-until about four decades ago, the electrical energy generation systems were developed based on the traditional nonrenewable sources - oil, gas, coal and nuclear - in order to attend the society needs, whose consumption was guided by the large-scale economy [1].

This scenario remained unchanged until the world energy crisis, 1973, when the employment of Renewable Energy Sources (RES) became a world trend [2]. Moreover, the policy for the pollutant emission and the environment impact reductions, strongly contributed for a complementary green energy matrix establishment [3].

From 2000 to now, many works referring grid-connected systems were published, relating aspects as power converters design, control strategies, PV and WT maximum

© Springer International Publishing AG 2018
M. Hatti (ed.), *Artificial Intelligence in Renewable Energetic Systems*, Lecture Notes in Networks and Systems 35, https://doi.org/10.1007/978-3-319-73192-6_31

power point tracker (MPPT), reactive compensation, active filtering and so on. In addition, important institutions as IEEE and IAS have developed standards and recommendations to guide the designers.

Solar power source is one of the most promising renewable power generation technology [4, 5]. FCs also shows great potential to be green power sources of the near future because of many advances they have (such as low emission of pollutant gases, high efficiency, and flexible modular structure) [6]. However, each source has its own drawbacks. For instance, solar power is highly dependent on climate while FCs needs hydrogen-rich fuel. FCs are good energy sources to provide reliable power at a steady rate, but they cannot respond to the electrical load transients as fast as desired. This is mainly due to their slow internal electrochemical and thermodynamic responses [7–9].

The power generated by a PV system is highly dependent on weather conditions. For example, during cloudy periods and at night, a PV system would not generate any power. In addition, it is difficult to store the power generated by a PV system for future use. To overcome this problem, a PV system can be integrated with other alternate power sources and/or storage systems, such as electrolyzer, hydrogen storage tank, FC systems and Battery bank [11].

Use of fuel cells (FC) in combination with a PV generator may ensure an uninterruptible power supply as long as the fuel cell power can meet the power deficit. Fuel cells show a particular promise as they can operate on hydrogen with zero emissions, have a relative high efficiency (30–60%), and have a limited number of moving parts with a flexible modular structure [12, 13].

In this paper the case with fuel cell supplied from a hydrogen container is considered.

In this paper, a hybrid alternative energy system consisting of PV, FC is proposed. The remainder of the paper is structured as follows: the Sect. 2 details the PV system modelling. In Sect. 3, a Solid oxide fuel cell system modelling and presented with your simulation. In Sect. 4, the modelling of battery characteristics, the test bench results for the proposed system are presented. Finally, this paper ends with concluding remarks for further study in Sect. 7.

2   PV System Modelling

The Fig. 1 shows a simplified scheme of a standalone PV system with DC–DC buck converter.

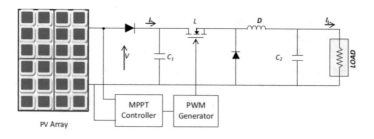

Fig. 1. A PV system with a DC–DC buck converter

This section is devoted to PV module modelling which is a matrix of elementary cells that are the heart of PV systems. The modelling of PV systems starts from the model of the elementary PV cell that is derived from that of the P–N junction [16].

2.1 Ideal Photovoltaic Cell

The PV cell combines the behavior of either voltage or current sources according to the operating point. This behavior can be obtained by connecting a sunlight-sensitive current source with a P–N junction of a semiconductor material being sensitive to sunlight and temperature. The dot-line square in Fig. 2 shows the model of the ideal PV cell. The DC current generated by the PV cell is expressed as follows

$$I = I_{PV,Cell} - I_{s,Cell}\left(e^{\frac{V}{aVt}} - 1\right) \tag{1}$$

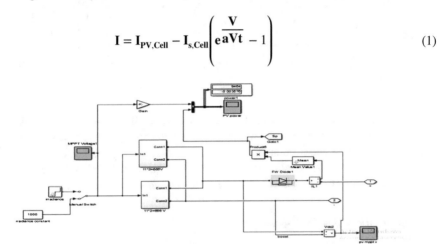

Fig. 2. Simulation model of PV module

The first term in Eq. (1), that is Ipv, cell, is proportional to the irradiance intensity whereas the second term, the diode current, expresses the non-linear relationship between the PV cell current and voltage. A practical PV cell, shown in Fig. 1, includes series and parallel resistances [17]. The series resistance represents the contact resistance of the elements constituting the PV cell while the parallel resistance models the leakage current of the P–N junction.

2.2 Simulation Model of PV

3 Solid Oxide Fuel Cell

Solid oxide fuel cell (SOFC) is an electrochemical conversion device that produces electricity directly from oxidizing a fuel. Fuel cells are characterized by their electrolyte material; the SOFC has a solid oxide or ceramic, electrolyte. Advantages of this class of fuel cells include high efficiency, long-term stability, fuel flexibility, low emissions, and relatively low cost. The largest disadvantage is the high operating temperature which results in longer start-up times and mechanical and chemical compatibility issues [18] (Fig. 3).

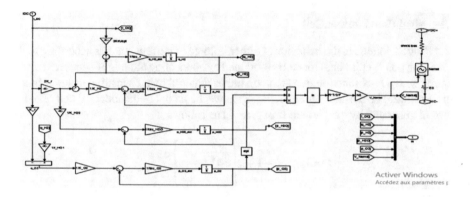

Fig. 3. Simulation SOFC model

SOFCs operate at extremely high temperature (typically above 800 °C). This high temperature give them extremely high electrical efficiencies and fuel flexibility, both of which contribute to better economics, but it also creates engineering challenges. By solving these engineering challenges with breakthroughs in materials science and revolutionary new design, target SOFC will be a cost-effective technology.

3.1 Simulation of SOFC Model

4 Batteries

Battery bank storage is sized to meet the load demand during non-availability period of renewable energy sources. At any time t, the charged quantity of the battery bank is subject to the following two constraints:

$$SOC_{min} \leq SOC(t) \leq SOC_{max} \, Cu_{bat,max}(t) \leq Cu_{max}$$

In the above relations, SOCmin (0.3) and SOCmax are the minimum and maximum SOC of the battery, respectively, SOC(t) is the battery SOC at each hour of the year, and Cumax is the maximum charge current which is determined as a battery specification by its manufacture. In the present study, the maximum value of the SOC (SOCmax) is 1, and 0.3 is utilized as the value of the SOCmin according to the battery specifications. Depending on the PV and wind energy production and the load power requirements.

5   Proposed Hybrid Power System

The architecture in Fig. 1 is proposed for the hybrid power system which is based on the centralized DC-bus system [18]. In the centralized DC-bus configuration, all the sources and storages are coupled to a common DC bus before they are connected to the load. The load in the figure stands for either a DC-load or a voltage source inverter depending on the application.

The PV generator is coupled to the DC bus via a buck based DC/DC maximum power point tracker (MPPT) to track the maximum power point where the PV plant generates the maximum possible power output for a given irradiance, ambient temperature and loading condition.

This enables to utilize the renewable energy to the maximum.

The low and highly variable load dependent voltage of the PEM fuel cell stack is boosted to the DC-bus level via a phase shifted PWM (PSPWM) transformer isolated DC/DC converter. The converter is current controlled to shape the fuel cell output to safe magnitude and rate.

A low volt-ampere (VA) rated buck DC/DC converter steadily charges the Ni-MH battery during light loading when the bus voltage becomes higher than the battery voltage. When the bus voltage goes below the battery voltage which is indication of a heavier loading in surplus of the power output from the fuel cell and PV combined, the diode becomes forward biased discharging the battery. During the longer period of normal loading, the low VA DC/DC converter having smaller inductor charges the battery slowly at steady-state (Fig. 4).

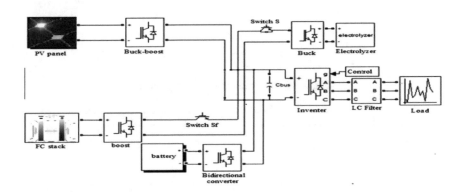

Fig. 4. Architecture proposed for PV/Fuel cell systems

6   Simulation Results

The developed PV-SOFC-Battery based standalone hybrid system during this work. The analysis of the developed model is done PV ARRAY (PVA) & SOFC both, HYBRID SOLAR SOFC (Figs. 5, 6 and 7).

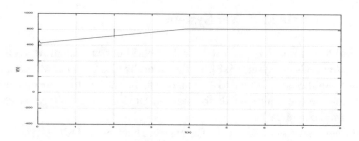

Fig. 5. MPPT of PV output power against time

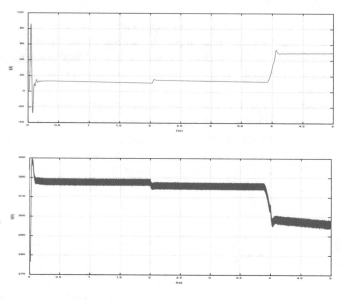

Fig. 6. Battery current and voltage output

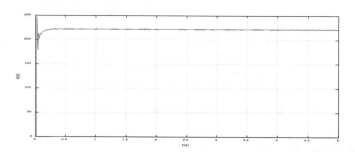

Fig. 7. MPPT buck converter output voltage

The simulation results of PV based on MPPT controller with changing irradiation from high to low level, from 700 W/m² to 600 W/m² and 500 W/m². Simulation starts with constant temperature of 25 °C and 700 W/m².

In this paper the Maximum Power Point Tracking (MPPT) controller for boost converter based on Incremental Conductance (IC) algorithms controller is run and developed (Figs. 8, 9 and 10).

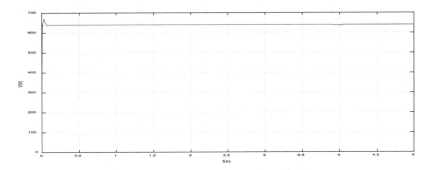

Fig. 8. Vdc output voltage for system

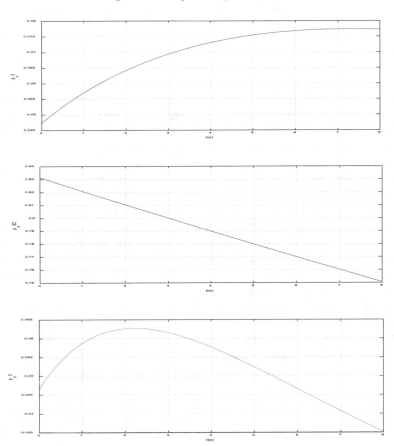

Fig. 9. H_2, H_2O and O_2 pressure curves of SOFC

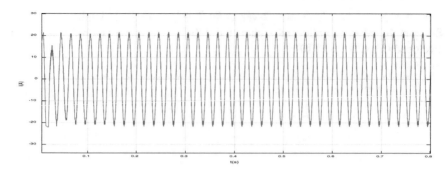

Fig. 10. Line currents for load

SOFC pressure for hydrogen (PH_2), water vapor (PH_2O) and oxygen (O_2) are displayed. It is seen that for SOFC off case the pressure of hydrogen (PH_2) and oxygen (O_2) is increase while the pressure of water vapor (PH_2O) is decrease. This is obvious that during SOFC it stores the energy so that during on situation energy can be released.

7   Conclusion

Modeling and operation of PV/Fuel cell with battery resources is analyzed in this paper. Rapidly changing irradiance and fuel cell performance are considered in this paper. In this paper a PVA model is designed in a MATLAB/SIMULINK environment for maximum power point search and a dynamic Incremental conductance (IC) algorithm is used to detect the maximum power point output of the module.

The results showing this are presented in this paper.

An effective load sharing and control strategy has been presented for a PV/FC/ Battery hybrid power system that optimizes availability, performance, fuel economy and safety. The simulation results show that the control strategy is effective in meeting high degree of power availability, and reduced cycling of a battery. The battery is also relieved from steep charging currents by slightly delaying the current reference improving charging efficiency. A near full controllability of battery and fuel cell enables operation of the units within safe limits in addition to making it possible to shape the trajectories of their power responses.

Solar photovoltaic system is considered to be the potential energy sources with a battery bank and a hydrogen storage fuel cell system as the prospective storage media. The incorporation of IC controller enables maximum power extraction from PV energy systems and accurately regulates the flow of power among the various system components to ensure optimum utilization of energy efficiently the whole day through.

Acknowledgment. This work was financially supported by the ENERGARID and SGRE laboratory and Tahri Mohammed University Béchar 08000 ALGERIA, under the scientific Programme Project Electro-energetique Industrial Option Renewable Energy, 2015/2016.

References

1. Dunn, S.: Micropower: The Next Electrical Era - Worldwatch Paper l51, pp. 5–94 (2000)
2. Xavier, M.Y., Borotni, E.D.C., Haddad, J.: Distrubeted generation in Brazil: oportunities and barriers (in Portuguese). Energy Brasilian Mag. **11**(2), 1–11
3. Coelho, R.F., Schimtz, L., Martins, D.C.: Grid-Connected PV -Wind-Fuel Cell Hybrid System Employing a Supercapacitor Bank as Storage Device to Supply a Critical DC Load. IEEE (2011)
4. Pan, C.-T., Cheng, M.-C., Lai, C.-M., Chen, P.-Y.: Current-ripple-free module integrated converter with more precise maximum power tracking control for PV energy harvesting. IEEE Trans. Ind. Appl. **51**(1), 271–278 (2015)
5. Saleh, S.A., Aljankawey, A.S., Alsayid, B., Abu-Khaizaran, M.S.: Influences of power electronic converters on voltage-current behaviors during faults in DGUs—part ii: photovoltaic systems. IEEE Trans. Ind. Appl. **51**(4), 2832–2845 (2015)
6. Sundarabalan, C.K., Selvi, K.: Compensation of voltage disturbances using PEMFC supported dynamic voltage. Int. J. Electric. Power & Energy Syst. **71**, 77–92 (2015)
7. Bizon, N.: Improving the PEMFC energy efficiency by optimizing the fueling rates based on extremum seeking algorithm. Int. J. Hydrogen Energy **39**(20), 10641–10654 (2014)
8. Thounthong, P., Piegari, L., Pierfederici, S., Davat, B.: Nonlinear intelligent DC grid stabilization for fuel cell vehicle applications with a supercapacitor storage device. Int. J. Electr. Power & Energy Syst. **64**, 723–733 (2015)
9. Vijayaraghavan, K., DeVaal, J., Narimani, M.: Dynamic model of oxygen starved proton exchange membrane fuel-cell using hybrid analytical-numerical method. J. Power Sources **285**, 291–302 (2015)
10. Zeng, J., Qiao, W., Liyan, Q.: An isolated three-port bidirectional DC–DC converter for photovoltaic systems with energy storage. IEEE Trans. Ind. Appl. **51**(4), 3493–3503 (2015)
11. Tani, A., Camara, M.B., Dakyo, B.: Energy management in the decentralized generation systems based on renewable energy— ultracapacitors and battery to compensate the wind/ load power fluctuations. IEEE Trans. Ind. Appl. **51**(2), 1817–1827 (2015)
12. Wang, C.: Modeling and control of hybrid wind/PV/FC distributed generation systems, PhD thesis- Montana State University, July 2006
13. Larminie, J., Dicks, A.: Fuel Cell Systems Explained, 2nd edn. Wiley (2003). Samson Gebre, T, Undeland, T.M., Ulleberg, Ø., Vie, P.J.S.: Optimal load sharing strategy in a hybrid power system based on PV/Fuel Cell/ Battery/Supercapacitor. IEEE Trans. Ind. Appl. (2009). IEEE
14. Coelho, R.F., Concer, F., Martins, D.C.: A MPPT approach based on temperature measurements applied in PV systems. In: IEEE International Conference on Sustainable Energy Technologies (ICSET), pp. 1–6 (2010)
15. Chao, P.C.P., Chen, W.D., Chang, C.K: Maximum power tracking of a generic photovoltaic system via a fuzzy controller and a two-stage DC–DC converter. Microsyst. Technol. **18**(9–10), 1267–1281 (2012) (2011)
16. Villalva, M.G., Gazoli, J.R.: Comprehensive approach to modeling and simulation of photovoltaic arrays. IEEE Trans. Power Electron. **24**, 1198–1208 (2009)
17. Bennett, T., Zilouchian, A., Messenger, R.: Photovoltaic model and converter topology considerations for MPPT purposes. Sol. Energy **86**, 2029–2040 (2012)
18. Ortjohann, E.: A Simulation Model for Expandable Hybrid Power Systems. University of applied Sciences Sudwestfalen

Using Probabilistic Neural Network (PNN) for Extracting Acoustic Microwaves (BAW) in Piezoelectric Material

Hichem Hafdaoui[(⊠)], Cherifa Mehadjebia, and Djamel Benatia

Electronics Department, Faculty of Technology,
University of Batna 2, Batna, Algeria
hichemhafdaoui@yahoo.fr

Abstract. In this paper, we propose a new method for Bulk waves detection of an acoustic microwave signal during the propagation of acoustic microwaves in a piezoelectric substrate (Lithium Niobate $LiNbO_3$). We have used the classification by probabilistic neural network (PNN) as a means of numerical analysis in which we classify all the values of the real part and the imaginary part of the coefficient attenuation with the acoustic velocity in order to build a model from which we note the Bulk waves easily. These singularities inform us of presence of Bulk waves in piezoelectric materials.

By which we obtain accurate values for each of the coefficient attenuation and acoustic velocity for Bulk waves. This study will be very interesting in modeling and realization of acoustic microwaves devices (ultrasound) based on the propagation of acoustic microwaves.

Keywords: Piezoelectric material · Probabilistic neural network (PNN)
Classification · Acoustic microwaves · Bulk waves · The attenuation coefficient

1 Introduction

The phenomenon of surface acoustic wave (SAW) propagation was first reported on by Lord Rayleigh in 1885. The interest in the use of piezoelectric materials as the wave propagation medium, lies in the propagation of appearance. These waves, in this case, spread in a resilient part (*or acoustic*) and power; hence the name electro elastic or electroacoustic [1]. This coupling allows the piezoelectric waves that propagate acoustic velocities to be very stable with frequencies. The resulting material will be considered a delay structure and allows an efficient signal processing [2].

So the processed signal is applied to the electrodes of the transducer which in turn leads to deformation and a piezoelectric wave arises is propagating in the Z direction [2, 3].

Theory on generation and propagation of BAW (Bulk acoustic waves) in a SAW (Surface acoustic waves) device with IDTS (Inter-Digitized transducers) is well explained in [1, 4].

Some materials, when deformed, become electrically polarized. This effect is known as piezoelectricity.

© Springer International Publishing AG 2018
M. Hatti (ed.), *Artificial Intelligence in Renewable Energetic Systems*, Lecture Notes in Networks and Systems 35, https://doi.org/10.1007/978-3-319-73192-6_32

In this work, we searched for all Acoustic velocity and Attenuation coefficients on the level of Piezoelectric substrate(LiNbO3) relying on fundamental relations and we classified all the values so that we can detect bulk wave easily.

2 Fundamental Relations

The electrical induction and the electric field E in the piezoelectric materials are bound by [5]:

$$D = \varepsilon.E + e^T.S \tag{1}$$

Where ε: is the permittivity tensor (F/m)

e: piezoelectric tensor (c/m)

S: strain tensor ($S_{ij} = 1/2[\partial U_i/\partial X_j + \partial U_j/\partial X_i]$)

The electric polarization of the medium under the effect of deformation, also involves the creation of stresses under the effect of an external electric field. This constraint is [6]:

$$T = C.S - e.E \tag{2}$$

Where C: represents the elastic tensor (N/m^2).

3 Piezoelectric Tensor Equations

In one form the stress tensor and the electric induction are defined as follows [6, 7]:

$$T_{ij} = C_{ijkl}.S_{kl} - e_{kij}.E_k \tag{3}$$

$$D_i = e_{jkl}.S_{kl} + \varepsilon_{ik}.E_k \tag{4}$$

With i, j, k, l = 1, 2, 3.

In the quasi-static approximation, the Maxwell equations reduce to the Poisson equation [6–8]:

$$\text{div}.\vec{D} = \frac{\partial D_i}{\partial X_i} = 0 \tag{5}$$

The movement of the particles under the action of stress is described by Newton's equation:

$$\nabla T = \rho.\frac{\partial U^2}{\partial t^2} \tag{6}$$

Substituting Eqs. (3) and (4) in Eqs. (5) and (6) we obtain the piezoelectric tensor phenomenological equations:

$$C_{ijkl} \frac{\partial^2 U_k}{\partial.X_i \partial.X_l} + e_{lij} \frac{\partial^2 U_4}{\partial.X_k \partial.X_i} = \rho \frac{\partial^2 U_j}{\partial.t^2} \tag{7}$$

$$e_{ikl} \frac{\partial^2 U_k}{\partial.X_i \partial.X_l} - \varepsilon_{ik} \frac{\partial^2 U_4}{\partial.X_k \partial.X_i} = 0 \tag{8}$$

4 General Form of the Solution

The general solution for the displacement and the electric potential in the medium is:

$$U_i = u_i \exp(j\beta.\alpha_i X_3).\exp - j(\omega.t - \beta(1 + j\gamma)X_1) \tag{9}$$

Where $u_i (i = 1, 2, 3)$ are displacement amplitudes, $u_i(i = 4)$ represents the amplitude of the electric potential, β is the propagation constant, α_i are the attenuation coefficients of the wave within the piezoelectric crystal, axis X3 and ω is the angular pulsation.

Solutions of this type of wave correspond to waves propagating with or without attenuation along the direction X1. The elastic displacements Ui and the electric potential U4 may vary from the normal direction to the flat surface (X3), but following invariant X2.

Substituting Eq. (11) into Eqs. (9) and (10) we obtain the system:

$$[A].[U] = [0] \tag{10}$$

Where [A] is a symmetric matrix 4×4.
$[U] = [u1, u2, u3, u4]^T$ Components determine.
Development of the determinant of [A] of the general, an eighth order polynomial:

$$Det(A) = \sum_{i=0}^{8} B_i.\alpha^i = 0 \tag{11}$$

This polynomial is called dispersion equation or secular equation.

The solution of Eq. (11) gives for each β ($\beta = 2.\pi.f/Vs$: constant propagation) eight roots. These roots are based on Vs (acoustic velocity). Each root generates three displacement components of the particle and an electric potential. Thus, the general solution is a combination of eight roots (8 secondary wave) given by the expression:

$$U_i = \sum_{n=1}^{8} C_n.D_i^n. \exp\{j\beta.\alpha^n.X_3 + j\beta.X_1(1 + j.\gamma) - j.\omega.t\} \tag{12}$$

$i = 1, 4$

Where D_i^n are the components of the eigenvector of the system (10) associated with the eigenvalue α^n

C_n: constant to be determined by the boundary conditions.

$$\alpha^{(1)} = \alpha^{(1)}_{re} + j\alpha^{(1)}_{im} \text{ et } \alpha^{(2)} = \alpha^{(2)}_{re} - j\alpha^{(2)}_{im}$$

Where $\alpha^{(j)}_{re} = \alpha^{(j+1)}_{re}$ et $\alpha^{(j)}_{im} = \alpha^{(j+1)}_{im}$ with $j = 1, 3, 5, 7$.

if $\alpha^{(i)}_{re} = 0$ et $\alpha^{(i)}_{im} \rangle 0$, the bulk waves will be obtained [6–8].

5   Application on LiNbO3 (Lithium Niobate) Cut YZ (Z = X1 and Y = X3)

In this example, we will show the variations of the real and the imaginary parts of attenuation coefficients ($\alpha 4$) depending on the acoustic velocity Vs (Fig. 1), we will indicate the presence of bulk waves (zero real part and imaginary part is positive) and for a good detection of bulk waves we should have (real part equals zero and the maximum positive value of imaginary part).

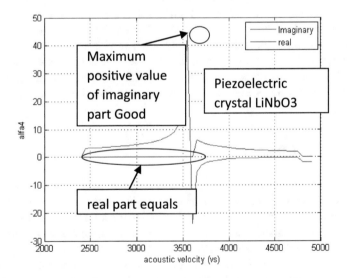

Fig. 1. Attenuation coefficient α 4 depending on the acoustic velocity Vs

We set the value of the imaginary part, which we will look for it in the classification (Fig. 2).

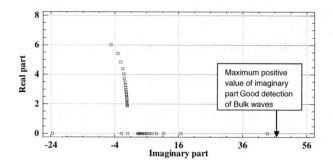

Fig. 2. Plot - Real part vs Imaginary part

6 Probabilistic Neural Network

A probabilistic neural network (PNN) is defined as an implementation of statistical algorithm called kernel discriminate analysis in which the operations are organized into multilayer feed forward network with four layers: input layer, pattern layer, summation layer and output layer. A pnn is predominantly a classifier since it can map any input pattern to a number of classifications [9] among the main advantages that discriminate pnn is: fast training process, an inherently parallel structure, guaranteed to converge to an optimal classifier as the size of the representative training set increases and training samples can be added or removed without extensive retraining. Accordingly, a PNN learns more quickly than many neural networks model and have had success on a variety of applications. based on these facts and advantages, PNN can be viewed as a supervised neural network that is capable of using it in system classification and pattern recognition [9]. The PNN consists of nodes allocated in three layers after the inputs (Fig. 3) [10].

7 Results and Discussion

To know the accurate values of each of the real part and the imaginary part which represent coefficient attenuation $\alpha 4$ and Acoustic velocity, Selection variable stays always Acoustic velocity and we change the Classification factor where we find:

In the Fig. 4, we have classification where we find classification factor is real part and selection variable is acoustic velocity and number of cases in training set 51, in the maximum positive value of imaginary part we find that the real part equals 0.

In the Fig. 5, we have classification where we find classification factor is imaginary part and selection variable is acoustic velocity and number of cases in training set: 51, We find that the Maximum positive value of imaginary part equals 43.7084.

In the Fig. 6, we have classification where we find classification factor: acoustic velocity and selection variable: acoustic velocity and number of cases in training set: 51, in the maximum positive value of imaginary part (43.7084) we find that the acoustic velocity equals 3550 m/s (Table 1).

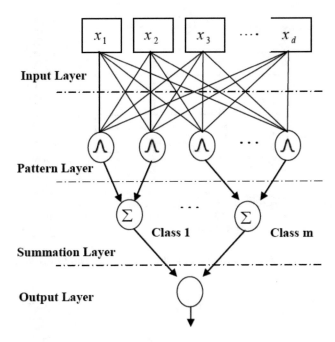

Fig. 3. The architecture of PNN [10]

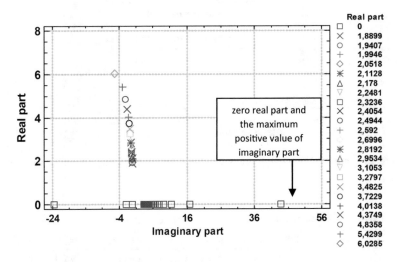

Fig. 4. Neural Network Bayesian Classifier - Real part (Acoustic velocity)

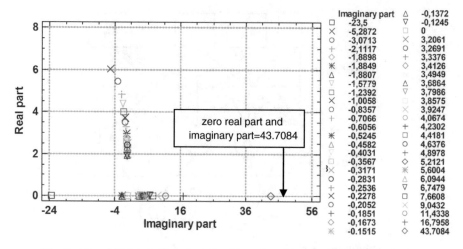

Fig. 5. Neural Network Bayesian Classifier - Imaginary part (Acoustic velocity)

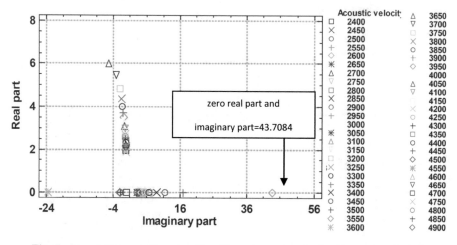

Fig. 6. Neural Network Bayesian Classifier - Acoustic velocity (Acoustic velocity)

Table 1. Summarizes the full work where we can notice bulk waves detection

Coefficient Attenuation	Imaginary part	Real part	Acoustic velocity	Bulk waves (BAW)
α_1	All values	All values	All values	No detection
α_2	All values	All values	All values	No detection
α_3	All values	All values	All values	No detection
α_4	**43,7043**	**0**	**3550 m/s**	**Good Detection**

8 Conclusion

In this article, we explained that the phenomenon of bulk waves are relying on numerical results at the level of attenuation coefficients. Changes in real and imaginary parts of the coefficients based on the acoustic velocity to detect these waves. It is quite clear that these waves are propagent into the piezoelectric material considered. Our method is applicable for any piezoelectric material, you just have to change the characteristics of the material parameters. The positive feature of this work gives us accurate values.

References

1. Royer, D., Dieulesaint, E.: Elastic Waves in Solids II – Generation, Acoustic-optic Interaction, Applications. Springer, New York (1999)
2. Wang, G., Pan, G.: Full wave analysis of microstrip line structures by wavelet expansion method. IEEE Trans. Microwave Theory Tech. **43**, 131–142 (1995)
3. Avramov, I.D.: High Q metal strip SSBW resonator using a SAW design. IEEE Trans. Ferroelectr. Freq. Control **37**, 530–534 (1990)
4. Deng, M.: Simulation of generation of bulk acoustic waves by interdigital transducers. In: IEEE Ultrasonics Symposium, pp. 855–858 (2001)
5. Benatia, D., Benslama, M.: Analysis of leaky and bulk acoustic microwaves by wavelet technique. J. Commun. Numer. Meth. Eng. **16**, 165–175 (2000)
6. Benatia, D.: Modélisation des ondes de volume rampantes sous la surface (SSBW) par une technique mixte ondelettes et fractales, Thèse de Doctorat d'Etat, Université de Constantine, Institut d'électronique (1999)
7. Hafdaoui, H., Benatia, D.: Identification of acoustics microwaves (bulk acoustic waves) in piezoelectric substrate (LiNbO$_3$ Cut Y-Z) by classification using neural network. J. Nano-electronics Optoelectron. **10**(3), 314–319 (2015)
8. Hafdaoui, H.: Détection et génération des micro-ondes acoustiques de volume dans les structures piézoélectriques, Thèse de Doctorat, Université de Batna 2 (2016)
9. Rao, P.N., Uma Devi, T., Kaladhar, D., Sridhar, G., Rao, A.A.: A probabilistic neural network approach for protein superfamily classification. J. Theor. Appl. Inf. Technol. **6**(1), 101–105 (2009)
10. Lotfi, A., Tlemsani, R., Benyettou, A.: Un Nouvel Algorithme d'Apprentissage pour les Réseaux Probabilistes. In: 11th African Conference on Research in Computer Science and Applied Mathematics, pp. 208–215 (2012)

Direct Torque Control Based on Artificial Neural Network of a Five-Phase PMSM Drive

Tounsi Kamel[1]($\boxtimes$), Djahbar Abdelkader[1], Barkat Said[2], and Atif Iqbal[3]

[1] Department of Electrical Engineering, LGEER Laboratory,
U.H.B.B-Chlef University, Chlef 02000, Algeria
t_kamel@outlook.com, a_djahbar@yahoo.fr
[2] Laboratoire de Génie Électrique, Faculté de Technologie, Université de M'sila,
28000 M'sila, Algérie
sa_barkati@yahoo.fr
[3] Department of Electrical Engineering, Qatar University, Doha, Qatar
atif.iqbal@qu.edu.qa

Abstract. Direct torque control (DTC) based on artificial neural network (ANN) of a five-phase permanent magnet synchronous motor drive (PMSM) is presented in this paper. Using the mathematical model of the five-phase motor, DTC control strategy is developed, and the corresponding controllers are properly designed in order to provide independent torque and flux control. In order to improve the performance of the DTC, a neural network based DTC scheme is adopted instead of the DTC based on the look-up table. The employed neural network uses the Levenberg-Marquardt back propagation algorithm for the adjustment of weights to increase the learning process accuracy. The efficacy of the proposed method is verified by simulation for various dynamic operating conditions, and the system's performance is compared with conventional DTC.

Keywords: Five-phase PMSM · Direct torque control
Artificial neural network

1 Introduction

Permanent magnet synchronous motor offers many advantages compared to induction motor, such as the elimination of rotor losses, high power density, small inertia, and more robust construction of the rotor [1]. Nowadays PMSM has become a leading in the industrial applications due to high power density, high torque/inertia ratio simple rugged structure, and high maintainability and economy. In addition to that it has high efficiency, lightweight, small volume, and it is also robust and immune to heavy overloading. Its small dimension compared with DC motor allows PMSM motor to be used widely in high performance drives such as electrical vehicles, industrial robots, machine tools, and servo applications [2, 3].

Nowadays, the advance and flexible use of high power semiconductor devices opens the door to new scopes in electrical drives. The multi-leg inverter maybe is one of the newest ideas in this field. This has led to an increasing interest in multiphase

© Springer International Publishing AG 2018
M. Hatti (ed.), *Artificial Intelligence in Renewable Energetic Systems*, Lecture Notes in Networks and Systems 35, https://doi.org/10.1007/978-3-319-73192-6_33

(greater than three-phase) drives. Here and now, in some industrial applications such as high power ship propulsion, electric aircraft, electric traction the use of this kind of drives is very suitable due to their tangible benefits over the conventional three-phase drives. Their advantages include reducing amplitude of torque and current pulsation, increasing the frequency of torque pulsation, reducing the stator current per phase without increasing the stator voltage per phase, lowering the DC link current harmonics, higher reliability, and fault tolerance capability [4–7].

In the recent years, many solutions have been suggested to control the permanent magnet synchronous machine, among them are the field oriented control (FOC), and direct torque control (DTC) methods [5, 8, 9].The DTC is based on the direct determination of the best control sequence applied to the switches of a voltage source inverter according to the difference between the reference and the real values of the of torque and flux linkage by using hysteresis controllers [1–3, 10].

Like an every control method has some advantages and disadvantages, the DTC method has too. Some of the advantages are lower parameters dependency, making the system more robust, fast dynamic response, and a pretty simple control [11]. However, the disadvantages are difficult to control flux and torque at low speed, variable switching frequency, a high sampling frequency needed for digital implementation of hysteresis controllers, and high torque ripple responsible for the noise and vibrations generation [12, 13]. The main reasons behind the high current and torque ripples in the conventional DTC is the presence of hysteresis comparators and the limited number of available voltage vectors [14]. As mentioned before, in the DTC the switching frequency of the inverter is not constant and changes with rotor speed, load torque, and bandwidth of the two hysteresis controllers [11]. The aforementioned drawbacks can be overcome by using artificial neural network (ANN) in DTC implementation. The advantages of ANN controller are multiple. The parallel distributed structure of ANN has the ability to learn the desired mapping between the inputs and the outputs of a given system, and has the capability to handle its nonlinearities as well. Also the ANN is known by its robustness since its conception does not rely on the mathematic model of the system. The ANN based controller is more robust as the parameter dependency is absent. These advantages support the ANN in playing a major role in solving uncertainty problems in motor drive systems, especially when DTC is used [3].

A quick glance at the literature informs that the DTC based on ANN can improve the performance of three-phase drives in low speed operations, and minimizes the torque ripples during short transients [14, 15]. The same concept can be extended for multiphase motor drive for DTC operation.

In this paper, an artificial neural network based DTC scheme of five-phase PMSM is presented. The neural network is used as intelligent selector to replace the hysteresis comparators and the switching table in order to lead the flux and the torque towards their reference values.

2 Five-Phase PMSM Modeling

The model of the five-phase PMSM is presented in a rotating d-q-x-y frame as [7, 16, 17]:

$$\frac{di_d}{dt} = (-r_s i_d + \omega L_q i_q + v_d)/L_d$$

$$\frac{di_q}{dt} = (-r_s i_q - \omega L_d i_d - \omega \Phi_f + v_q)/L_q$$

$$\frac{di_x}{dt} = (-r_s i_x + v_x)/L_{ls}$$ \quad (1)

$$\frac{di_y}{dt} = (-r_s i_y + v_y)/L_{ls}$$

Where v_d, v_q, v_x and v_y are the stator voltages in the d, q, x and y axis, respectively, i_d, i_q, i_x and i_y are the stator currents in d, q, x and y axis respectively, r_s is the stator resistance. L_d, L_q and L_{ls} are inductances in the rotating frame.

The mechanical and torque equations can be written as:

$$J\frac{d\Omega}{dt} + f_r\Omega = T_{em} - T_l$$

$$T_{em} = \frac{5}{2}p(\Phi_d i_q - \Phi_q i_d)$$ \quad (2)

With J is the inertia coefficient, f_r is the friction coefficient, p is the number of poles pairs, and T_l is the external load torque.

3   Five-Leg Inverter Modeling

The scheme of the proposed drive is shown in Fig. 1. The load is taken as star-connected, and the inverter output phase voltages are denoted as illustrated Fig. 1. Each leg switching function (s_a, s_b, s_c, s_d, s_e) can take either 1 or 0 value based on the state of the upper or lower switch.

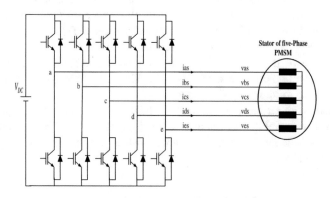

Fig. 1. Schematic diagram of a five-phase voltage source inverter.

The inverter output phase-to-neutral voltages can be obtained from the switching states $s_i, i = a, b, c, d, e$ and the DC-link voltage v_{dc} as follows:

$$v_{as} = \frac{v_{dc}}{5}(4s_a - s_b - s_c - s_d - s_e)$$

$$v_{bs} = \frac{v_{dc}}{5}(-s_a + 4s_b - s_c - s_d - s_e)$$

$$v_{cs} = \frac{v_{dc}}{5}(-s_a - s_b + 4s_c - s_d - s_e) \qquad (3)$$

$$v_{ds} = \frac{v_{dc}}{5}(-s_a - s_b - s_c + 4s_d - s_e)$$

$$v_{es} = \frac{v_{dc}}{5}(-s_a - s_b - s_c - s_d + 4s_e)$$

The five-phase inverter has total thirty two space voltage vectors, thirty non-zero active voltage vectors and two zero voltage vectors. These space vectors can be projected on α-β subspace and x-y subspace as shown in Fig. 2 by using the following two space vectors transformation:

$$v_{\alpha\beta} = \frac{2}{5}v_{dc}\left(s_a + s_b e^{j\frac{2\pi}{5}} + s_c e^{j\frac{4\pi}{5}} + s_d e^{j\frac{6\pi}{5}} + s_e e^{j\frac{8\pi}{5}}\right)$$

$$v_{xy} = \frac{2}{5}v_{dc}\left(s_a + s_b e^{j\frac{6\pi}{5}} + s_c e^{j\frac{2\pi}{5}} + s_d e^{j\frac{8\pi}{5}} + s_e e^{j\frac{4\pi}{5}}\right) \qquad (4)$$

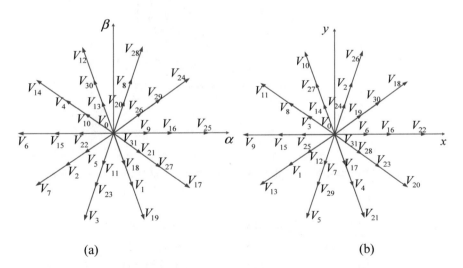

Fig. 2. Inverter switching voltage vectors in stationary reference frames: (a) α-β subspace, (b) x-y subspace.

From Fig. 2, the active switching vectors can be divided into three groups in accordance with their magnitudes small, medium and large space vector groups. The magnitudes are identified with indices s, m, and l and are given as

$|V_s| = (4/5)\cos(2\pi/5)v_{dc}$, $|V_m| = (2/5)v_{dc}$, $|V_l| = (4/5)\cos(\pi/5)v_{dc}$ respectively [18]. It can be observed from Fig. 2 that medium length space vectors of the α-β plane are mapped into medium length vectors in the x-y plane, and large vectors of the α-βplane are mapped into small vectors in the x-y plane, and vice-versa.

4   Conventional DTC Using Large Voltage Vectors Only

The DTC involves direct control of the electromagnetic torque and the flux developed in the motor. The principle of controlling the flux in DTC is to keep its magnitude within a predefined hysteresis band [19]. A comparison between the estimated and the actual values of the motor torque and flux are the inputs to two hysteresis controllers. With the knowledge of the torque and flux errors and the sector number in which the stator flux space vector is located as shown in Fig. 3, the switching signals are adequately generated as shown in Table 1.

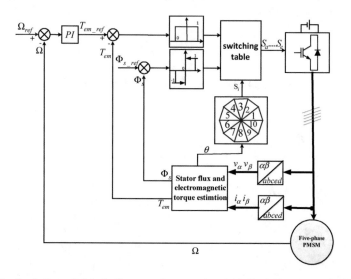

Fig. 3. Schematic block diagram of conventional DTC of five-phase PMSM

Table 1. Optimum active voltage vector look-up Table

d_Φ	d_{Tem}	S_1	S_2	S_3	S_4	S_5	S_6	S_7	S_8	S_9	S_{10}
1	1	V_{24}	V_{28}	V_{12}	V_{14}	V_6	V_7	V_3	V_{19}	V_{17}	V_{25}
1	0	V_0	V_{31}	V_0	V_{31}	V_0	V_{31}	V_0	V_{31}	V_0	V_{31}
1	−1	V_{17}	V_{25}	V_{24}	V_{28}	V_{12}	V_{14}	V_6	V_7	V_3	V_{19}
0	1	V_{14}	V_6	V_7	V_3	V_{19}	V_{17}	V_{25}	V_{24}	V_{28}	V_{12}
0	0	V_{31}	V_0	V_{31}	V_0	V_{31}	V_0	V_{31}	V_0	V_{31}	V_0
0	−1	V_7	V_3	V_{19}	V_{17}	V_{25}	V_{24}	V_{28}	V_{12}	V_{14}	V_6

In each sampling time, the control scheme estimates the electromagnetic torque and stator flux linkage. The stator flux components in the stationary reference frame can be estimated by:

$$\Phi_\alpha = \int (v_\alpha - r_s i_\alpha)dt$$
$$\Phi_\beta = \int (v_\beta - r_s i_\beta)dt \tag{5}$$

The vector voltage components v_α and v_β are estimated using the Eq. (4). The magnitude of the stator flux and its position are calculated as follows:

$$\Phi_s = \sqrt{\Phi_\alpha^2 + \Phi_\beta^2}$$
$$\theta_s = \tan^{-1} \frac{\Phi_\beta}{\Phi_\alpha} \tag{6}$$

The electromagnetic torque is estimated by:

$$T_{em} = \frac{5}{2} p(\Phi_\alpha i_\beta - \Phi_\beta i_\alpha) \tag{7}$$

5 DTC-ANN Approach

The idea here is to use neural network to replace the switching table in the conventional DTC. The back propagation algorithm is used to train the neural network. The training algorithm adopted is Levenberg Marquardt back propagation; it updates weight and bias values. The network consists of three layers: input layer, the hidden layer, and the output layer. Each of the inputs is connected to each neuron in the hidden layer, and in turn, each of the hidden layer neurons is connected to the output.

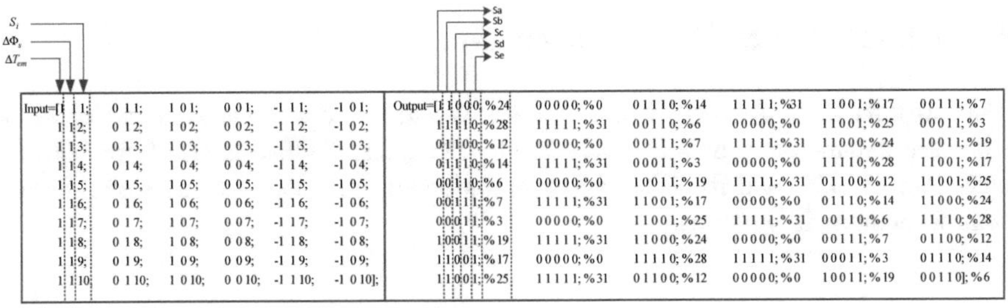

Fig. 4. Coding of the inputs and the outputs used for learning

Figures 4 and 5 give an idea how the DTC-ANN is implemented. The ANN inputs are the position of the stator flux vector represented by the corresponding sector number (1 to 10), the normalized deviation (0 or 1) between the estimated stator flux value and its reference and the standardized deviation (−1, 0 or 1) between the estimated electromagnetic torque and its reference, are three neurons in the input layer. For the output layer, it also consists of five neurons, each representing the switching state of the five-phase inverter.

As soon as the training procedure is over, the neural network gives almost the same output pattern for the same or nearby values of input.

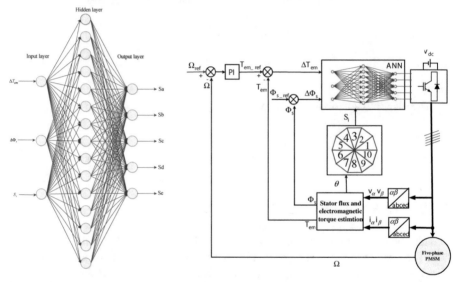

Fig. 5. Structure Neural Network for impulsion vector

Fig. 6. Schematic block diagram of DTC-ANN of five-phase PMSM

6 Simulation Results and Discussion

The performance of the control strategy illustrated in Fig. 6 was analyzed by simulation using the parameters shown in Table 2. The five-phase PMSM is tested for both the conventional DTC and the DTC-ANN, and a comparison is drawn between the two control methods is presented.

Table 2. Five-phase PMSM parameters

P	L_d	L_q	L_{ls}	Φ_f	J	r_s	f_r
2	8.5 mH	8 mH	0.2 mH	0.175 T	0.004 kg.m^2	1 Ω	0

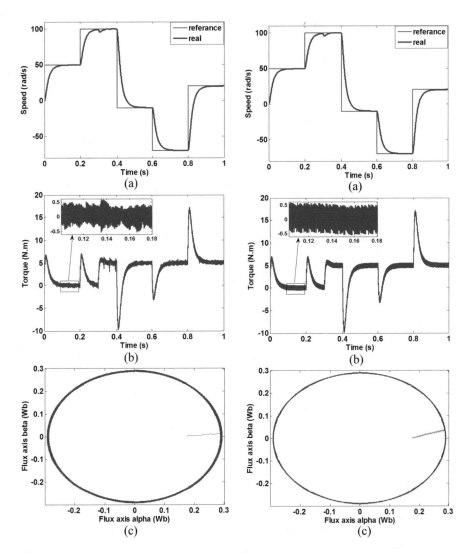

Fig. 7. Dynamic responses of five-phase PMSM controlled by DTC-ANN under variable load torque and reference speed profile test: (a) Reference and actual speeds, (b) Electromagnetic torque, (c) α-β stator flux.

Fig. 8. Dynamic responses of five-phase PMSM controlled by conventional DTC under variable load torque and reference speed profile test: (a) Reference and actual speeds, (b) Electromagnetic torque, (c) α-β stator flux.

The simulation results of DTC for the five-phase PMSM based on the artificial neural networks ANN are compared with those obtained using conventional DTC under the same operating conditions. The responses of the five-phase PMSM shown in Figs. 7 and 8 are done under a sudden change in load torque from 0 Nm to 5 Nm at t = 0.3 s. In this test, the reference speed is step changed as follows: 50 100-10-70 and 20 rad/s at t = 0.2, 0.4, 0.6, 0.8 s respectively.

The speed responses are shown in Figs. 7(a) and 8(a). The torque responses for a step change of the load torque are shown in Figs. 7(b) and 8(b). The α-β stator fluxin the DTC-ANN and the conventional DTC are shown in Fig. 7(c) and Fig. 8(c), respectively.

From Figs. 7 and 8, it is observed that both the torque and the speed of the five-phase PMSM drive are very close to their reference values. The torque changes as the speed of the motor is changed and the electromagnetic torque settles at 5 Nm compensating the load torque. From the speed responses it can be observed that there is no significant difference between the two DTC methods in term of tracking performance.

Figures 7(c) and 8(c) describe the stator flux vector trajectories which are almost circular. In these figures, it can be seen that neural network controller offers the better and fast response.

It can be seen also that the torque ripples with DTC-ANN in steady state are significantly reduced compared to conventional DTC.

7 Conclusion

In this paper, the performance of the direct torque control based base on an artificial neural network of five-phase PMSM has been highlighted. Simulations results show the effectiveness of the adopted control scheme. Indeed, a high performance and quick torque response are obtained in different tests such as load torque application and speed reference reversion. Moreover, the proposed DTC presents reduced flux and torque ripples compared to conventional one.

References

1. Guiti, M., Naitseghir, A.: Direct direct torque control with ANN hysteresis controller for PMSM. In: The 4th International Conference on Electrical Engineering (ICEE) 2015, Boumerdes, Algeria, 13–15 December, pp. 1–6 (2015)
2. El Idrissi, A.E.J., Zahid, N., Jedra, M.: Optimized DTC by genetic speed controller and inverter based neural networks SVM for PMSM. In: Second International Conference on Innovative Computing Technology (INTECH), Casablanca, 18–20 September, pp. 1–4 (2012)
3. Gulez, K., Adam, A.A.: Adaptive neural network based controller for direct torque control of PMSM with minimum torque ripples. In: SICE Annual Conference 2007, Takamatsu, Japan, 17–20 September, pp. 174–179 (2007)
4. Kim, N., Kim, M.: Modified direct torque control system of five phase induction motor. J. Electr. Eng. Technol. 4(2), 266–271 (2009)
5. Raj, L.R.L.V., Jidin, A., Che Wan Mohd Zalani, C.W.M.F., Abdul Karim, K., Wee Yen, G., Jopri, M.H.: Improved performance of DTC of five-phase induction machines. In: 2013 IEEE 7th International Power Engineering and Optimization Conference, Langkawi, Malaysia, 3–4 June, pp. 613–618 (2013)

6. Hosseyni, A., Trabelsi, R., Mimouni, M.F., Iqbal, A.: Vector controlled five-phase permanent magnet synchronous motor drive. In: 23rd International Symposium on Industrial Electronics (ISIE), Istanbul, pp. 2122–2127 (2014)
7. Sneessens, C., Labbe, T., Baudart, F., Matagne, E.: Position sensorless control for five-phase permanent-magnet synchronous motors. In: International Conference on Advanced Intelligent Mechatronics, Besançon, France, July 8–11, pp. 794–799 (2014)
8. Moghbeli, H., Zarei, M., Mirhoseini, S.S.: Transient and steady states analysis of traction motor drive with regenerative braking and using modified direct torque control. In: Proceedings of PEDSTC, pp. 615–620 (2015)
9. Jannati, M., Anbaran, S., Zaheri, D.M., Idris, N.R.N., Aziz, M.J.A.: A new speed sensorless SVM-DTC in induction motor by using EKF. In: 2013 IEEE Student Conference on Research and Development, Putrajaya, December 16–17, pp. 94–99 (2013)
10. Zolfaghari, M., Taher, S.A., Munuz, D.V.: Neural network-based sensorless direct power control of permanent magnet synchronous motor. Ain Shams Eng. J. 7(2), 729–740 (2017)
11. Bakouri, A., Mahmoudi, H., Abbou, A., Moutchou, M.: Optimizing the wind power capture by using DTC technique based on artificial neural network for a DFIG variable speed wind turbine. In: The 10th International Conference on Intelligent Systems: Theories and Applications (SITA), Rabat, Morocco, 20-21 October, pp. 1–7 (2015)
12. Barikand, S.K., Jaladi, K.K.: Five-phase induction motor DTC-SVM scheme with PI Controller and ANN controller. In: Global Colloquium in Recent Advancement and Effectual Researches in Engineering, Science and Technology (RAEREST 2016), vol. 25, pp. 816–823 (2016)
13. Krishnaveni, D., Sivaprakasam, A., Manigandan, T.: Optimum voltage vector selection in direct torque controlled PMSM using intelligent controller. In: International Conference on Green Computing Communication and Electrical Engineering (ICGCCEE), Coimbatore, India, 6–8 March, pp. 1–6 (2014)
14. Gottapu, K., Kiran, U.S., Raju, U.S., Nagasai, P., Prasad, S., Rao, P.T.: Designand analysis of artificial neural network based controller for speed control of induction motor using DTC. Int. J. Eng. Res. Appl. 4(4), 259–264 (2014)
15. Bossoufi, B., Karim, M., Ionita, S., Lagrioui, A.: DTC control based artificial neural network for high performance PMSM drive. J. Theor. Appl. Inf. Technol. 33(2), 165–176 (2011)
16. Xue, S., Wen, X.: Simulation analysis of two novel multiphase SVPWM strategies. In: IEEE International Conference on Industrial Technology, ICIT 2005, Hong Kong, pp. 1337–1342 (2005)
17. Yu, F., Zhang, X., Qiao, M., Du, C.: The direct torque control of multiphase permanent magnet synchronous motor based on low harmonic space vector PWM. In: International Conference on Industrial Technology, pp. 1–5 (2008)
18. Jones, M., Dordevic, O., Bodo, N., Levi, E.: PWM algorithms for multilevel inverter supplied multiphase variable-speed drives. Electronics 16(1), 22–31 (2012)
19. Parsaand, L., Toliyat, H.A.: Sensorless direct torque control of five-phase interior permanent magnet motor drives. In: 39th IAS Annual Meeting on Industry Applications Conference 2004, 3–7 October, vol. 2, pp. 992– 999 (2004)

Wind Maximum Power Point Prediction and Tracking Using Artificial Neural Network and Maximum Rotation Speed Method

Aoued Meharrar[1]($\boxtimes$) and Mustapha Hatti[2]

[1] Sciences and Technology Department, Tissemsilt University Center,
Tissemsilt, Algeria
Meharrar_aoued@yahoo.fr
[2] EPST, UDES, Unité de Développement des Equipements Solaires,
11th National Road, Bou Ismail, Tipaza, Algeria
musthatti@gmail.com

Abstract. The power characteristic of a wind turbine is naturally nonlinear, because the position of the maximum power varies with the wind speed, for each wind speed, it is necessary that the system finds the maximum power. To approach this goal, a specific command must be used. In this paper, a variable speed wind generator maximum power point tracking (MPPT) based on artificial neural network (ANN) is presented.

Which at first allows prediction of the maximum rotation speed by using the ANN, which is then determined the optimal speed rotation by using the new maximum rotation speed method (MRS).

The generator used in this study is a synchronous permanent magnet (PMSG), controlled by an electronic converter with pulse width modulation (PWM); this last uses of a vector and an MPPT (Maximum power point tracking) controller to check the electromechanical magnitudes such as the torque or the rotational speed of the generator in order to extract the maximum wind energy. The simulation results show the effectiveness and robustness of the proposed control.

Keywords: Wind turbine · Neural network
Maximum power point tracking (MPPT)

1 Introduction

Today, the production of electric power is primarily based on the use of fossil fuels such as oil, coal and natural gas, or the use of nuclear fission. The use of fossil fuels reduces the cost of electricity production, but leads to emissions of greenhouse gases. Energy from nuclear fission does not reject greenhouse gases, but the risk of accidents related to their use (as the case of the Chernobyl disaster, April 26, 1986 in Ukraine), or due to natural phenomena such as the earthquake tsunami (for example, Fukushima nuclear accident in Japan, March 11, 2011), are thus very probable and responsible of radioactive emissions that have undesirable consequences on humanity and the environment. Then, the treatment of nuclear waste is very expensive and their radioactivity remains high for many years. To reduce the risk associated with the use of these

© Springer International Publishing AG 2018
M. Hatti (ed.), *Artificial Intelligence in Renewable Energetic Systems*, Lecture Notes in Networks and Systems 35, https://doi.org/10.1007/978-3-319-73192-6_34

technologies, many countries are beginning to produce an amount of their electrical power from the renewable sources (as the sun and the wind). Wind power does not reject carbon dioxide and can therefore meet the requirements of green electricity production. However, the mechanical characteristic of a wind turbine is naturally nonlinear and are imperative to adopt a research strategy of maximum power points (MPPT). For this, several control algorithms have been developed, such as that based on fuzzy logic [1, 2], artificial neural network [3, 4] or both (Neuro - Fuzzy) [5]. In our case the MPPT proposed a device based on the combination of artificial neural network (ANN) and the new maximum rotation speed method (MRS), which at first allows prediction of the maximum rotation speed by using the ANN, which is then determined the optimal speed rotation by using the maximum rotation speed method (MRS). The control of the (PMS) generator used is vector type made using a type of electronic converter (PWM). MATLAB Simulink-software is used for the modeling and simulation of the sets of component of the conversion chain.

2 Wind Energy Conversion Subsystem (WECS)

The wind power conversion chain used in this work is given by the following Fig. 1:

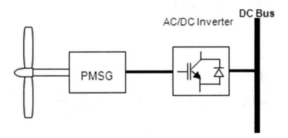

Fig. 1. Wind conversion chain.

The mechanical power of the wind turbine can be written as [6, 7]:

$$P_t = \frac{1}{2}\rho A C_p v^3 \tag{1}$$

Where:
v: Wind speed (m/s).
ρ: Air density (kg/m^3).
Cp: Power coefficient.
A: Area swept by the rotor blades (m^2).

The Cp is a dimensionless parameter expressing the efficiency of the wing in the conversion of kinetic energy of wind into mechanical energy. This coefficient depends on the wind speed, the number of blades, their radius of their pitch angle and speed of rotation; it is generally given in terms of reduced speed λ defined by:

$$\lambda = \frac{\Omega R}{v} \tag{2}$$

Where:

Ω: Rotational speed (rad/s).

R: Blade radius (m).

The electromechanical equation of the turbine and the generator is given by the following equation [8, 9]:

$$J\frac{d\Omega}{dt} = T_m - T_{em} - f\Omega \tag{3}$$

Where:

J: Inertia of turbine and generator (Kg.m^2).

Ω: Rotational speed (rad/s).

f: friction coefficient (N.m.s.rad^{-1}).

T_{em}:   Electromagnetic torque (N.m).

T_m: Mechanical torque (N.m).

2.1   PMSG Model

The theory of the space vector gives the dynamic equation of the stator currents as follows [9, 10]:

$$\frac{dI_{sd}}{dt} = \frac{1}{L_d}(V_{sd} - R_s I_{sd} + p\Omega\Omega_q I_{sq}) \tag{4}$$

$$\frac{dI_{sq}}{dt} = \frac{1}{L_q}(V_{sq} - R_s I_{sq} - pL_{sd}I_{sd}\Omega - p\Omega\Phi_m) \tag{5}$$

Where:

R_s: Stator resistance (Ω).

L_d and L_q: Inductances of the stator (H).

Φ_m: Flux of the permanent magnet (wb).

V_{sd}, V_{sq}:   Stator voltages (V).

I_{sd}, I_{sq}:   Stator currents (A).

P: Number of pairs of poles.

Ω: Rotation speed (rad/s)

The electromagnetic torque is given by the following equation:

$$T_{em} = p(\Phi_m I_{sq} + (L_d - L_q)I_{sd}I_{sq}) \tag{6}$$

3 MPPT Control Strategy of the WECS

The power characteristic of a wind turbine is naturally nonlinear, for each wind speed, it is necessary that the system finds the maximum power. To approach this goal, a specific command must be used. In this work, our method of research of maximum power point tracking (MPPT) is performed as follows:

In the fist The ANN model is used to predict the maximum rotation speed using the variation of the wind speed and the blade radius as the inputs. Then from the maximum rotation speed, the maximum rotation speed method computes the optimal speed rotation and thus the aerodynamic torque Tmref. The optimal torque gives the q-axis reference current Isqref. The d- and q- axis reference currents applied to two PI regulators and two decoupling stages give the d- and q-axis reference voltages Vsdref and Vsqref. These voltages applied to two modulators provide the switching functions of the rectifier which gives the modulated current I. The control strategy of the WECS, previously described, is illustrated in Fig. 2.

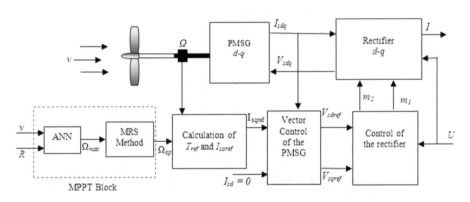

Fig. 2. MPPT control strategy of the WECS

4 Artificial Neural Network

The use of artificial neural networks has developed in many disciplines (such as: economics, ecology, environment, biology, medicine and energy …).

They are particular applied to solve the problems of classification, prediction, categorization, optimization, pattern recognition and associative memory [11].

An ANN is considered as a very good approximation method for the nonlinear systems, particularly useful when these systems are difficult to model using conventional statistical methods.

One artificial neuron may have many inputs X_i, each of them associated to a weight function, synapses W_{ij}. The outputs of all added synapses n_i are submitted to an activation function $h(n_i)$, in order to restrict the output signal amplitude. Any collection of input dates will generate a certain output n_i, as a Boolean output, for example. One neural network is a neuron linkage arranged in interconnected layers [12] (Fig. 3).

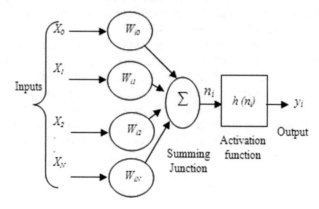

Fig. 3. Model of neural network

Where:

$$n_i = \sum_{j=0}^{N} W_{ij} X_i \tag{7}$$

$$y_i = h(n_i) \tag{8}$$

A neural network (see Fig. 4), is organized in three successive layers: an input layer, an output layer and, between the two, one or more intermediate layers, called hidden layers [13, 14].

The connections between the units of different layers of the network are weights and biases. The ANN is trained to learn the functional mapping of inputs to outputs using input/output training pairs. The goal is to train the network until the output of the neural network is suitably close to the target output [15].

To be able to produce the correct output data, the network was trained with an improved version of the back propagation Levenberg–Marquardt algorithm (LM-BP). During the learning process the error function was minimized with an increasing number of training epochs [16], as shown in Fig. 5. An epoch is a cycle that is finished when all the available training input patterns have been presented to the network once.

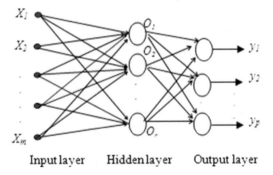

Input layer   Hidden layer   Output layer

Fig. 4. Network networks architecture.

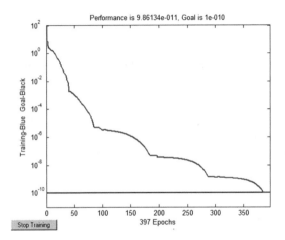

Fig. 5. Levenberg–Marquardt–BP training error during the learning process.

5 Maximum Rotation Speed Method (MRS)

In this work we developed a new maximum rotation speed method; this method is based on the fact that, the ratio between the optimal speed rotation and the maximum rotation speed is equal to constant K.

$$k = \frac{\Omega_{op}}{\Omega_{mux}} \approx 0.63 \tag{9}$$

The value of k it can be also verified by the following Figs. 6 and 7:

Therefore, for any maximum rotation speed the optimum rotational speed is also calculated using Eq. 9.

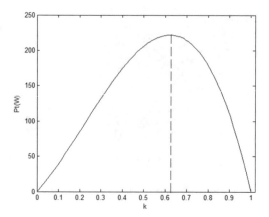

Fig. 6. Mechanical power/speed ration characteristic for wind speed: v = 6 m/s.

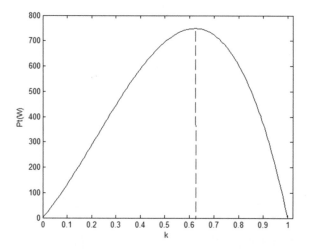

Fig. 7. Mechanical power/speed ration characteristic for wind speed: v = 9 m/s

6 Results

In this work, a small horizontal turbine with three blades was used, where, the power extracted by the turbine depending on the speed of rotation Ω, for different wind values is shown in the following Fig. 8:

For a rapid change in wind speed (see Fig. 9), the mechanical power variation of the wind with the use of an MPPT controller is presented in Fig. 10.

For a real variation of the wind speed (see Fig. 11), Fig. 12; give the rotor speed optimized by maximum power point prediction and tracking using neural network device compared with MPPT calculated.

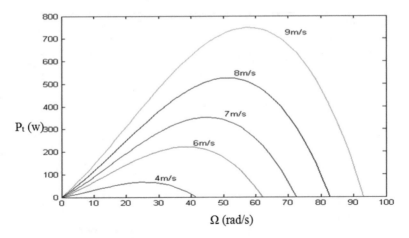

Fig. 8. Mechanical power/rotation speed characteristic

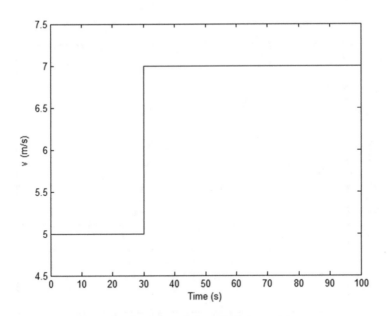

Fig. 9. Step variation of wind speed

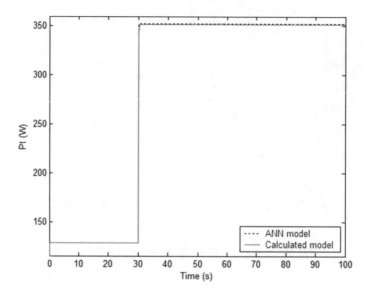

Fig. 10. Output power response

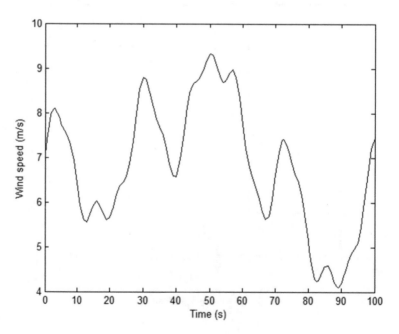

Fig. 11. Wind speed.

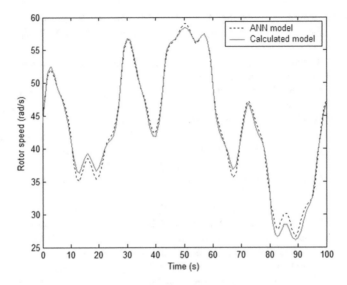

Fig. 12. Rotor speed

7 Conclusions

In this paper, the WECS was modeled using d-q rotor reference frame.

The feasibility of using artificial neural network and maximum rotation speed method for prediction and tracking of the maximum power available in both wind energy system are demonstrated.

The simulation results showed the effectiveness of the ANN control strategy.

References

1. Belmokhtar, K., Doumbia, M.L., Agbossou, K.: Novel fuzzy logic based sensor less maximum power point tracking strategy for wind turbine systems driven DFIG (doubly-fed induction generator). Int. J. Energy **76**(1), 679–693 (2014). http://www.sciencedirect.com/science/article/pii/S0360544214010196
2. Yin, X., Lin, Y., Li, W., Gu, Y., Lei, P., Liu, H.: Sliding mode voltage control strategy for capturing maximum wind energy based on fuzzy logic control. Int. J. Electr. Power Energy Syst. **70**, 45–51 (2015)
3. Ata, R.: Artificial neural networks applications in wind energy systems: a review. Int. J. Renew. Sustain. Energy Rev. **49**, 534–562 (2015)
4. Lopez, F.J., Kenne, G., Lagarrigue, F.L.: A novel online training neural network- based algorithm for wind speed estimation and adaptive control of PMSG wind turbine system for maximum power extraction. Renew. Energy **86**, 38–48 (2016)
5. Meharrar, A., Tioursi, M., Hatti, M., Stambouli, A.B.: A variable speed wind generator maximum power tracking based on adaptive Neuro-fuzzy inference system. Int. J. Expert Syst. Appl. **38**(6), 7659–7664 (2011)

6. Kamal, E., Koutb, M., Sobaih, A.A., Abozalam, B.: An intelligent maximum power extraction algorithm for hybrid wind – diesel - storage system. Int. J. Electr. Power Energy Syst. **32**(3), 170–177 (2010)
7. Cherif, H., Belhadj, J.: Energy output estimation of hybrid Wind-Photovoltaic power system using statistical distributions. Int. J. Electr. Syst. **10**(2), 117–132 (2014)
8. Errami, Y., Ouassaid, M., Maaroufi, M.: Modelling and optimal power control for permanent magnet synchronous generator wind turbine system connected to utility grid with fault conditions. World J. Model. Simul. **11**(2), 123–135 (2015)
9. Lalouni, S., Rekioua, D., Idjdarene, K., Tounzi, K.M.: An improved MPPT algorithm for wind energy conversion system. J. Electr. Syst. **10**(4), 484–494 (2014)
10. Bu, F., Hu, Y., et al.: Wide-speed-range-operation dual stator-wind induction generator DC generating system for wind power applications. IEEE Trans. Power Electron. **30**(2), 561–573 (2015)
11. Aleksander, H.M.: An Introduction to Neural Computing, 2nd edn.
12. Veiga, J.L., Carvalho, A.A., da Silva, I.C., Rebello, J.M.: The use of artificial neural network in the classification of pulse-echo and TOFD ultra-sonic signals. J. Braz. Soc. Mech. Sci. Eng. **27**(4), 394–398 (2005). Technical Editor: Atila P. Silva Freire
13. Efe, M.O., Abadoglu, E., Kaynak, O.: A novel analysis and design of a neural network assisted nonlinear controller for a bioreactor. Int. J. Robust Nonlinear Control **9**, 799–815 (1999)
14. Saadi, A., Moussi, A.: Neural network use in the MPPT of photovoltaic pumping system. Rev. Energ. Ren., 39–45 (2003). ICPWE
15. Kalogirou, S.A.: Artificial neural networks in renewable energy systems applications: a review. Renew Energy Rev., 373–401 (2001)
16. Sun, T., Cao, G., Zhu, X.: Nonlinear modeling of PEMFC based on neural networks identification. J. Zhejiang Univ. **6**(5), 365–370 (2005)

Modeling of an Improved Liquid Desiccant Solar Cooling System by Artificial Neural Network

Tayeb Benhamza[(⊠)], Maamar Laidi, and Salah Hanini

Department of GPE, Faculty of Technology,
Yahia Farès Univ Medea, Medea, Algeria
benhamzatayeb@gmail.com, maamarw@yahoo.fr,
s_hanini2002@yahoo.fr

Abstract. This paper proposes the use of an artificial neural network (ANN) to model an improved liquid desiccant solar system. This system consists of six major components: an air dehumidifier or absorber, a solution regenerator or desorber, a heat and mass exchanger, and an air-to-air heat exchanger. The experimental data were collected from literature during four successive typical days (5–8 September 2010) from a large volume of data. By developing a MatLab program an ANN was used with the main goal of predicting the coefficient of performance (COP) based on the lowest input variables. The optimal topology of the static neural network is (8-8-1): {8 neurons in the input layer; 8 neurons in the hidden layer and 1 neuron in output (COP)}. The optimized neural network reproduces the experimental data from the literature with a great accuracy: an average mean square error of 1.93% and determination coefficient $R^2 > 0.90$ which can be considered very satisfactory.

Keywords: Liquid desiccant solar system · Modeling · Neural network

1 Introduction

With such abundant renewable clean energy −5.9 billion KWh is generated by solar energy and 2.7 billion kWh is produced by photovoltaic generators (Allouhi et al. 2015) and due to energy saving policies and concerns over climate change. New innovative works in solar energy to produce domestic hot water and to cover the heating and cooling demands in single homes or office buildings are growing.

In the last few years, Cooling applications like air-conditioning and refrigeration are becoming basics of everyday life. For that reason developing solar cooling techniques is of a great importance since cooling is needed most of the time when solar radiation is available.

Gommed and Grossman (2012) for example made an improved Liquid Desiccant System (LDS) which is the base for our actual work. Where they used an innovative Heat and Mass Exchanger (HME) to correct many drawbacks of the classical cooling model. Including long idling time, high pressure drop in the previous heat exchanger, level control problems, heat losses and pressure drops due to solution exchange between absorber and desorber.

© Springer International Publishing AG 2018
M. Hatti (ed.), *Artificial Intelligence in Renewable Energetic Systems*, Lecture Notes in Networks and Systems 35, https://doi.org/10.1007/978-3-319-73192-6_35

The actual performance of the system in field level depends on many factors, such as cooling demand, cooling water temperature availability, sources of heat input, and it's potential. Hence, detailed modeling is required to predict the performance of the chiller considering all the above factors.

In the last decades Artificial neural networks have successfully passed the research stages and found real time applications in various fields of aerospace, defense, automotive, mathematics, engineering, meteorology, medicine, and many others. Some of the most important ones are weather and market trends forecasting, prediction of mineral exploration sites, electrical and thermal load prediction, adaptive and robotic control and many others.

(Kalogirou et al. 2014) various researchers have proven that building energy systems are exactly the types of problems and issues for which the artificial neural network (ANN) approach appear to be most applicable. It can be more reliable at predicting energy consumption and performance in these buildings than any other traditional statistical approach. So many comprehensive overviews of ANN application for thermal engineering and renewable energy system have been made:

Kalogirou (2001), Kalogirou (2007), Yang (2008), Mellit and Kalogirou (2008), Mellit et al. (2009), Mohanraj et al. (2012).

A number of researchers have already demonstrated the application of ANNs in traditional and solar cooling systems.

Chow et al. (2002) has proposed a new concept of control system for absorption chiller integrating artificial network with genetic algorithms.

Yang et al. (2003) presented an ANN application system to control energy consumption in buildings;

Sözen and Akçayol (2004) analysed the performance of an absorption cooling cycle with a solar energy activated ejector using ANN approach.

Manohar et al. (2006) modeled a double effect absorption chiller applying the ANN method:

Sencan (2006) used an ANN. to simplify the performance analysis of an ammonia-water absorption chiller: Şencan et al. (2007) used in their work, amongst other approaches ANNs, for modeling an absorption heat transformer; Hernandez et al. (2008) has proposed a predictive ANN model to determine the performance of a water purification process integrated into an absorption heat transformer.

Rosiek and Batlles (2010), Rosiek and Batlles (2011) used a neural network to model the thermal behavior of a solar-assisted air-conditioning system that consists mainly of an absorption chiller, a solar collector array and a cooling tower: Labus (2011) used experimental data obtained on a test bench, to obtain stationary ANN model for low-capacity absorption refrigeration machines: Even though there are so many works already available related to ANN modeling of energy systems in general and some for sorption chillers, but -apparently- there is no known study, of an ANN predictive COP model for the improved liquid desiccant solar cooling system.

The main purpose of this study is to predict and evaluate the behavior of the improved liquid desiccant solar cooling system installed in the Faculty of Mechanical Engineering, Technion in Israel Institute of Technology, Haifa presented by Gommed and Grossman (2012).

The main goal of this work is to use a neural network to estimate the coefficient of performance of the total solar cooling installation.

2 Description of Liquid Desiccant System Language

In this work, we use data registered in the solar-assisted air conditioning system well explained in Gommed and Grossman (2012). Authors in their paper have used a novel HME designed to serve as desiccant solution for the desorber and absorber. The HME helps eliminating the need for external recuperative heat exchanger by enabling mass transfer between them and minimize transfer losses. This HME associated with an improved solution eliminate heat losses due to circulation by assuring adiabatic absorption/desorption. After varying operating conditions and studying characteristic performance it was found that the new LDS has improved many drawbacks compared to classical systems. The new HME has helped to reduce time constant of the system, correct idling and level control problems and ensures maximum solution concentration on absorber (Gommed and Grossman 2012). The system employs a flat plate solar collector of conventional design, selected with 20 m^2 as collector area. Figure 1 presents the view of the system installed. Four successive typical days (5–8 September 2010) were selected from this paper to be used in the modeling stage. These are days during which the system operated rather smoothly, representing what may be expected ultimately after all the various operational and control problems have been fixed. Normally, data was taken at 1-min intervals. Readings of temperatures, humidities, flow rates, etc. at various parts of the system.

Fig. 1. Photograph of the experimental liquid desiccant system: 1 absorber/dehumidifier; 2 desorber/regenerator; 3 air ducts; 4 fan; 5 rotary air/air heat exchanger; 6 HME; 7 water/solution heat exchanger; 8 solution distribution header (Gommed and Grossman 2012)

3 Methodology

3.1 ANN Model Description

A neural network can be defined as a system of simple processing elements, neurons that are connected into a network by a set of (synaptic) weights. The function of the network is determined by the architecture of the network, the magnitude of the weights and the processing element's mode operation. The neuron, or node as it is also called, is a processing element that takes a number of inputs, weights them, sums them up, and uses the result as the argument for a singular valued function, namely the activation function.

The displacement, $w_{i,o}$ is called the bias and can be interpreted as a weight applied to a pseudo input that is clamped to the constant value of 1. Essentially, the activation function f_i can take any form. Units can be combined into a network in numerous fashions. The most common of these is the multilayer perceptron (MLP) network. The basic MLP network is constructed by ordering the units in layers, letting each unit in a layer take as input only the outputs of units in the previous layer or external inputs. Due to the structure, this type of network is often referred to as a feed forward network. Figure 2 shows the topology of this network.

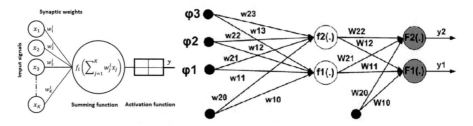

Fig. 2. Schematic of a neural network and a fully connected two-layer feed forward network with three inputs, two hidden, and two outputs.

The second layer in Fig. 2 is called the output layer, referring to the fact that it produces the output of the network. The first layer is known as the hidden layer because in some sense, it is hidden between the external inputs ($\varphi 1$, $\varphi 2$, $\varphi 3$) and the output layer. The mathematical formula expressing the activity of the MLP-network takes the form:

$$y_1(t) = g_1[\varphi, \theta] = F_1\left[\sum_{j-1}^{n_h} W_{1,j}f_j\left(\sum_{l-1}^{n_\varphi} W_{j,l}\varphi_1 + W_{j,0}\right) + W_{1,0}\right] \quad (1)$$

Θ specifies the parameter vector, which contains all the adjustable parameters of the network, i.e., the weights and biases $\{W_j, 1; W_{i,j}\}$. To determine the weighting values, one must have a set of examples of how the outputs, y_i, should relate to the inputs. The task of determining the weights from these examples is called training or learning. That is, the weights are estimated from the examples in such a way that the network, according to some metric, models the true relationship as accurately as possible. The neural network selected here is a multilayer feed forward perceptron (MLP) with one hidden layer. A tan-sigmoid transfer function was used as the activation function for the hidden layer, and a linear transfer function was used for the output layer. The Levenberg Marquardt (LM) algorithm was applied as the method for achieving fast optimization.

3.2   ANN Model Development and Results

To determinate the ANN liquid desiccant solar cooling System's model, a set of 1028 data points with a 1 min sampling period was used. Table 1 presents the input and output parameters used for training the ANN absorption system's model. The parameters were monitored by the data acquisition system of four successive typical days (5–8 September 2010) and. In order to carry out the network training, 719 data patterns were used, and the remaining 309 patterns were used as the test data set.

Table 1.   Input and output parameters used for ANN absorption system's model.

Description	Temperature (°C)	Flow rate (kg/s)	Humidity (g/kg)
Hot water inlet	65–100	0.15–0.36	
Hot water outlet	53.6–88.6	0.15–0.36	
Desorber solution before heating	34.3–94.3	0.35–0.75	
Desorber solution after heating	41.8–101.8	0.35–0.75	
Cooling water inlet	22–27	0.25–0.52	
Cooling water outlet	27.9–32.9	0.25–0.52	34.62–37.41
Absorber solution before cooling	21.7–59.7	0.1–0.22	
Absorber solution after cooling		0.1–0.22	6.81–9.30
Solution after desorption	11.5–49.5	0.35–0.75	15.27–16.81
Solution after absorption	34.5–94.5	0.1–0.22	15.27–16.81
Exhaust air before heat recovery		0.1–0.3	
Supply (dehumidified) air	61.2–71.5	0.1–0.5	34.62–37.41
Ambient (fresh) air			
Desorber inlet air after heat recovery	30.3–40.3	0.1–0.3	
Desorbers exhaust air after heat recovery	24–35 55.3–65.3 30.1–40.1	0.1–0.3	
Coefficient of performance (COP)	0.34–0.86		

The training and testing data were normalized between 0 and 1, using Eq. (1)

$$X_{scaled} = \frac{x - x_{min}}{x_{max} - x_{min}} \tag{2}$$

Where x_{max} and x_{min} are equal to the maximum and minimum recorded values for each variable x. In order to determine the ANN system's model, the neural network toolbox under the Matlab environment was used. When a large number of variables are eligible to be included in a model, selecting optimal inputs becomes a critical step prior to the model development itself, since computational cost can be considerably reduced and because not all the variables considered are always available. In this study, We attempt to predict the coefficient of performance of the liquid desiccant solar system driven with the main aim of lowering the initial input parameters.

In order to assess the accuracy of the neural models, we analyze the results in terms of the Root Mean Square Error (RMSE) and Mean Bias Error (MBE) expressed as a percentage of the measured mean. The RMSE gives the dispersion of the experimental data and is defined as:

$$RMSE = \sqrt{\frac{\sum (x_{estimated} - x_{measured})^z}{N}} \qquad (3)$$

Where $X_{estimated}$ is the predicted value, $X_{measured}$ is the measured value, and N is the number of data patterns. The MBE gives the tendency above the underestimation of experimental data and is expressed by the following equation:

$$MBE = \sqrt{\frac{\sum (x_{estimated} - x_{measured})}{N}} \qquad (4)$$

To verify the influence of the above mentioned input parameters, we considered the Root Mean Square Error (RMSE) as a percentage of the mean measured values. In Table 2, one can see the statistical results of RMSE errors of coefficient of performance obtained during selection of inputs variables for the ANN liquid desiccant solar system.

Table 2. Errors of coefficient of performance

RMSE [%]	RMSE COP
$T_{C,IN}$, $\dot{m}_p$, $\omega_{a,in}$, $\omega_{a,out}$, $\dot{m}_{e,air}$, $T_{e,a}$, T_{amb}, T_s, $\dot{m}_{h,w}$, $T_{h,in}$, $T_{h,out}$	0.898
$T_{C,IN}$, $\dot{m}_p$, $\omega_{a,in}$, $\omega_{a,out}$, $\dot{m}_{e,air}$, $T_{e,a}$, T_{amb}, T_s, $\dot{m}_{h,w}$, $T_{h,out}$	1.33
$\dot{m}_p$, $\omega_{a,in}$, $\omega_{a,out}$, $\dot{m}_{e,air}$, $T_{e,a}$, T_{amb}, T_s, $\dot{m}_{h,w}$, $T_{h,out}$	1.79
$T_{C,IN}$, $\omega_{a,in}$, $\omega_{a,out}$, $T_{e,a}$, T_{amb}, T_s, $T_{h,in}$, $T_{h,out}$, $\dot{m}_{h,w}$	2.38
$T_{C,IN}$, $\omega_{a,in}$, $\omega_{a,out}$, $T_{e,a}$, T_{amb}, T_s, $T_{h,in}$, $T_{h,out}$, $\dot{m}_{h,w}$	3.26
$T_{C,IN}$, $\omega_{a,in}$, $\omega_{a,out}$, $T_{e,a}$, T_{amb}, T_s, $T_{h,out}$, $\dot{m}_{h,w}$	1.89
$T_{C,IN}$, $\omega_{a,in}$, $\omega_{a,out}$, $T_{e,a}$, T_{amb}, T_s, $T_{h,in}$, $\dot{m}_{h,w}$	2.49
$\omega_{a,in}$, $\omega_{a,out}$, $T_{e,a}$, T_{amb}, T_s, $\dot{m}_p$, $\dot{m}_{e,air}$	1.79

The main goal of this work is to determine the unique ANN model with the minimal number of input patterns able to estimate the output variable (COP). Through the analysis of the results presented in Table 2, we select the more favorable configuration of ANN model input variables. To carry out the network training, 719 data patterns were used, where the number of input variables was varied. The selection of the input variables was started with the configuration of 10 variables (process air mass flow rate, air humidity going in absorber, air humidity going out of absorber, exhaust air mass flow rate, exhaust air temperature, ambient air temperature, supply air temperature, hot water mass flow rate, hot water inlet temperature, hot water outlet temperature) and we progressively decreased the number of variables by taking into account their importance.

Finally, we chose: cold water inlet temperature, hot water inlet temperature, hot water mass flow rate, exhaust air temperature, ambient air temperature, air humidity going in absorber, air humidity going out absorber as the more favorable configuration of the network inputs. Table 2 show that the statistical results of RMSE errors of coefficient of performance for this network configuration was 1.89%.

After the input and output model variables were fixed, the next step consists of determining the network architecture. Several MLP's networks with different numbers of hidden neurons (Nh) were trained. Figures 3 and 4 illustrate RMSE and MBE evaluation versus the increasing number of hidden neurons, respectively. As can be seen, for numbers of hidden neurons higher than 8, the RMSE became almost constant, and the maximum and minimum deviations were found are 0.04 and −0.08, respectively. Finally, the architecture 8-8-1 (8 inputs, 8 hidden and 4 output neurons) appears to be the most optimal topology.

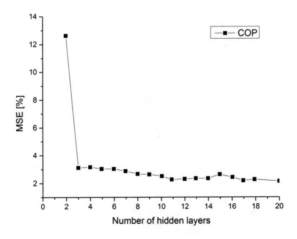

Fig. 3. RMSE evolution vs. the increase of hidden units

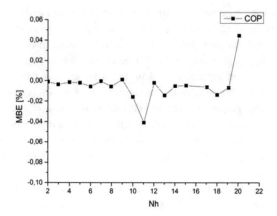

Fig. 4. MBE evolution vs. the increase of hidden units

Figure 5 presents the configuration of the two-layer back propagation network selected in this work. The input layer includes: cold water inlet temperature $T_{C,IN}$, hot water inlet temperature $T_{H,IN}$, hot water mass flow rate $\dot{m}_h$, exhaust air temperature $T_{e,a}$, ambient air temperature T_{amb}, air humidity going in absorber $\omega_{a,in}$, air humidity going out absorber $\omega_{a,out}$ as the more favorable configuration. The hidden layer has eight nodes, and the output layer includes coefficient of performance (COP).

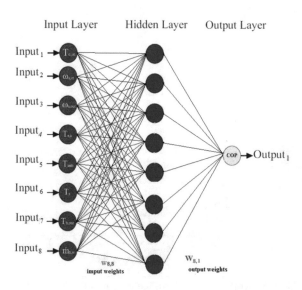

Fig. 5. ANN architecture used for the absorption system.

Table 3. Hidden layer's weight coefficients values used for the ANN absorption system model.

	$T_{C,IN}$	$\omega_{a,in}$	$\omega_{a,out,}$	$T_{e,a}$	T_{amb}	T_s	$T_{h,out}$	$\dot{m}_{h,w}$	COP
1	−1.527	0.7926	−2.3514	2.4059	−0.2787	−0.64607	0.039	−1.527	2.8762
2	0.5546	−0.1828	−3.221	3.3173	−0.3655	−0.20298	0.171	−0.182	−0.028259
3	−11.83	8.748	0.4765	−0.8177	−0.1445	1.6177	−1.949	0.4765	1.4433
4	−0.309	−1.1464	−1.3636	1.4366	−0.0370	2.1581	3.850	1.4366	0.051989
5	5.399	−5.0261	−0.9125	0.8948	−0.0166	0.06449	−0.284	−0.016	−4.678
6	−0.021	−0.0513	1.2055	−1.2423	−1.3933	0.062088	−0.063	0.0620	−0.07957
7	0.0131	−0.0169	0.6748	−0.6830	−0.0063	0.0030	−0.009	−0.009	0.9464
8	0.4398	1.7373	−2.3709	2.3785	−0.2035	1.9116	1.283	0.0084	0.0084452

Table 3 presents the hidden layers and output layers weight coefficients values used for the ANN liquid desiccant solar cooling system. Those coefficients can be used to derive the mathematical formulations to calculate the COP. Once the training process was finished, we proceeded to compare the predicted values from ANN model with actual values. We used a set of 309 patterns as the test data. The accuracy of the ANN model was evaluated on the basis of the regression analysis of estimated versus measured values of the determination coefficient R. Figure 6 present the comparison between the actual and predicted values of COP for liquid desiccant system.

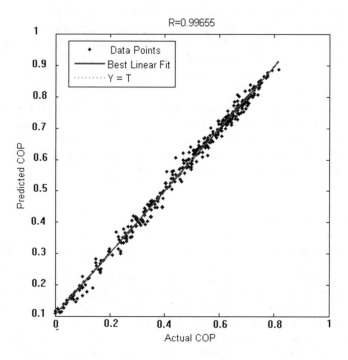

Fig. 6. Comparison of actual and ANN-predicted values of COP for the improved liquid desiccant system

The actual COP can be calculated mathematically but were taken directly from Gommed and Grossmen results. The aforementioned figure shows that the majority of the experimental points are located over the perfect adjust line 1:1, illustrating minimal dispersion. The RMSE error caused by the network in every case is less than 1.9%, and the deviation is practically null. It is noted that R values are similar and reach around 0.996, and all the slope values are equal to 1 while the intercept values are very small. We can observe that the training values resulting in a good match to the experimental values.

4   Conclusion

The main goal of this study was to describe the approach followed to set up an artificial neural network model that was found to be practically applicable to an improved liquid desiccant system. In this work, real data of an existing operating solar liquid desiccant system was used to derive an ANN system's model to predict the coefficient of performance of the system.

We attempt to outline a feasible path for the model's input parameters and architecture identification. The architecture 8-8-1 (8 inputs, 8 hidden and 1 output neurons) was the most successful from those tried. The main aim of the present study was to determine the unique ANN model with a minimal number of input variables. Results demonstrate accurate predictions from the ANN model, yielding a Root Mean Square Error (RMSE) less than 1.9% and practically null deviation, which can be considered very satisfactory. The ANN model applied in this paper was developed for an improved liquid desiccant using LiCl as a desiccant but it can be easily adapted to other liquid desiccant systems. It can be used to predict the coefficient of performance of the whole system, considering only the key variables of absorber's humidities and desorbers's exhaust air temperature, ambient temperature, hot water in and out temperature and masse flow rate and cold water temperature. Considering the obtained acceptable results, we point out that future study in this field should focus on the use of the artificial neural network to predict the cooling capacity of the dehumidifier and the thermal coefficient of performance 1, 2 of the system working at various, operation modes, considering the energy provided from all heat sources.

References

Allouhi, A., et al.: Solar driven cooling systems: an updated review. Renew. Sustain. Energy Rev. **44**, 159–181 (2015)

Ameel, T.A., Gee, K.G., Wood, B.D.: Performance predictions of alternative, low cost absorbents for open-cycle absorption solar cooling. Sol. Energy **54**(2), 65–73 (1995)

Chow, T.T., et al.: Global optimization of absorption chiller system by genetic algorithm and neural network. Energy Build. **34**(1), 103–109 (2002)

Gommed, K., Grossman, G.: Investigation of an improved solar-powered open absorption system for cooling, dehumidification and air conditioning. Int. J. Refrig. **35**(3), 676–684 (2012). https://doi.org/10.1016/j.ijrefrig.2011.10.001

Henning, H.-M.: Solar assisted air conditioning of buildings – an overview. Appl. Therm. Eng. **27**(10), 1734–1749 (2007)

Hernández, J.A., et al.: COP prediction for the integration of a water purification process in a heat transformer: with and without energy recycling. Desalination **219**(1–3), 66–80 (2008). http://linkinghub.elsevier.com/retrieve/pii/S0011916407005644. Accessed 29 July 2016

Hernández, J.A., et al.: Optimal COP prediction of a solar intermittent refrigeration system for ice production by means of direct and inverse artificial neural networks. Solar Energy **86**(4), 1108–1117 (2012). http://www.sciencedirect.com/science/article/pii/S0038092X11004622

Hernández, J.A., et al.: Optimum operating conditions for a water purification process integrated to a heat transformer with energy recycling using neural network inverse. Renew. Energy **34**(4), 1084–1091 (2009)

Jani, D.B., Mishra, M., Sahoo, P.K.: Performance prediction of rotary solid desiccant dehumidifier in hybrid air-conditioning system using artificial neural network. Appl. Thermal Eng. **98**, 1091–1103 (2016). http://dx.doi.org/10.1016/j.applthermaleng.2015.12.112

Kalogirou, S.: Artificial Intelligence in Energy and Renewable Energy Systems. Nova Publishers (2007). https://books.google.dz/books?id=PpynLn1aXw4C

Kalogirou, S.A., Mathioulakis, E., Belessiotis, V.: Artificial neural networks for the performance prediction of large solar systems. Renew. Energy **63**, 90–97 (2014). http://www.sciencedirect.com/science/article/pii/S0960148113004655

Kalogirou, S.A.S.: Artificial neural networks in renewable energy systems applications: a review. Renew. Sustain. Energy Rev. **5**(4), 373–401 (2001). <Go to ISI>://WOS:000171215600003. Accessed 29 July 2016

Kumar, R., Aggarwal, R.K., Sharma, J.D.: Energy analysis of a building using artificial neural network: a review (2013). http://www.sciencedirect.com/science/article/pii/S0378778813003459

Labus, J.: Modelling of Small Capacity Absorption Chillers Driven by Modelling of small capacity absorption chillers driven by solar thermal energy or waste heat A thesis submitted in partial fulfillment of the requirements of, Universitat Rovira i Virgili (2011). http://www.tdx.cat/handle/10803/51878. Accessed 29 July 2016

Lof, G.O.G.: Cooling with solar energy. In: Congress on Solar Energy, pp. 171–189 (1955)

Manohar, H.J., Saravanan, R., Renganarayanan, S.: Modelling of steam fired double effect vapour absorption chiller using neural network. Energy Convers. Manag. **47**(15–16), 2202–2210 (2006)

Mellit, A., et al.: Artificial intelligence techniques for sizing photovoltaic systems: a review. Renew. Sustain. Energy Rev. **13**(2), 406–419 (2009)

Mellit, A., Kalogirou, S.A.: Artificial intelligence techniques for photovoltaic applications: a review. Prog. Energy Combust. Sci. **34**(5), 574–632 (2008)

Mohanraj, M., Jayaraj, S., Muraleedharan, C.: Applications of artificial neural networks for refrigeration, air-conditioning and heat pump systems - a review. Renew. Sustain. Energy Rev. **16**(2), 1340–1358 (2012)

Rosiek, S., Batlles, F.J.: Modelling a solar-assisted air-conditioning system installed in CIESOL building using an artificial neural network. Renew. Energy **35**(12), 2894–2901 (2010). http://www.sciencedirect.com/science/article/pii/S0960148110001850

Rosiek, S., Batlles, F.J.: Performance study of solar-assisted air-conditioning system provided with storage tanks using artificial neural networks. Int. J. Refrig. **34**(6), 1446–1454 (2011). http://www.sciencedirect.com/science/article/pii/S0140700711001095

Sencan, A.: Performance of ammonia-water refrigeration systems using artificial neural networks. Renew. Energy **32**(2), 314–328 (2006). http://www.sciencedirect.com/science/article/pii/S096014810600022X

Şencan, A., et al.: Different methods for modeling absorption heat transformer powered by solar pond. Energy Convers. Manag. **48**(3), 724–735 (2007). http://www.sciencedirect.com/science/article/pii/S0196890406002925

Sözen, A., Akçayol, A.M.: Modelling (using artificial neural-networks) the performance parameters of a solar-driven ejector-absorption cycle. Appl. Energy **79**(3), 309–325 (2004)

Yang, I.H., Yeo, M.S., Kim, K.W.: Application of artificial neural network to predict the optimal start time for heating system in building. Energy Convers. Manag. **44**(17), 2791–2809 (2003)

Yang, K.-T.: Artificial Neural Networks (ANNs): a new paradigm for thermal science and engineering. J. Heat Transf. **130**(9), 93001 (2008). http://www.scopus.com/inward/record.url?eid=2-s2.0-56449112462&partnerID=tZOtx3y1. Accessed 31 July 2016

Artificial Neural Networks Modeling of an Active Magnetic Refrigeration Cycle

Younes Chiba[1]([⊠]), Yacine Marif[2], Noureddine Henini[3], and Abdelhalim Tlemcani[3]

[1] Mechanical Engineering Department, Faculty of Technology, University of Medea, Medea, Algeria
Chiba.younes@univ-medea.dz
[2] LENREZA, Faculté des Mathématiques et Sciences de la Matière, Université de Ouargla, Ouargla, Algeria
[3] LREA, Electrical Engineering Department, Faculty of Technology, University of Medea, Medea, Algeria

Abstract. The aim of this work is to use multi-layered perceptron artificial neural networks and multiple linear regressions models to predict the efficiency of the magnetic refrigeration cycle device operating near room temperature. For this purpose, the experimental data collection was used in order to predict coefficient of performance and temperature span for active magnetic refrigeration device. In addition, the operating parameters of active magnetic refrigerator cycle are used for solid magnetocaloric material under application 1.5 T magnetic fields. The obtained results including temperature span and coefficient of performance are presented and discussed.

Keywords: Gadolinium · Magnetocaloric effect · Neural networks
Numerical simulation

1 Introduction

In order to deal with the drawbacks of the global warming and the negative impact of synthetic refrigerant chlorofluoro carbon (CFC), hydro chlorofluoro carbon (HCFC) and hydro fluoro carbon (HFC) on the environment [1], several alternative cooling technologies were proposed [2, 3]. In this context, the magnetic refrigeration based on the magnetocaloric effect is currently considered as one of the best alternative for conventional systems [1]. The magnetocaloric effect can be defined as the thermal response of a magnetic material when subjected to an external magnetic field which is due interplay between the magnetic moments and the phonons lattice [3]. Aiming to predict the performance of active magnetic refrigerator thermodynamic cycles, a one and multi-dimensional numerical models was proposed [4]. On the other hand, new designs and innovative prototypes working with active magnetic refrigerator cycles were recently reported in the literature [5]. Neural networks have broad applicability to scientific problems. In fact, they have already been applied successfully in many

© Springer International Publishing AG 2018
M. Hatti (ed.), *Artificial Intelligence in Renewable Energetic Systems*, Lecture Notes in Networks and Systems 35, https://doi.org/10.1007/978-3-319-73192-6_36

domains. Since a neural network has been considered a good method for identifying models or trends in data, they are well suited for simulation or prediction needs including [6, 7]:

- Performance predictions
- Industrial process control
- Optimization systems
- Data validation

The aim of this paper is to develop two models with good performances based on Artificial Neural Networks and to predict coefficient of performance and temperature span values of active magnetic refrigerator cycle based on gadolinium under magnetic field 1.5 T.

2 Neural Networks Discription

Typically, the artificial neural networks were organized in layers form. All layers are made up of a certain number of nodes that contains the activation function. In the presented model, the artificial neural network can be considered as a dialog box between output variables and input variables where, the input variables are communicated with hidden variables. The principle of artificial neural networks has been shown in the Fig. 1 [6, 7].

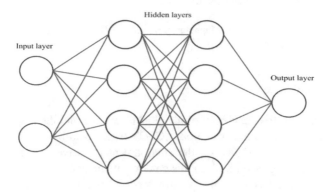

Fig. 1. Basics of neural networks diagram.

- The role of the input variable represents the information that is fed into the artificial neural network.
- The role of each hidden variable is determined by the activities of the input variables and the weights on the connections between the input and the hidden variables.
- The output variable depends on the activity of the hidden variable and the weights between the hidden and output variable.

Neural networks are considered universal interpolator, and they work best if the system you are using them to model has a high tolerance for error. However, they work very well for:

- The relation is difficult to describing the problem by using conventional approaches.
- Variables number or diversity of the points is very good.
- The relation between variables is vaguely understood.

3 Magnetic Refrigeration Cycle

The design of the here used AMR thermodynamic cycle is reported in Fig. 2. Each AMR cycle breaks up into four steps [1–3]:

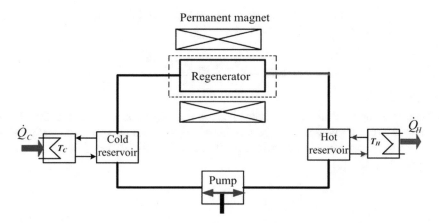

Fig. 2. Schematic diagram of an active magnetic refrigerator (AMR)

1. The magnetocaloric material is adiabatically magnetised by applying an external magnetic field.
2. Flow of the heat transfer fluid through the magnetized regenerator to evacuate the resulting heat.
3. The magnetocaloric material is adiabatically demagnetized by removing the applied magnetic field leading to the decrease of its temperature.
4. Flow of the heat transfer fluid in the opposite direction to evacuate fridges.

4 Mathematical Model

The collection of database used for the development of Artificial Neural Networks models for coefficient of performance and temperature span were reported in reference [1]. The general detail of an architecture used for an active magnetic refrigerator cycle

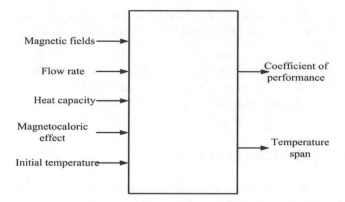

Fig. 3. Neural networks architecture used for an active magnetic refrigerator cycle

models is presented in Fig. 3. The Mean Square Error (MSE) generated during learning can be calculated by using relationship:

$$MSE = \sum_{k=1}^{n} \frac{\left(D_{Exp} - D_{Pre}\right)^2}{n} \tag{1}$$

where, n is the total number of data, D_{Exp} is the experimental data and D_{Pre} is the predicted data for coefficient of performance and temperature span. The error values between the experimental data and the predicted data Artificial Neural Networks can be expressed by Eq. (2).

$$Error(\%) = \left| \frac{D_{Exp} - D_{Pre}}{D_{Exp}} \right| \times 100 \tag{2}$$

5   Results and Discussions

A correlation coefficient of R = 0.988 illustrated in Fig. 4 was obtained between the predicted and the experimental values according to data presented in Table 1, which acts to study the effect flow rate on the COP of active magnetic refrigerator stabilised during time. The total average error value obtained between experimental and predicted results of *COP* equals to 0.90%. The value of this error is very low and indicates that the capacity of the proposed model is very good for the prediction of *COP* for various operating parameters of active magnetic refrigerator.

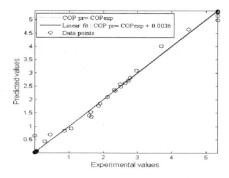

Fig. 4. Correlation between experimental and predicted data for COP of an active magnetic refrigerator cycle

Table 1. Training parameters values used in neural networks model for *COP* prediction

Neural networks parameters	Values and nomination in Matlab
Number of input layer	02
Number of output layer	01
Train function	TRAINLM
Transfer function	TANSIG
Performance function	MSE
Error after learning	0.001
Train epochs	1000

A correlation coefficient of R = 0.999 illustrated in Fig. 5 was obtained between the predicted and the experimental values according to data presented in Table 2, which acts to study the effect local time on the temperature span of active magnetic refrigerator stabilised within time about 35 °C. The total average error value obtained between experimental and predicted results of temperature span equals to 0.95%. The value of this error is very low and indicates that the capacity of the proposed model is very higher to predict of temperature span for various operating parameters of active magnetic refrigerator.

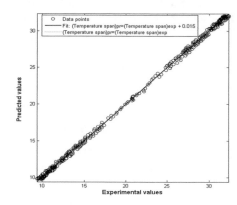

Fig. 5. Regression analysis plot for the optimum model between output and target of temperature span of an AMR cycle

Table 2. Training parameters values used in neural networks model for temperature span

Neural networks parameters	Values and nomination in matlab
Number of input layer	05
Number of output layer	01
Train function	TRAINLM
Transfer function	TANSIG
Performance function	MSE
Error after learning	0.001
Train epochs	1000

6 Conclusion

The study was made in order to develop artificial neural networks models used for predicting performances of an active magnetic refrigerator. The study carried out in this work showed the feasibility of using a simple neural network to predict the Span temperature and coefficient of performance at different operating parameters. A gadolinium magnetocaloric material has been proposed as active regenerator for the AMR cycle applications. The following conclusion can be noted is the good performance of artificial neural networks model for the prediction of temperature span and coefficient of performance with correlation coefficient about 0.999 and 0.988 respectively corresponding to regression experimental values.

References

1. Chiba, Y., Smaili, A., Mahmed, C., Balli, M., Sari, O.: Thermal investigations of an experimental active magnetic regenerative refrigerator operating near room temperature. Int. J. Refrig. **37**, 36–42 (2014)
2. Chiba, Y., Smaili, A., Sari, O.: Enhancements of thermal performances of an active magnetic refrigeration device based on nanofluids. Mechanics **23**(1), 31–38 (2017)
3. Chiba, Y., Sari, O., Smaili, A., Mahmed, C., Nikkola, N.: Experimental study of a multilayer active magnetic regenerator refrigerator-demonstrator. In: Progress in Clean Energy, vol. 1, pp. 225–233. Springer, Cham (2015)
4. Kitanovski, A., Tusek, J., Tomc, U., Plaznik, U., Ozbolt, M., Poredos, A.: Magnetocaloric Energy Conversion: From Theory to Applications. Springer, Cham (2015)
5. Balli, M., Jandl, S., Fournier, P., Kedous-Lebouc, A.: Advanced materials for magnetic cooling: fundamentals and practical aspects. Appl. Phys. Rev. **4**, 021305 (2017)
6. Christopher, M.B.: Neural Networks for Pattern Recognition. Clarendon Press, Oxford (1995)
7. Haykin, S.: Neural Networks: A Comprehensive and Foundation, 2nd edn. Prentice Hall International, Inc., New Jersey (1999)

Storage and Optimization

Analysis of Novel Flywheel Energy Storage System Based on Dual Stator Induction Machine Incorporated in Wind Energy Systems Using Intelligent Approach

Meriem Bouras[✉] and Katia Kouzi[✉]

Laboratoire des Semi-conducteurs et Matériaux Fonctionnels,
Université Amar Telidji Laghouat, Laghouat, Algeria
meriembouras7@gmail.com, kouzi.univ@gmail.com

Abstract. The important feature of wind energy is the fluctuation of the power produced over time. The stability of the network is based on the balance between production and consumption. For this, the idea of storage has been exploited. Owing of this, in first stage, an ANFIS controller is proposed for speed control in order to ensure the real-time tracking of the optimum operating point and MPPT giving online a maximum production of electric power for different wind speeds. In second stage, we present a new solution for the wind energy storage based on short term storage. This solution is based on the use of the intelligent flywheel based on fuzzy logic. The new Fuzzy FESS can be used not only to minimize wind power fluctuations, but also to adjust the frequency and the voltage of the grid during operating conditions. Simulation testes on a 1.5 MW DSIG system are given to illustrate the feature of control method and the large interest of energy storage in such WECS.

Keywords: Dual stator induction generator · Vector control method
Intelligent MPPT · Flywheel energy storage system (FESS)
Fuzzy logic control

1 Introduction

In recent years, there has been a shift in electricity generation from wind power. This energy source has been developed taking into account the diversity of exploitable areas and the relatively attractive cost. It is the fastest growing source of electricity in the world (Wu et al. 2011). The dual star induction machine (DSIG) seems a good solution for variable speed wind energy system. The wind energy conversion systems (WECS) based on DSIG is usually adjusted in the way to maximize the generated power using the Maximum Power Point Tracking (MPPT) strategy. However, wind generators are generators whose primary source of energy is wind. It is well known that wind has very fluctuating and unpredictable characteristics and it is impossible to predict its value for a given moment. This poses many problems for the energy system managers for two reasons: the balance between the generated power and the consumed power must be ensured; on the other hand, the power consumed is difficult to predict and its variable.

© Springer International Publishing AG 2018
M. Hatti (ed.), *Artificial Intelligence in Renewable Energetic Systems*, Lecture Notes
in Networks and Systems 35, https://doi.org/10.1007/978-3-319-73192-6_37

Due to these restrictions, the current wind generators cannot operate without being associated with a conventional source of energy (Khaterchi et al. 2009), (Davigny 2007), (Leclerq 2004). Various studies have been developed (Davigny 2007), (Cimuca 2005) for power generated control in the way to ensure the equilibrium between production and consumption by exploiting the storing energy idea that the base operating is to store power under kinetic energy form in the flywheel. The energy stored will be reconverted into the electrical form for use in case of deficit. There are two categories of energy storage methods: –Short-term storage: For this category, storage time is less than 10 min. –Long-term storage: Storage time is more than 10 min (Cimuca 2005). Today, the short-term energy storage systems have been attracting great attention due to the rapid response, long lives, and low costs. In the light of the above, we propose in this study new fuzzy FESS coupled to DSIM integrated to WECS. This work concern the analysis of fuzzy FESS coupled to DSIM incorporated in Wind Energy Systems: Sect. 2 provides the mathematical model of the wind generator. Section 3 presents the field oriented control of DSIG. Section 4 focuses on the conception of ANFIS for DSIG speed control. The model and fuzzy control of FESS are discussed in Sect. 5. Section 6 illustrates the performance of the suggested algorithm control through simulation results. At the end, some concluding remarks are summarized in Sect. 7.

2 Mathematical Model of the Wind Generator

2.1 Model of the Wind Turbine and Gearbox

In Wind turbine mechanical power is defined by (Abo-Khalil et al. 2004), (Chekkal et al. 2011):

$$P_t = C_p(\lambda)\rho SV^3/2 \tag{1}$$

Where C_p the power coefficient, V is the wind velocity. The turbine torque is given by (Abo-Khalil et al. 2004):

$$T_t = \frac{P_t}{\Omega_t} \tag{2}$$

The turbine is integrated to the generator shaft through the gearbox whose gear ratio G is chosen to regulate the generator shaft speed with respect of reference speed range. The torque and speed of the wind turbine are deducted by:

$$T_g = \frac{T_t}{G}, \Omega_t = \frac{\Omega_r}{G} \tag{3}$$

With the T_g driving is generator torque and Ω_r is the generator speed. Although this equation seems simple, C_p is based on the ratio λ between the turbine angular velocity Ω_t and the wind speed V. This ratio is given by (Amimeur et al. 2012):

$$\lambda = \frac{\Omega_t R}{V} \tag{4}$$

The wind torque is as follows (Abo-Khalil et al. 2004):

$$T_t = \frac{P_t}{\Omega_t} = C_p(\lambda)S\rho V^3/2\Omega_t \tag{5}$$

2.2 The Model of DSIG

The model of DSIM, given in asynchronous frame (d, q) and defined in state-space by this form (Amimeur et al. 2012):

$$\dot{X} = AX + BU \tag{6}$$

With: $X = \left[\phi_{ds1}\phi_{qs1}\phi_{ds2}\phi_{qs2}\phi_{dr}\phi_{qr}\right]^T$; $U = \left[v_{ds1}v_{qs1}v_{ds2}v_{qs2}00\right]^T$;
The system matrices are expressed by:

$$A = \begin{bmatrix}
\frac{L_a-L_{s1}}{T_{s1}L_{s1}} & \omega_s & \frac{L_a}{T_{s1}L_{s2}} & 0 & \frac{L_a}{T_{s1}L_r} & 0 \\
-\omega_s & \frac{L_a-L_{s1}}{T_{s1}L_{s1}} & 0 & \frac{L_a}{T_{s1}L_{s2}} & 0 & \frac{L_a}{T_{s1}L_r} \\
\frac{L_a}{T_{s2}L_{s1}} & 0 & \frac{L_a-L_{s2}}{T_{s2}L_{s2}} & \omega_s & \frac{L_a}{T_{s2}L_r} & 0 \\
0 & \frac{L_a}{T_{s2}L_{s1}} & -\omega_s & \frac{L_a-L_{s2}}{T_{s2}L_{s2}} & 0 & \frac{L_a}{T_{s2}L_r} \\
\frac{L_a}{T_r L_{s1}} & 0 & \frac{L_a}{T_r L_{s2}} & 0 & \frac{L_a-L_r}{T_r L_r} & \omega_{gl} \\
0 & \frac{L_a}{T_r L_{s1}} & 0 & \frac{L_a}{T_r L_{s2}} & -\omega_{gl} & \frac{L_a-L_r}{T_r L_r}
\end{bmatrix} \quad \text{And,}$$

$$B = \begin{bmatrix}
1 & 0 & 0 & 0 \\
0 & 1 & 0 & 0 \\
0 & 0 & 1 & 0 \\
0 & 0 & 0 & 1 \\
0 & 0 & 0 & 0 \\
0 & 0 & 0 & 0
\end{bmatrix} \quad \text{With} \quad T_r = \frac{L_r}{R_r}, \ T_{s(1,2)} = \frac{L_{s(1,2)}}{R_{s(1,2)}}$$

The mechanical equation is as follows by:

$$J\frac{d\Omega_r}{dt} = T_r - T_{em} - J\Omega_r \tag{7}$$

3 Field Oriented Control of DSIG

The idea of vector control lies in the orientation of the flux in the machine to the stator, to the rotor, or to the air gap according to one of the two axes d or q. The goal of vector control is the decoupled control of the rotor flux and electromagnetic torque of the generator to obtain high dynamic and static performance. Rotor flux orientation is obtained by aligning the d-axis of the synchronous reference frame with the rotor flux vector, the resultant d-q-axis rotor flux components are $\psi_{rq} = 0$ and $\psi_{rd} = \psi_r$, the system model can be given by simplified equation as follow (Pant and Singh 1999), (Vas 1998);

$$i_{dr} = \frac{\varphi_r^*}{L_m + L_r} - \frac{L_m}{L_m + L_r}(i_{ds1} + i_{ds2}) \tag{8}$$

$$i_{qr} = -\frac{L_m}{L_m + L_r}(i_{qs1} + i_{qs2}) \tag{9}$$

$$\omega_{gl}^* = \frac{r_r L_m}{(L_m + L_r)} \frac{(i_{qs1} + i_{qs2})}{\varphi_r^*} \tag{10}$$

Besides the electromagnetic torque is given by:

$$T_{em}^* = P \frac{L_m}{L_m + L_r}(i_{qs1} + i_{qs2}) \cdot \varphi_r^* \tag{11}$$

4 Conception of ANFIS for Dual Star Induction Generator Speed Control

The presented ANFIS regulator has two inputs: speed error e and its variation Δe, and one output is the electromagnetic reference torque change. At the input layer, the e and Δe are sampled and fuzzified according to pre-defined fuzzy rules. The training method of ANFIS is based on FLC (see Fig. 1). The ANFIS parameters are as follows: Takagi Sugeno Type; Number of membership functions is 7; Number of iteration is 50; Error tolerance is 10–3.

5 Model and Regulation of FESS

In order to involve the variable speed wind turbine in the system services, FESS is envisaged. The FESS is mechanically coupled to DSIM and driven by power converter as shown in Fig. 2.

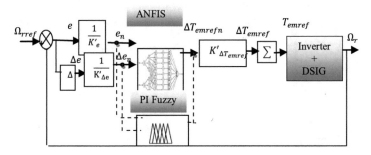

Fig. 1. Basic structure of an ANFIS controller based on an indirect field oriented control of DSIG.

The energy stored by the FES system is in the form of kinetic energy, which can be represented by this equation:

$$E_c = \frac{1}{2} J_{vol} \Omega_{vol}^2 \tag{12}$$

The Eq. (12) shows that for the same energy, high speed operation reduces considerably the value of inertia. The relation of the latter is expressed by the following relationship:

$$J_{vol} = K m_{vol} R_{vol}^2$$

5.1 Dynamic Equation of the Electrical Machine of the (FESSs)

The dynamic equation makes the relation between the mechanical section and the electromagnetic section of the electrical machine, such as:

$$J_{vol} \frac{d\Omega_{vol}}{dt} = C_{emIM} - f_{vol}\Omega_{vol} - C_s \tag{14}$$

5.2 Principle of Control of the (FESSs) Associated with the Wind Generator

The power provided by a generator is always variable, because of variations in the speed of the wind. On the other hand, it is the consumer which must obtain a stable power. Figure 3 shows the principle of control of a (FESSs) combines has a wind generator. Having the power produced by the wind generator (P_{ge}), and knowing the power that he must deliver to the network (P_r), the desired power for the (FESSs) (P_w) can be determined as follows:

$$P_r^* = P_r^* - P_{eg} \tag{15}$$

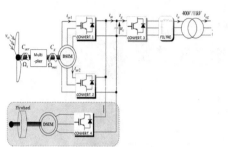

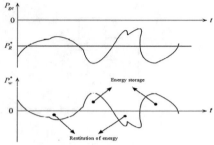

Fig. 2. Wind power production with storage system.

Fig. 3. Control of FESS integrated in the wind generator.

If the desired power is positive, it is that there is a surplus of energy that must be stored. If the desired power is negative, there is a deficit of energy which should be given by the energy stored.

6   Design of Fuzzy of FESS Speed

The used fuzzy controller has two inputs linguistic variables: the speed error e and its variation Δe and the electromagnetic torque change ΔT_{em} is considered as the output linguistic variable. The inputs fuzzy variables are given by (Kouzi et al. 2004):

$$e(k) = K_e \left(\widehat{i}_s(k) - i_s(k) \right) \tag{16}$$

and its variation is:

$$\Delta e(k) = K_{\Delta e}(e(k) - e(k-1)) \tag{17}$$

The five fuzzy sets are namely: NB: Negative Big; NS: Negative Small; EZ: Zero; PS: Positive Small; PB: Positive Big. Hence, 25 fuzzy rules were created (see Table 1).

Table 1. Inference rules

$\Delta T_{refn}(k+1)$		e(k)				
		NB	NS	EZ	PS	PB
$\Delta e(k)$	NB	NB	NB	NB	NS	EZ
	NS	NB	NB	NS	EZ	PS
	EZ	NB	NS	EZ	PS	PB
	PS	NS	EZ	PS	PG	PG
	PG	EZ	PS	PG	PG	PB

7   Simulation Results and Discussion

Figure 5 illustrate the active power of fuzzy FSSS and its reference. This latter show that the flow of this power varies with the power generated by the wind generator. In the state where this latter is greater than the power that must be provided to the grid, the difference between the latter two is stored in the FSSS ($P\_w > 0$), otherwise, the power is restored (from storage $P\_w < 0$) to compensate for the power deficit generated by the wind generator. The curve of Fuzzy FESS torque and its reference is represented in Fig. 6, it can be seen that it is positive during the storage phases (the DSIM drives the flywheel, so it operates as a motor), and during the phases of energy restitution it is negative (the flywheel drives the DSIM, generator). The speed of the FSSS is shown in Fig. 7. The speed decreases during the modes of operation of the flywheel in energy

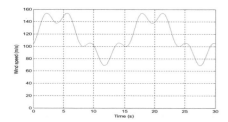

Fig. 4. Random wind profile.

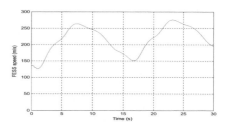

Fig. 7. Fuzzy FESS speed.

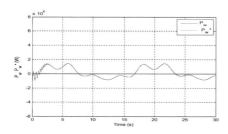

Fig. 5. Active power of the (Fuzzy FESS) and its reference.

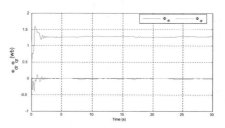

Fig. 8. Direct and quadratic rotor flux of the Fuzzy FESS

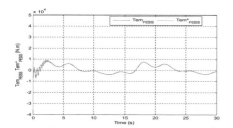

Fig. 6. Fuzzy FESS Torque and its reference.

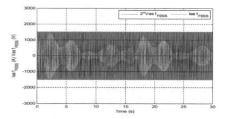

Fig. 9. Stator voltage and current for stator 1 of the Fuzzy FESS.

removal, and it increases during the energy storage operating modes. The properties of vector control of fuzzy FSSS is given in Fig. 8. Figure 9 shows the voltage and current characteristics of stator phase of the (DSIM). From the stator voltage and current waveforms, it is clearly that, the stator works nearly at unity power factor. The DC link voltage is maintained at a constant level (1130 V) (see Figs. 4, 10 and 11).

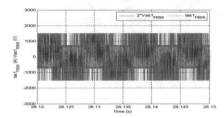

Fig. 10. Stator voltage and current for stator 1 of the (FESS) zooming

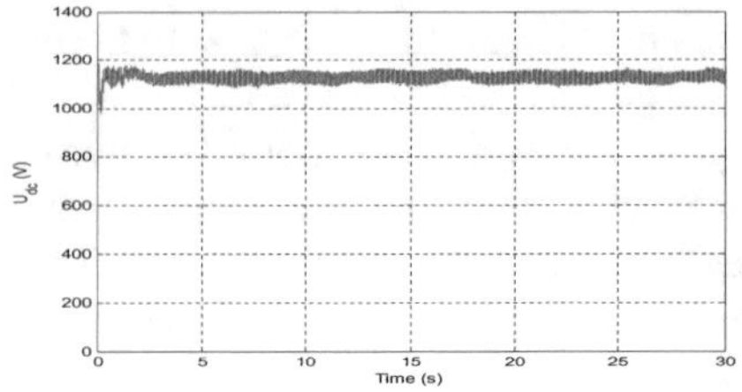

Fig. 11. DC link voltage

8   Conclusion

In this paper, to maintain a constant power transits and to contribute in wind energy system services, it has suggested fuzzy FESS coupled to DSIM integrated to WECS. The presented control scheme is used for WECS based on DSIG. The performance of the fuzzy FESS associate to wind generator has been tested under several variations of wind. From simulation results, it was remarked that the dynamic and static behavior of the whole wind power generation system (wind generator and fuzzy FSSS) is very acceptable.

Appendix. A Parameters

DSIG: 1.5 MW,   400 V,   50 Hz,   2 pole pairs, $R_{s1} = R_{s2} = 0.008\,\text{X}$, $L_1 = L_2 = 0.134\,\text{mH}$, $L_m = 0.0045\,\text{H}$, $R_r = 0.007\,\text{X}$, $L_r = 0.067\,\text{mH}$, $J = 104\,\text{kg m}^2$ (turbine + Machine), $f_r = 2.5\,\text{N m S/rd}$: (turbine + Machine).

Turbine : Radius = 35 m, blades = 3, Hub height = 85 m.

References

Wu, B., Lang, Y., Zargari, N., Kouro, S.: Power Conversion and Control of Wind Energy Systems. IEEE Press Series on Power Engineering. Wiley, Hoboken (2011)

Khaterchi, M., Elhadj, J., Elleuch, M.: DPC for three-phase inverter to improve the integration of wind turbine associated to flywheel energy storage system into the grid. In: Proceedings of the 6th Annual International Multi Conference on System, Signals and Devices (2009)

Davigny, A.: Participation aux services systèmes de fermes d'éoliennes à vitesse variable intégrant du stockage inertial d'énergie. Ph.D thesis. Lillle University (2007)

Leclerq, L.: Apport du stockage inrtiel associé à des éoliennes dans un réseau électrique en vue d'assurer des services systèmes. Ph.D thesis, Lillle University (2004)

Cimuca, G.O.: Système Inertiel de Stockage D'énergie Associé à des générateurs Eoliens '. Ph.D thesis, Lille University (2005)

Abo-Khalil, A., Lee, D.C., Seok, J.K.: Variable speed wind power generation system based on fuzzy logic control for maximum output power tracking. In: Proceedings of the 35th Annual IEEE Conference on Power Electronics Specialists, pp. 2039–2043, June 2004

Chekkal, S., Aouzellag, D., Ghedamsi, K., Amimeur, H.: New control strategy of wind generator based on the dual-stator induction generator. In: July of the 10th Annual IEEE Conference (EEEIC), pp. 68–71, May 2011

Amimeur, H., Aouzellag, D., Abdessemed, R., Ghedamsi, K.: Sliding mode control of a dual-stator induction generator for wind energy conversion systems. Trans. Electr. Power Energ. Syst. **42**(1), 60–70 (2012). Elsevier

Pant, V., Siugh, G.K., Singh, S.N.: Modeling of a multi-phase induction machine under fault condition. In: July IEEE Conference Power Electronics and Drive Systems, pp. 92–97, July 1999

Vas, P.: Sensorless Vector and Direct Torque, p. 267. Oxford University Press, U.K (1998). Chapter 4

Kouzi, K., Mokrani, L., Naït-Saïd, M-S.: High performances of fuzzy self-tuning scaling factor of PI fuzzy logic controller based on direct vector control for induction motor drive without flux measurements. In: Proceedings of IEEE International Conference on Industrial Technology (ICIT), pp. 1106–1111, December 2004

Dynamic Modeling and Optimal Control Strategy of Energy Storage Elements in Hybrid Electrical Vehicle "Fuel Cell and Ultracapacitor"

O. Heddad[1,3(✉)], L. Ziet[1(✉)], C. Gana[2], W. Dana[3], and K. Chettoueh[3]

[1] Electronic Department, University of Setif, Sétif, Algeria
haddedw@yahoo.fr, lahcene.ziet@yahoo.fr
[2] Electronic Department, University of Bejaia, Béjaïa, Algeria
[3] Electronic Department, University of BBA, Bordj Bou Arreridj, Algeria

Abstract. This paper presents Proton exchange membrane fuel cells (PEM-FC) modeling, ultracapacitor modeling, analysis of the parameter influences for performance evaluation and optimal control of energy storage elements. Theoretical results are verified by simulation models for a PEM-FC and Ultracapacitor (UC). This paper introduces the electrochemical model of fuel cell (FC). The Simulink model based on it is suitable. The results of the model are used to control the output of energy storage elements.

Keywords: Fuel cell · Modeling · Hybrid vehicle · Ultracapacitor
Renewable energy · Optimal control

1 Introduction

Increased demand for oil, future oil supply shortage concerns and climate change concerns, have led to the fast development of renewable energy firms. In spite of production costs of these alternative energies are still high, the sector accomplished has accomplished remarkable progress and attracted attention to clean energy, both at the industry level and at the academic side [1]. Globally, many countries face several challenges in their energy sectors to produce the renewable energy, green energy or clean energy, such as the fuel cells. The principal idea is to combine FC energy source with a fast power source such as UC carrying out a hybrid power source [2, 3]. This solution offers both high power and high energy densities, Different control strategies have been proposed in the literature for hybrid system power management. The main of these strategies which produce good results in terms of energy management is optimal control [4]. In this context, an efficient characterization of UC and FC has been proposed and successfully performed using simulation models (Sect. 2). The performance of the optimal control is presented in Sect. 3.

© Springer International Publishing AG 2018
M. Hatti (ed.), *Artificial Intelligence in Renewable Energetic Systems*, Lecture Notes in Networks and Systems 35, https://doi.org/10.1007/978-3-319-73192-6_38

2 Hybrid System: Fuel Cell, Ultracapacitor and Load Model

2.1 Fuel Cell System

A fuel cell is an electrochemical cell that converts the chemical energy from a fuel directly to the electrical energy through an electrochemical reaction of hydrogen-containing fuel [5] with oxygen or another oxidizing agent (Fig. 1).

The most common classification of fuel cells is by the type of electrolyte used in the cells. One of the types is Polymer Electrolyte membrane or Proton Exchange Membrane Fuel Cell (PEMFC). Development of this alternative energy is very important because it has a number of advantages. The chemical reactions for the PEMFC system can be expressed as follows:

$$\textit{Anode Reaction}: \quad H_2 \rightarrow 2H^+ + 2e^- \qquad\qquad \textit{Cathode Reaction}: \quad 1/2\,O_2 + 2H^+ + 2e^- \rightarrow H_2O$$
$$\textit{Overall Cell Reaction}: \quad H_2 + 1/2\,O_2 \rightarrow H_2O + heat + electricity$$

$$(1)$$

The FC systems have been showing up as a promising and very favorable alternative energy source due their high efficiency to the environment conservation, excellent dynamic response, and superior reliability and durability [6, 7]. There are many studies related to the modeling of fuel cells. This paper presents an analysis of the dynamic response of variation temperature, surface, humidification membrane and resistance (Figs. 13, 14, 15 and 16) of fuel cell stack to step changes in load (Fig. 6), which are characteristic of automotive fuel cell system applications (Fig. 12). The goal is a better understanding of the electrical and electrochemical processes when accounting for the characteristic cell voltage response during transients [8]. The analysis is based on a fuel cell (PEMFC) stack (Fig. 2), which is similar to those used in several.

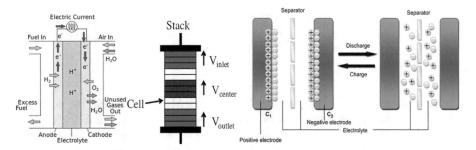

Fig. 1. Scheme of a proton conducting FC [9] **Fig. 2.** Scheme of a FC stack **Fig. 3.** Scheme of ultracapacitor

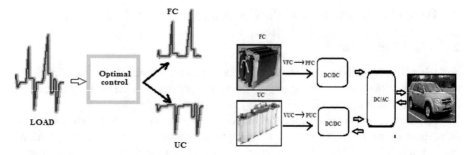

Fig. 4. Bloc diagram of differents power   **Fig. 5.** Bloc diagram of hybrid group

2.2   Fuel Cell Model and Fuel Cell Voltage

The voltage fuel cell is defined by the following equation:

$$V_{cell} = E_{cell} - V_{act} - V_{ohm} - V_{conc} \tag{2}$$

V_{act}: Activation voltage drop, V_{ohm}: Ohmic voltage drop and V_{conc}: Concentration voltage drop.

Finally the model used for the fuel cell system is given by:

$$
\begin{aligned}
V_{PAC} = &\left[0,2817 - 0.85 \cdot 10^{-3} \cdot (T - 298,15) + 4.31 \cdot 10^{-5} \cdot T \cdot [\ln\left(\frac{3}{4} \cdot P_{anode}\right) + \frac{1}{2}\ln\left(\frac{1}{2} \cdot P_{cath}\right)] \right] \\
&+ \left[2,86 \cdot 10^{-3} + 2 \cdot 10^{-4} \cdot \ln(A) + \left(4,3 \cdot 10^{-5}\right)\ln\left(\frac{\frac{3}{4} \cdot P_{anode}}{1,09 \cdot 10^6 \cdot \exp\left(\frac{77}{T}\right)}\right) \right] \cdot T \\
&+ 7,6 \cdot 10^{-5} \cdot T \cdot \ln\left(\frac{\frac{1}{2} \cdot P_{cath}}{5.08 \cdot 10^6 \cdot e^{\frac{-498}{T}}}\right) - 1,93 \cdot 10^{-4} \cdot T \cdot \ln(i_{PAC}) \\
&- i_{PAC}\left[\left[\frac{181,6\left[1 + 0,03\left(\frac{i_{PAC}}{A}\right) + 0,062\left(\frac{T}{303}\right)^2 \cdot \left(\frac{i_{PAC}}{A}\right)^{2,5}\right]}{\left[\lambda_{H_2O/SO_3^-} - 0,634 - 3\left(\frac{i_{PAC}}{A}\right)\right] \cdot \exp\left[4,18\left(\frac{T-303}{T}\right)\right]} \cdot \frac{1}{A} \cdot R_C \right] \right] + B\left(1 - \frac{J}{J_{max}}\right)
\end{aligned}
\tag{3}
$$

The voltage Ecell is the "Nernst voltage" of a hydrogen fuel cell. The activation loss *Vact* is due to the movement of electrons between the anode and the cathode. Ohmic losses *Vohm* are caused by the resistance of the polymer membrane to the transfer of protons. The concentration loss *Vconc* results from the imbalance in the supply of reactant and the product production through the reaction.

2.3   Ultracapacitor Model and UC Power

In Electric Vehicle, the fuel cell forms the primary energy storage. Sometimes available power from this fuel cell may not be sufficient to meet peak load demands. So, a secondary storage like ultracapacitor can be used in parallel with FC source to meet the

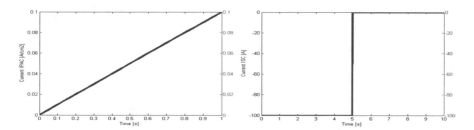

Fig. 6. Current of fuel cell   **Fig. 7.** Current of ultracapacitor

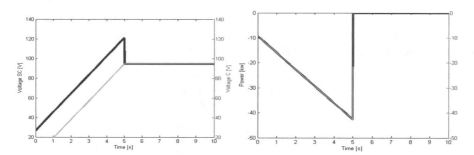

Fig. 8. Voltage of ultracapacitor   **Fig. 9.** Response of power

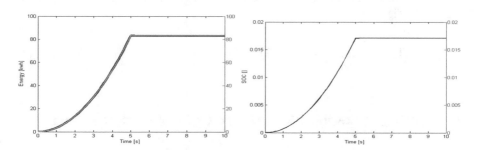

Fig. 10. Response of energy   **Fig. 11.** State of charge

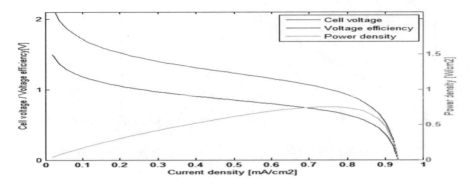

Fig. 12. Caractiristic of fuel cell (Voltage Cell, Voltage Efficiency, Power Density)

power demand, where in the high frequency current requirements are met by ultracapacitor [10] (Fig. 3). The regeneration current transients can be taken by ultracapacitor, preventing battery recharge with high current transients (Fig. 7). Therefore, the UC variable (Figs. 8, 9, 10 and 11) are obtained by the next system

$$E_{SC}(t) = E_{SC}(0) - \frac{1}{C_{SC}} \int_0^t I_{SC}(\tau).d\tau \tag{4}$$

$$V_{SC}(E_{SC}, I_{SC}) = E_{SC} - R_{SC}.I_{SC} \tag{5}$$

$$X_{SC}(E_{SC}) = \frac{1}{2} C_{SC}.E_{SC}^2 \tag{6}$$

2.4   Load Model and Traction Power

The power load specification is mainly due to speed variations [11], this power can be expressed (Fig. 18):

$$P_{LOAD}(T) = [\frac{1}{2}.P_{AIR}.S_F.C_X.V_{VH}^2(T) + M_{VH}.G.C_R + M_{VH}\left(\frac{DV_{VH}(T)}{DT}\right)]V_{VH}(T) \tag{7}$$

3   Modeling of Storage Elements and Optimal Control System

The model used for the control system is given by Eq. 8:

$$\begin{cases} \dot{Q}_{SC} = -I_{SC} \\ U_{SC} = \frac{1}{C_{SC}} Q_{SC} - R_{SC}I_{SC} \\ P_{SC} = U_{SC}I_{SC} \\ P_{PAC} = P_{VH}(t) - P_{SC} \end{cases} \tag{8}$$

4 Simulation Results

the data plotted in Figs. 6, 12, 13, 14 and 15 are obtained from a fuel cell simulation, it shows the effect of different parameters, resistance and temperature on the fuel cell voltage, the effect of area and membrane humidity are included. The calculation of parameters in Eq. 3 requires the knowledge of cathode pressure, anode pressure fuel cell and temperature. For the current study, fixed stack temperature is assumed. Even though the fuel cell has a static response, a dynamic content is imposed to this response by the dynamics of each subsystem. This is achieved through the influence of subsystems state variables, ($T°FC$, $PH2$, $PO2$, ...) on the cell voltage response. For this reason we are interested at modeling the aspect of the PEMFC and representing it by 3D. In particular, this characterization has been proposed. Simulated results show the cell temperature, the resistance, the area and the membrane humidification rate effects on the electrical outputs and the power economy. Also, we have considered the electrical phenomenon of the ultracapacitor, the results appear to be coherent with the inputs change and the ultracapacitor operating principle. Figure 8 shows voltage of UC. Response of power, response of energy and SOC are represented in Figs. 9, 10 and 11 respectively. The simulation results provide an effective improvement reference for the modeling of fuel cell and ultracapacitor.

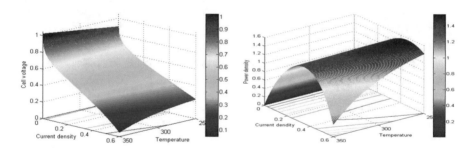

Fig. 13. The dynamic response of temperature variation (voltage cell, power cell)

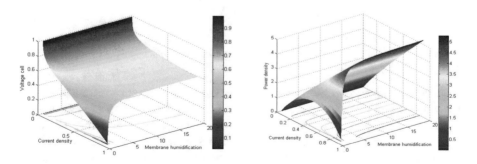

Fig. 14. The dynamic response of membrane humidification variation (voltage cell, power cell)

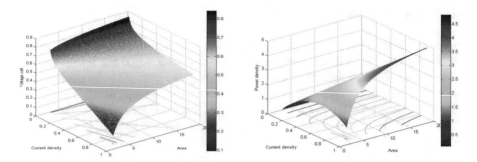

Fig. 15. The dynamic response of area variation (voltage cell, power cell)

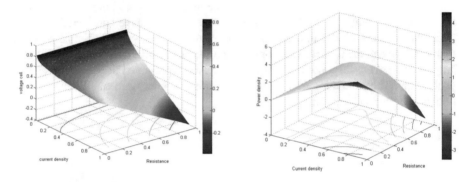

Fig. 16. The dynamic response of resistance variation (voltage cell, power cell)

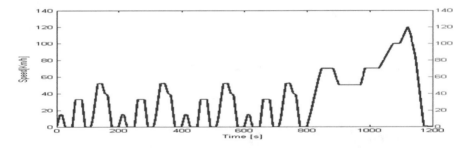

Fig. 17. New European driving cycle

The results are taken by considering new European driving cycle (NEDC) which leads to a specific traction power/current (Fig. 17) to study and to evaluate the performance of the optimal control. This control strategy permits to define the control inputs directly from the desired output trajectories. However, a stabilization adopting feedback loops are required to ensure robustness to the system. The current demanded to the fuel cell is converted in air flow that should be delivered by the compressor. In this paper, the optimal control strategy is applied for hybrid fuel cell and ultracapacitor power source of the instantaneous power split between sources (Fig. 4). A simple and effective has been successfully tested using Mathlab/Simulink to assign load power component to the appropriate source. The low power part is supplied by the fuel cell and hygh power part is absorbed by the ultracapacitor (Fig. 5). Simulation results clearly show that the system performance are improved. A non linear modeling was applied to the system. A many important variables are the main objectives of this modeling: resistance, membrane humidification, area and pressure. A non linear control strategy "optimal control" is applied to the non linear system presented previously. This results presents good study of this hybrid system and justify the performance of this modeling and of the power train control scheme of the fuel cell (Fig. 19).

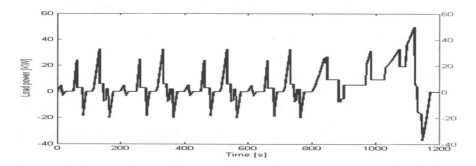

Fig. 18. Demanded power traction

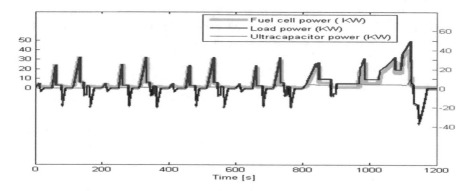

Fig. 19. Hybrid system response (loadPower, Fuel cell Power and Ultracapacitor Power)

References

1. Madaleno, M., Pereira, A.M.: Clean energy firms' stock prices, technology, oil prices and carbon prices. In: Proceedings of the 12th International Conference on the European Energy Market EEM, pp. 1–5 (2015)
2. Lachazze, J.: Etudedes stratégies et des structures de commande pour pilotage des systèmes énergétiques à pile combustible (PAC) destinée à la traction, Thèse, de l'Institut National Polytechnique de Toulouse (2004)
3. Hattab, M.: Optimisationet gestion d'énergie pour un système hybride: association pile à combustible et super-condensateur, Thèse, de l'université de technologie Belfort Montbéliard (2008)
4. Dubray, A.: Adaptation des lois de gestion d'énergie des véhicules hybrides suivant le profil de mission suivi, Thèse, de l'école Doctorale, Electronique, Electrotechnique, Automatique, Télécommunication, Signal (2002)
5. Haddad, A.: Modélisation dynamique non linéaire de la pile à combustible tu type PEM: application à la régulation de l'humidité dans la membrane électrolytique, Thèse, de l'université de technologie de Belfort-Montbéliard (2009)
6. Ghnnoum, R.: Pile à combustible, Projet de fin d'étude, de l'université Libanaise (2012)
7. Vilchez ynca, E.: Etude comparative d'algorithmes de gestion de l'énergie pour véhicule hybride à pile à hydrogène, Projet de fin d'étude, de l'université du Québec-Ecole de technologie supérieure (2014)
8. Gaoua, Y.: Modèles mathématiques et techniques d'optimisation non linéaire et combina-toire pour la gestion d'énergie d'un système multi-sources: vers une implantation temps réel pour différentes structures électriques de véhicules hybrides, Thèse, de l'université de Toulouse (2004)
9. Breeze, P.: Fuel Cell. LAP LAMBERT Academic Publishing, Saarbrucken (2017). Paperback book
10. Venet, P.: Les super-condensateurs, Conférence, à l'université de Lyon 1 (2012)
11. Destiny, L., Bounzeki, M.: Modélisation, conception et expérimentation d'un véhicule léger pour usage urbains, Thèse, de l'université de Franche-Comté (2012)

Application of Simulated Annealing Optimization Algorithm in Selective Harmonic Elimination Problem

Fayçal Chabni[1], Rachid Taleb[1(✉)], and Mustapha Hatti[2]

[1] Laboratoire Génie Electrique et Energies Renouvelables (LGEER),
Electrical Engineering Department, Hassiba Benbouali University, Chlef, Algeria
chabni.fay@gmail.com, rac.taleb@yahoo.fr
[2] Unité de Développement des Equipements Solaires (UDES)/EPST, Bou Ismail, Tipaza, Algeria
mustapha.hatti@ieee.org

Abstract. Harmonic pollution is a very common issue in the field of power electronics, Harmonics can cause multiple problems for power converters and electrical loads alike, this paper introduces a modulation method called selective harmonic elimination pulse width modulation (SHEPWM), this method allows the elimination of a specific order of harmonics and also control the amplitude of the fundamental component of the output voltage. In this work SHEPWM strategy is applied to a five level cascade inverter. The objective of this study is to demonstrate the total control provided by the SHEPWM strategy over any rank of harmonics using the simulated annealing optimization algorithm and also control the amplitude of the fundamental component at any desired value. Simulation and experimental results are presented in this work.

Keywords: Multilevel inverter · Harmonic elimination · Simulated annealing Optimization

1 Introduction

Electronic motor controllers play a major role in our daily life, these devices can be found anywhere especially in industrial applications. There are multiple types of power converters such as AC to DC, DC to DC and DC to AC. The Direct to Alternative Current converters (DC to AC) are the most used type of power converters for the control of alternating current motors. Cascade DC to AC multilevel inverters are suitable for high power applications they can withstand a huge amount of voltage stress; they are also very easy to make and to maintain due to their modular structure. The conventional multilevel cascade configuration can be achieved by connecting multiple H-bridge modules in series; this configuration will be briefly covered in this work. The harmonic content in an AC voltage waveform generated by an inverter can affect significantly the performance of AC machines. For example harmonics can raise the temperature of an AC motor which decreases the lifetime of the insulation and consequently the lifetime of the motor itself. One way to fight this problem is by choosing the right modulation strategy. Several modulation strategies have been proposed and studied for the control of multilevel inverters such as Sinusoidal Pulse width modulation (SPWM) [1] and space

© Springer International Publishing AG 2018
M. Hatti (ed.), *Artificial Intelligence in Renewable Energetic Systems*, Lecture Notes in Networks and Systems 35, https://doi.org/10.1007/978-3-319-73192-6_39

vector pulse width modulation (SVPWM) [2]. A more efficient method called selective harmonic elimination pulse width modulation (SHE-PWM) is also used; the method offers a lot of advantages such as operating the inverters switching devices at a low frequency which extends the lifetime of the switching devices. The main disadvantage of selective harmonic elimination method is that a set of non-linear equations extracted from the targeted system model must be solved to obtain the optimal switching angles to apply this strategy. Multiple computational methods have been used to calculate the optimal switching angles such as Newton-Raphson (N-R) [3], this method dependents on initial guess of the angle values in such a way that they are sufficiently close to the global minimum (desired solution). And if the chosen initial values are far from the global minimum, non-convergence can occur. Selecting a good initial angle, especially for a large number of switching angles can be very difficult. Another approach is to use optimization algorithms such as genetic algorithm (GA) [4], firefly algorithm (FFA) [5] and particle swarm optimization (PSO) [6] and differential evolution (DE) [7]. The main advantage of these methods is that they are free from the requirement of good initial guess. This article discusses the possibility of using the simulated annealing algorithm to solve the selective harmonic elimination problem, and also to demonstrate the possibility of eliminating any undesired harmonic of any rank. The simulated annealing algorithm was first introduced in 1979 by Armen G. Khachaturyan in and it was used in multiple applications such as solving facility layout problems [8], telecommunication network problems [9], optimal reactive power problem [10] and integrated circuits design [11]. In this work the SA algorithm is used to compute the optimal switching angles necessary for the SHEPWM method, in the case of a uniform step five level waveform, only one harmonic is eliminated and the fundamental component is controlled. This work is organized as follows the next section will present briefly the SHEPWM for multilevel inverters and the simulated annealing optimization method. The third section presents the obtained simulation results, simulation and experimental results are presented in the last section.

2   SHEPWM for Multilevel Inverts

2.1   Proposed Converter and the SHEPWM Strategy

The structure of the converter chosen in this study is presented in left side of Fig. 1, the converter consists of two H-bridges, each bridge is powered by its own isolated direct current power source $Vdc1$ and $Vdc2$ with $Vdc1 = 25$ V and $Vdc2 = 25$ V, this particular configuration can generate five voltage levels. In order to apply the SHEPWM strategy to this inverter the generated output voltage waveform has to be a simple stepped signal, the left side of Fig. 1 illustrates a generalized form of a uniform stepped voltage waveform with θ_1 and θ_2 are the optimal angles to be computed in order to eliminate the undesired harmonics and control the fundamental component simultaneously.

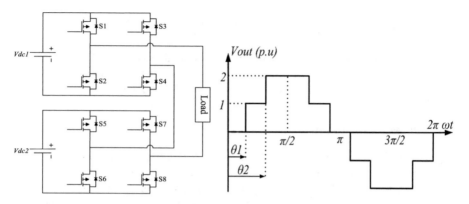

Fig. 1. Schematic of the proposed multilevel converter (Left) Generalized five level voltage waveform (Right)

The number of voltage levels that can be generated by Cascade multilevel inverters is generally presented by $2P + 1$ where P represents the number of voltage levels or switching angles in a quarter waveform of the signal, and $P - 1$ is the number of undesired harmonics that can be eliminated from the generated waveform. In a five level inverter with uniform step voltage waveform, the number of voltage levels generated in quarter waveform is two plus the zero level which means only one harmonic can be eliminated. To control the peak value of the output voltage and eliminate any harmonic, with quarter and half wave symmetry characteristics of the voltage waveform are taken in consideration, the Fourier series expansion is given as:

$$V(\omega t) = \sum_{n=1,3,5,\dots}^{\infty} \left[\frac{4V_{dc}}{n\pi} \sum_{i=1}^{P} \cos\left(n\theta_i\right) \right] \sin\left(n\omega t\right) \tag{1}$$

Where n is rank of harmonics, $n = 1, 3, 5, \dots$, and $p = (N - 1)/2$ is the number of switching angles per quarter waveform, and θ_i is the ith switching angle, and N is the number of voltage levels per half waveform. The optimal switching angles θ_1 and θ_2 can be determined by solving the following system of non-linear equations:

$$\begin{cases} H_1 = \cos\left(\theta_1\right) + \cos\left(\theta_2\right) = M \\ H_n = \cos\left(n\theta_1\right) + \cos\left(n\theta_2\right) = 0 \end{cases} \tag{2}$$

Where $M = (((N - 1)/2)r/4)$, r is the modulation index and. The obtained solutions must satisfy the following constraint:

$$0 < \theta_1 < \dots < \theta_p < \pi/2 \tag{3}$$

An objective function is necessary to perform the optimization operation, the function must be chosen in such way that allows the elimination of low order harmonics while maintaining the amplitude of the fundamental component at a desired value Therefore the objective function is defined as:

$$f(\theta_1, \theta_2) = (\sum\nolimits_{n1}^{2} (\cos(\theta_n) - M))^2 + (\sum\nolimits_{n1}^{p} (\cos(n\theta_n)))^2 \qquad (4)$$

The optimal switching angles are obtained by minimizing Eq. (4) subject to the constraint Eq. (3). The main problem is the non-linearity of the transcendental set of Eq. (2), the simulated annealing is used to overcome this problem.

2.2 Simulated Annealing

The simulated annealing is a stochastic global optimization method that can differentiate between multiple local optima points. The algorithm is inspired from the process of cooling metal after heating it to get a perfect crystal structure with minimum defects. While many optimization methods get stuck in a local minimum instead of converging to a global minimum, the simulated annealing solves this problem by performing a random search. Figure 2 presents a simplified flowchart of the simulated annealing algorithm.

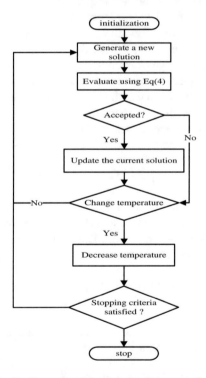

Fig. 2. Generalized five level voltage waveform

The algorithm deals with the minimization of an objective function using a parameter called temperature to evaluate the probability of accepting worst values to escape local minima. The algorithm starts by defining the values of parameters and algorithm operators, and also sets the temperature parameter T to an initial value with initial set of solutions. In this algorithm new random solutions are generated for each iteration, if the newly generated solution improves the objective function $f(x)$ expressed in (4) and gave better result than the previous one, then the proposed solution is accepted. Another technique to evaluate the improvement of the system, is to accept the new random solution with a likelihood according to a probability of $e^{(-\Delta f)}$, where Δf is the variation of the objective function, this variation can be expressed by the following equation.

$$\Delta f = f\left(x^{k}\right) - f\left(x^{k-1}\right) \tag{5}$$

Where k is the current iteration.

3   Simulation Results

In order to prove the theoretical predictions and to test the effectiveness of the proposed algorithm, the control method and the mathematical model of the proposed inverter were developed and simulated using MATLAB/SIMULINK scientific programming environment; the optimization program was executed on a computer with Intel(R) Core(TM) i3 CPU@ 2.13 GHz Processor and 4 GB of RAM, the optimization algorithm takes 127.43 s to complete the computation process.

The left side of Figs. 3, 4 and 5 show the generated waveforms in the case of eliminating the third, fifth and the seventh harmonics respectively, and for different modulation indices r where r1 = 0.7, r2 = 0.85 and r3 = 0.9, whereas the left side of the same figures show the FFT analysis of the generated waveforms for the above mentioned cases and also for different values of r. It can be noticed from the voltage waveforms that by decreasing the modulation index, the switching angles will have higher values and this will lead to a decrease in the amplitude of the fundamental component this effect can be clearly observed in the FFT analysis figures. And also it can be clearly seen from the FFT analysis that the undesired harmonics were successfully eliminated in each case, for example in Fig. 3 the third harmonic was eliminated while the fifth harmonic remained untouched, whereas in Fig. 4 the fifth harmonic was eliminated while the third and the seventh harmonics remained untouched.

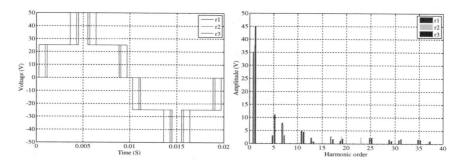

Fig. 3. Generated Voltage waveforms (Left) and FFT analysis (Right) in the case of eliminating the third harmonic

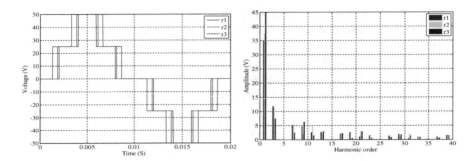

Fig. 4. Generated Voltage waveforms (Left) and FFT analysis (Right) in the case of eliminating the fifth harmonic

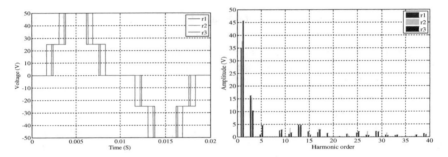

Fig. 5. Generated Voltage waveforms (Left) and FFT analysis (Right) in the case of eliminating the seventh harmonic

4 Experimental Results

A five level inverter prototype was built to validate the results obtained from the simulation process; Irf640 MOSFETS were used as switching devices for the proposed inverter, 4N25 optocouplers were used to protect the microcontroller used in this experiment, Siglent SDS 1000 oscilloscope with FFT capability was used to preview the voltage waveforms and to perform FFT analysis. The left side Figs. 6, 7 and 8 show the experimental waveforms in the case of eliminating the third, fifth and seventh harmonic respectively, and each waveform was generated for a particular value of modulation index r. The right side of same figures show the FFT analysis of generated waveforms for the cases mentioned previously, and it can be clearly seen that the third, fifth and the seventh harmonics were successfully eliminated, and also it can be noticed that there is a slight change in the value of the fundamental component.

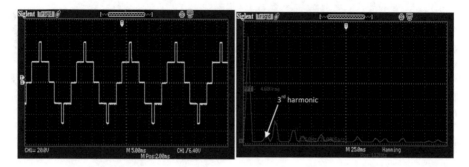

Fig. 6. Generated Voltage waveforms (Left) and FFT analysis (Right) in the case of eliminating the third harmonic r = 0.7.

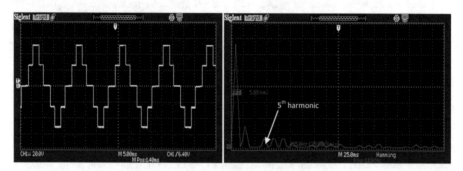

Fig. 7. Generated Voltage waveforms (Left) and FFT analysis (Right) in the case of eliminating the fifth harmonic r = 0.85.

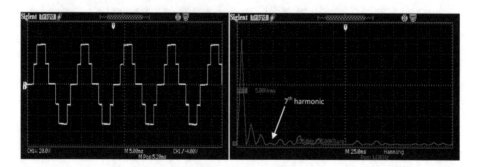

Fig. 8. Generated Voltage waveforms (Left) and FFT analysis (Right) in the case of eliminating the seventh harmonic for r = 0.9.

5   Conclusions

This paper demonstrated the ability of the selective harmonic elimination strategy for multilevel inverters of eliminating any undesired harmonics and maintain the fundamental component at a desired value, and also the possibility of using the simulated annealing algorithm to solve the optimal switching problem for multilevel inverters. The set of non-linear equations that describe the overall system are solved to obtain the optimal switching angles using the proposed optimization algorithm which belongs to the physics inspired optimization methods. The selective harmonic elimination strategy was tested in this paper for multiple situations and different values of modulation index r in order to prove the efficiency of this control method. Simulation and experimental results show a great agreement in this work.

References

1. Karami, B., Barzegarkhoo, R., Abrishamifar, A., Samizadeh, M.: A switched-capacitor multilevel inverter for high AC power systems with reduced ripple loss using SPWM technique. In: Power Electronics, Drives Systems & Technologies Conference, pp. 627–632 (2015)
2. Jana, K.C., Biswas, S.K.: Generalized switching scheme for a space vector pulse-width modulation-based N-level inverter with reduced switching frequency and harmonics. IET Power Electron. 8(12), 2377–2385 (2015)
3. Mistry, T., Bhatta, S.K., Senapati, A.K., Agarwal, A.: Performance improvement of induction motor by Selective Harmonic Elimination (SHE) using Newton Raphson (N-R) method. In: International Conference on Energy Systems and Applications, pp. 364–369, May 2015
4. Deniz, E., Aydogmus, O., Aydogmus, Z.: Implementation of ANN-based selective harmonic elimination PWM using hybrid genetic algorithm-based optimization. Measurement 85, 32–42 (2016)
5. Gnana Sundari, M., Rajaram, M., Balaraman, S.: Application of improved firefly algorithm for programmed PWM in multilevel inverter with adjustable DC sources. Appl. Soft Comput. 41, 169–179 (2016)
6. Letha, S.S., Thakur, T., Jagdish, K.: Harmonic elimination of a photo-voltaic based cascaded H-bridge multilevel inverter using PSO (particle swarm optimization) for induction motor drive. Energy 107, 335–346 (2016)
7. Chabni, F., Taleb, R., Helaimi, M.: Differential evolution based SHEPWM for seven-level inverter with non-equal DC sources. Int. J. Adv. Comput. Sci. Appl. (IJACSA) 7(9), 304–311 (2016)
8. Grobelny, J., Michalski, R.: A novel version of simulated annealing based on linguistic patterns for solving facility layout problems. Knowl. Based Syst. 124, 55–69 (2017)
9. Valdivieso, C., Novillo, F., Gomez, J., Dik, D.: Centralized channel assignment algorithm for WSN based on simulated annealing in dense urban scenarios. In: 8th IEEE Latin-American Conference on Communications, pp. 1–6 (2016)
10. Raha, S.B., Mandal, K.K., Chakraborty, N.: Parametric variation based simulated annealing for reactive power dispatch. In: IET Chennai Fourth International Conference on Sustainable Energy and Intelligent Systems, pp. 29–34 (2013)
11. Sait, S.M., Oughali, F.C., Al-Asli, M.: Design partitioning and layer assignment for 3D integrated circuits using tabu search and simulated annealing. J. Appl. Res. Technol. 14(1), 67–76 (2016)

Modeling and Control Thermal of Building for Improve Comfort Level by Using PID and On/Off Methods in the Case South-West Algeria

Merabti Soufiane[1,2(✉)], Bahra Imane[2], Rahoui Ikram[2], Bounaama Fatah[2], and Daoui Belkacem[1]

[1] Department of Mechanical Engineering, University of Tahri Mohamed, Béchar, Algeria
merabti.soufiane1@gmail.com, bdraoui@yahoo.com
[2] Department of Electrical Engineering, University of Tahri Mohamed, Béchar, Algeria
ikram99004@gmail.com, fbounaama2002@yahoo.fr

Abstract. Heating, ventilation and air conditioning (HVAC) systems in buildings aim to control the indoor climate in order to keep occupants comfortable by control the temperature and air flow specially in arid zone like south west Algeria when we have hot summer and cold winter. To achieve the comfort, it is necessary to have adaptable control systems that could deal with the parameters required to control the indoor climatic conditions. This paper, we have developed a nonlinear physical model of order seven by using lumped capacitance method. The experimental data measured in the laboratory we used for the validation phase of the models. The identification of the thermal parameters was did manually under the Matlab/Simulink environment. The control phase aims to provide user comfort. The choice was made on conventional controllers adapted to this kind of model. The PID and On/Off controllers are given good results.

Keywords: Thermal comfort · Building management systems · PID
On/Off controller · HVAC

1 Introduction

Algeria's total final consumption has been steadily increasing in recent years. The country has seen an increase of 22% in just three years. Taking a closer look at various sectors, the residential sector is the one which consumes the most energy 43%, followed by the transport sector 36% and the industry sector 21%. In electricity consumption the residential sector is the biggest consumer in Algeria, representing 38.1% of the nationally consumed energy. Other important sectors are the tertiary sector 20.93% and the manufacturing industry 17.83% [1]. For this reason we interested to control thermal in buildings for reduce energy consumption and keep thermal comfort.

There are a wide variety of software could offer the possibility to create thermodynamic models for buildings climatic conditions so that it would be possible to simulate them on graphs or some other data, in addition to control these conditions so that would be compatible with the occupants comfort. Simulink/Matlab is one of the effective software that has advanced possibility to design thermodynamic models for indoor climatic

M. Hatti (ed.), *Artificial Intelligence in Renewable Energetic Systems*, Lecture Notes in Networks and Systems 35, https://doi.org/10.1007/978-3-319-73192-6_40

conditions. Control building temperature is a major issue of control indoor conditions, this process has a problem of the severe and frequent change in the indoor temperature. This problem may be solved by modeling of the real parameters in order to be as close as possible to the ideal parameters.

1.1 Comfort Conditions

The comfort of a person is mainly dependent upon the following factors temperature, humidity, clothing, air flow and work rate. An evaporative cooler decreases temperature and increases air flow but increases humidity. The effect of the temperature reduction far outweighs the increase in humidity. Any additional air flow further improves the comfort level.

In the nineties, the buildings lacked natural ventilation and smart architecture etc. when the buildings needs for energy saving, because people spend the most time in buildings. The environmental comfort in a work place is strongly related to the occupant's satisfaction and productivity. On the other hand, as well known, energy consumption is also strongly and directly related to the operation cost of a building.

In the recent days, special emphasis has been given to the bioclimatic architecture of buildings. Bioclimatic architecture is geared towards energy savings and comfort by using shadowing and glazing systems, natural ventilation, solar spaces thermal mass, Trombe walls, cooling systems with radiation and evaporation, etc.

The quality of comfort in buildings is determined by three basic factors: Thermal comfort, indoor air quality and visual comfort [2]. When discussing thermal comfort, there are two main different models that can be used: the static model PMV (Predictive Mean Vote) PPD (Predicted Percentage Dissatisfied) and the adaptive model. The PMV/PPD model was developed by P.O. Fanger using heat balance equations and empirical studies about skin temperature to define comfort. PMV predicts the mean thermal sensation vote on a standard scale for a large group of persons. The American Society of Heating Refrigerating and Air Conditioning Engineers (ASHRAE) developed the thermal comfort index by using coding -3 for cold, -2 for cool, -1 for slightly cool, 0 for natural, $+1$ for slightly warm, $+2$ for warm, and $+3$ for hot.

Indoor Air Quality (IAQ) refers to the air quality within and around buildings and structures, especially as it relates to the health and comfort of building occupants. Understanding and controlling common pollutants indoors can help reduce your risk of indoor health concerns. Indoor air quality can be affected by gases (including carbon monoxide, radon, volatile organic compounds) [2]. IAQ is part of Indoor Environmental Quality (IEQ), which includes IAQ as well as other physical and psychological aspects of life indoors (lighting, visual quality, acoustics, and thermal comfort).

At present, there are two modes of outdoor air supply for controlling occupant microenvironment, including the constant quantity outdoor air supply and Demand Control Ventilation (DCV). As for the constant quantity outdoor air supply, it cannot meet the indoor air needs induced by pollutant source and supply outdoor air quality. Meanwhile, its energy saving potential is very limited. On the contrary, the DCV has advantages in these two aspects.

Domestic and foreign researchers have focused on the DCV mode, control strategy and its energy saving property [3, 4]. Yang and Xin [5] improved DCV mode by designing a testing platform and obtaining the function relation of carbon dioxide concentration analog and electric signals valve openings. Yand et al. [6] proposed a novel complex air supply model for indoor air quality control via the occupant micro-environment demand ventilation. Lu et al. [7] proposed a novel and dynamic demand controlled ventilation strategy for CO2 control and energy saving in buildings. Nassif, [8] in this work provides insight into the performance of a multi-zone VAV system under different operating and ventilation conditions, discusses the difficulties in the CO2-based DCV, and proposes a robust DCV strategy based on the supply air CO2 concentration. Tong et al. [9] measured the impact of traffic-related air pollution on the indoor air quality of a naturally ventilated building.

According to the small review above, we have many studies about thermal comfort and indoor air quality but is still lacking. In fact, the occupant micro-environment demand air supply is a complex system, which covers the aspects of the demand target, the conservation relations, and the influence of different factors. Meanwhile, the air supply characteristics change dynamically over time. In particular, the air diffusion difference will lead to the variations of effective supply outdoor airflow reaching occupant micro-environment. Therefore, it is necessary to determine reasonable outdoor air supply to solve this problem.

2   Material and Methods

To study the thermal and energy behavior of the building, the model chosen is defined in this article. The research laboratory at Tahri Mohammed Bechar University, with an area of 416 m². The temperature measurement is carried out using a PT100 probe (accuracy ±0.1°C) located at the middle of the building. Radiation using pyranometers (accuracy of ±5 W/m²) on the roof of the building, the wind speed per cup anemometer (start

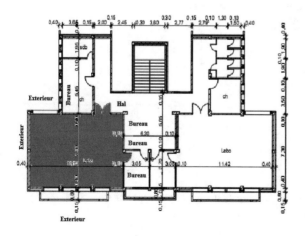

Fig. 1. Laboratory plan

threshold = 0.5 m/s) located on the roof of the building. All of these measurements were recorded in 10 min (Fig. 1).

3   The Thermoelectric RC Models of the Envelope Walls of the Building

The way of representing and modeling thermal of building, in some aspects, peculiar and dictated by the models that we know can then be implemented.

The heat conduction analogy with electrical conduction can be used to interpret the discretized structure obtained as a series of "RC" circuits where resistance and capacity are those of a slice of the wall. When the number of elements is small, this approach is known as the "grouped capacity" method. The individual RC cells may be in the form of "T", "TT" or more general structure [2] (Fig. 2).

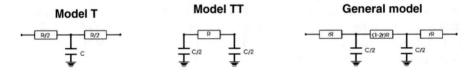

Fig. 2. Type of RC model [10].

For our research the model adopted for the envelope walls of the laboratory is TT type (Fig. 3).

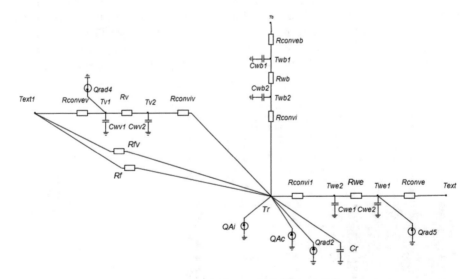

Fig. 3. Simplified model of building by using lumped capacitance method.

3.1 Equations of the Simplified Thermoelectric Model

We adopt the structure type 1R2C or TT we obtain 7 first order differential equations with 25 parameters.

$$\frac{dTwb_1}{dt} = \left(\frac{2}{Cwb_1}\right) \times \left(\frac{Tb - Twb_1}{Rconveb} + \frac{Twb_2 - Twb_1}{Rwb}\right) \tag{1}$$

$$\frac{dTwb_2}{dt} = \left(\frac{2}{Cwb_2}\right) \times \left(\frac{Tr - Twb_2}{Rconvi} + \frac{Twb_1 - Twb_2}{Rwb}\right) \tag{2}$$

$$\frac{dTv_1}{dt} = \left(\frac{2}{Cwv_1}\right) \times \left(\frac{Text - Tv_1}{Rconvev} + \frac{Tv_2 - Tv_1}{Rv} + (\alpha v \times Qrad \times Sv)\right) \tag{3}$$

$$\frac{dTv_2}{dt} = \left(\frac{2}{Cwv_2}\right) \times \left(\frac{Tr - Tv_2}{Rconvi} + \frac{Tv_1 - Tv_2}{Rv}\right) \tag{4}$$

$$\frac{dTwe_1}{dt} = \left(\frac{2}{Cwe_1}\right) \times \left(\frac{Text - Twe_1}{Rconve} + \frac{Twe_2 - Twe_1}{Rwe} + (\alpha m \times Qrad \times Si)\right) \tag{5}$$

$$\frac{dTwe_2}{dt} = \left(\frac{2}{Cwe_2}\right) \times \left(\frac{Tr - Twe_2}{Rconvi} + \frac{Twe_1 - Twe_2}{Rwe}\right) \tag{6}$$

$$\frac{dTr}{dt} = \left(\frac{2}{Cr}\right) \times \left(\frac{Text - Tr}{Rf} + \frac{Text - Tr}{Rfv} + \frac{Twb_2 - Tr}{Rconvi} + \frac{Tv_2 - Tr}{Rconvi} + \frac{Twe_2 - Tr}{Rconvi} + QAi + QAc + (\alpha r \times Qrad \times Sp)\right) \tag{7}$$

After choosing the RC structure for the envelope we obtained a non-linear physical model.

The use of the database collected on the site will be used in the phase of identification of the parameters and simulation of the models in the following work.

4 Validation of Thermal Model

To validate the physical model of building it is necessary to identify these parameters so that the calculated output temperature Tin of the system is identical to an error near the temperature measured by the measurement probe inside the building (Fig. 4).

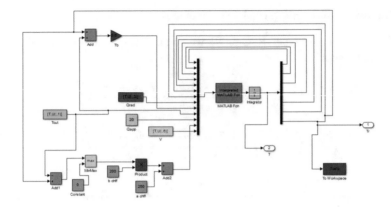

Fig. 4. Implementation of the thermoelectric model in Matlab/Simulink.

4.1 Simulation of Thermoelectric Physical Model of the Building

The temperatures calculated by the model and measured inside the building are given by the following figures:

The temperature calculated by the model in blue is quite close to the temperature measured in red as shown in Fig. 5.

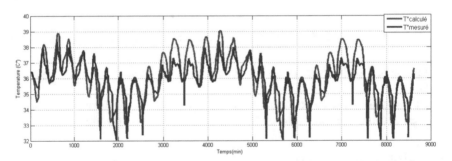

Fig. 5. Comparison between measured and calculated temperature PID (August 2015).

5   Indoor Climatic Management of the Building Model

5.1   By PID Controller

See (Figs. 6 and 7).

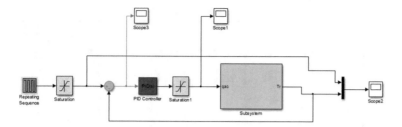

Fig. 6. Implementation of PID-controlled model in Matlab/Simulink.

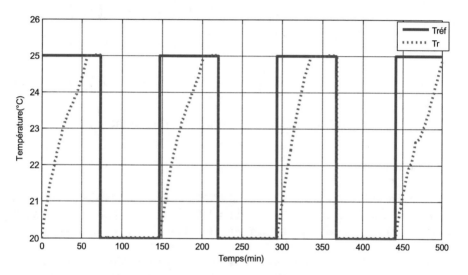

Fig. 7. The system response with PID (August 2015)

5.2 By On/Off Controller

Our goal is to achieve temperature control using the digital controller. This type of conventional controller sometimes gives good results on non-linear systems and is very simple to implant because it is a relay and a gain.

Air conditioning phase (with a "saturation" constraint on the control). For this phase of air conditioning we chose the self of August 2015 because it is a very warm period and the air conditioning gives good results (Fig. 8).

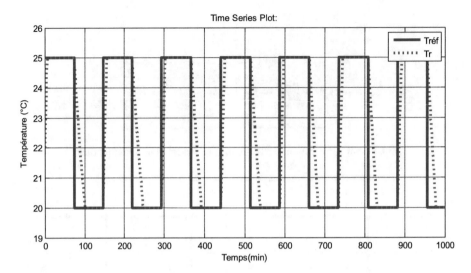

Fig. 8. The system response with On/Off (August 2015).

6   Conclusion

The objective of this article was twofold. The first was to select a model for predicting the thermal behavior of a laboratory building. It had to not only predict the air conditioning needs and the daily interior temperature, but also be controllable and stable in nature.

We have modeled a thermal system of the laboratory building, which has been identified by measurement data collected at the laboratory site itself. These models were used for the automatic internal climate control of the ENERGARID laboratory. The control of thermal models for buildings is a very important field of research and therefore in this brief, we focused on the thermal comfort of the users. The climate management of the internal environment of the laboratory was done by a PID and On/Off controller. Lastly, we can say that the domain of modeling and control of the sectors of the building and a vast and multidisciplinary field of where the necessity of the work of group of different specialties.

References

1. Merabti, S., Draoui, B., Bounaama, F.: A review of control systems for energy and comfort management in buildings. In: 2016 8th International Conference on Modelling, Identification and Control (ICMIC), pp. 478–486 (2016)
2. Standard, A.: Ventilation and Acceptable Indoor Air Quality in Low-Rise Residential Buildings (2003)
3. Fisk, W.J., De Almeida, A.T.: Sensor-based demand-controlled ventilation: a review. Energy Build. **29**, 35–45 (1998)

4. Emmerich, S.J., Mitchell, J.W., Beckman, W.A.: Demand-controlled ventilation in a multi-zone office building. Indoor Built Environ. **3**, 331–340 (1994)
5. Yang, H.X., Xin, S.: Study on new type demand control ventilation. Build Energy Environ (1999)
6. Yang, J., Zhou, B., Jin, M., Wang, J., Xiong, F.: A novel complex air supply model for indoor air quality control via the occupant micro-environment demand ventilation. Chaos, Solitons Fractals **89**, 474–484 (2016)
7. Lu, T., Lü, X., Viljanen, M.: A novel and dynamic demand-controlled ventilation strategy for CO_2 control and energy saving in buildings. Energy Build. **43**, 2499–2508 (2011)
8. Nassif, N.: A robust $CO2$-based demand-controlled ventilation control strategy for multi-zone HVAC systems. Energy Build. **45**, 72–81 (2012)
9. Tong, Z., Chen, Y., Malkawi, A., Adamkiewicz, G., Spengler, J.D.: Quantifying the impact of traffic-related air pollution on the indoor air quality of a naturally ventilated building. Environ. Int. **89–90**, 138–146 (2016)
10. Kummert, M.: Contribution to the Application of Modern Control Techniques to Solar Buildings (2000–2001)

Optimal Reconfiguration of an Algerian Distribution Network in Presence of a Wind Turbine Using Genetic Algorithm

Mustafa Mosbah[1,2(✉)], Salem Arif[2], Ridha Djamel Mohammedi[3], and Rabie Zine[4]

[1] Algerian Distribution Electricity and Gas Company, Blida, Algeria
mosbah.mustapha@gmail.com
[2] LACoSERE Laboratory, Department of Electrical Engineering,
Amar Telidji University of Laghouat, Laghouat, Algeria
s.arif@lagh-univ.dz
[3] Department of Electrical Engineering, University of Djelfa, Djelfa, Algeria
r.mohammedi@univ-djelfa.dz
[4] Department of Mathematics, Faculty of Sciences and Humanities,
Prince Sattam Bin Abdul-Aziz university, Aflaj, Saudi Arabia
rabie.zine@gmail.com

Abstract. In this paper a Genetic Algorithm (GA) method based on graphs theory is proposed to determine the distribution network reconfiguration in presence of wind turbine based DG considering all technical and topological constraints. The objective function considered in this study is the minimization of real power loss. A detailed performance analysis is applied on (33 bus, 69 bus and 84 bus networks) to illustrate the effectiveness of the proposed method. Then this method was validated on Algerian distribution network (116 bus).

Keywords: Distribution network · Optimal reconfiguration · Wind turbine

1 Introduction

Distribution networks is an important element in the power system. In recent years, the Algerian Distribution Electricity and Gas Company has pursued its constant policy of modernizing and automating the operation of its medium-voltage distribution network. The main objectives are to achieve quality compatible with current standards and requirements and less binding and more secure. Among the essential tasks in the management of distribution networks, is optimization network reconfiguration [1]. To determine the optimal reconfiguration, it is necessary to close and open some switches [2]. The distribution networks are usually meshed, but operating in radial structure. The objective of distribution network are reconfiguration is to find the topology of networks which give minimum real loss, and/or maximum voltage stability, and/or load balancing between distribution feeders, under technical and topological constraints [3]. The optimal distribution network reconfiguration is a nonlinear and discrete optimization nature, which belongs to class of NP-complete problems [4]. This requires a

© Springer International Publishing AG 2018
M. Hatti (ed.), *Artificial Intelligence in Renewable Energetic Systems*, Lecture Notes in Networks and Systems 35, https://doi.org/10.1007/978-3-319-73192-6_41

robust optimization technique. Different methods have been proposed in previous works, to determine the optimal distribution network reconfiguration, among these methods:

Deterministic optimization methods have been presented in literature to determine the optimal distribution networks reconfiguration, as an example, Simplex Method [5, 6], a Spanning Tree Method [7] and Mixed-Integer Convex Programming Method [8].

Many algorithms are based on Artificial intelligence and/or metaheuristic search algorithms have been used to solve distribution network reconfiguration for example, Genetic Algorithms [9], Modified Taboo Search [10], Hybrid Big Bang Big Crunch, Non-dominated Sorting Genetic Algorithm (NSGA-II), Fireworks Algorithm, Memetic Algorithms, Ant Colony Algorithm, Simulated Annealing Algorithm, Fuzzy Logic Multiobjective, Harmony Search Algorithm, Honeybee Mating Optimization, Particle Swarm Optimization, Artificial Neural Networks Algorithm, Non-Dominated Sorting Particle Swarm Optimization, Hybrid Fuzzy Bees Algorithm, Cuckoo Search Algorithm, Bacterial Foraging Optimization Algorithm, Hybrid The Minimum Spanning Tree and Improved Heuristic Rules Algorithm, Refined Genetic Algorithm, Backtracking Search Optimization Algorithm, Artificial Immune System, Binary Gravitational Search Algorithm, Differential Search Algorithm and Differential Evolutionary Algorithm. Runner-Root Algorithm [11], Stochastic Dominance Concepts Algorithm. Reference examined some of the most recent methods for distribution network reconfiguration.

The objective in this paper is the application of Genetic Algorithm (GA) method based on graphs theory to design an optimal distribution network reconfiguration considering wind turbine based DG. This reconfiguration, determine the adjustment of switches state, in order to minimize the total active power losses. This study, proposed to adapt the principles of GA method to the strategy case of branches permutation, this opens and closes the switches devices. The proposed method is tested on different types of IEEE distribution network (33 bus, 69 bus and 84 bus) and validated on Algerian distribution network (116 bus).

2 Problem Formulation

2.1 Objective Function

The optimal reconfiguration problem is formulated as an optimization problem by considering minimization of active power as objective while satisfying system constraints. In radial distribution networks, each receiving bus is fed by only one sending bus. From Fig. 1, the line losses between the receiving and sending end buses P_{loss}^T can be calculated using Eq. 1.

$$P_{loss}^T = I_{pq}^2 R_{pq} = \frac{S_{pq}^2}{V_p^2} R_{pq} = \frac{P_{pq}^2 + Q_{pq}^2}{V_p^2} R_{pq} \qquad (1)$$

$$F = Min \sum_{i=1}^{NB} P_{loss(i)} \qquad (2)$$

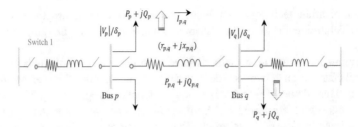

Fig. 1. One line diagram of a two-bus system.

where $|V_p|/\delta_p$, is voltage and angle voltage at bus p, r_{pq} and x_{pq} are resistance and reactance of line connecting bus p and bus q, respectively, P_{pq} and Q_{pq} are active and reactive power through the branch between bus p and bus q, NB is number of lines and N is number of buses.

2.2 Equality Constraints

The equality constraints are active/reactive power flow equations as:

$$P_G + P_{DG} = P_D + P_L$$
$$Q_G + Q_{DG} = Q_D + Q_L \tag{3}$$

where (P_G, Q_G), are the total active and reactive power of generator, respectively, (P_{DG}, Q_{DG}) are active and reactive power of wind turbine based DG, respectively. (P_D, Q_D) the total active and reactive power of load, respectively, (P_L, Q_L) is the total active and reactive power losses, respectively.

2.3 Inequality Constraints

Voltage limit:

$$V_{imin} \leq V_i \leq V_{imax} \quad for\, i = 1.\ldots.N \tag{4}$$

Line thermal limit:

$$S_k \leq S_{kmax} \quad for\, k = 1.\ldots.NB \tag{5}$$

Real power generation limit:

$$P_{Gimin} \leq P_{Gi} \leq P_{Gimax} \quad for\, i = 1.\ldots NG \tag{6}$$

The DG source limit:

$$0 \leq \sum_{i=1}^{NDG} P_{DGi} \leq 0.3 * \sum_{i=1}^{Nbus} P_{Di}\, for\, i = 1.\ldots.NDG \tag{7}$$

2.4 Radiality and Connectivity Constraints

The network topology should always be radial; the topology is radial if it satisfies the two following conditions:

1. No load out of service.
2. The network topology must be radial (no-loop).

2.5 Preserving Solution Feasibility

Note that the control variables are generated in their permissible limits using strategist preservation feasibility (perform a random value between the minimum and maximum value), while for the state variables, including the voltages of load bus, the power flowing in distribution lines, it appealed to penalties functions that penalize solutions that violate these constraints. The introduction of penalty in the objective function, transforms the optimization problem with constraints in an optimization problem without constraints [12–15], so it is easier to deal, in this case the Eq. 2 shall be replaced by:

$$F_p = Min \sum_{i=1}^{NB} P_{loss(i)} + P_f \tag{8}$$

$$P_f = k_v.\Delta V + k_s.\Delta S + k_m.N_m + k_i.N_i \tag{9}$$

$$\Delta V = \sum_{i=1}^{NL} \left(V_{Li} - V_{Li}^{lim}\right)^2 \tag{10}$$

$$\Delta S = \sum_{i=1}^{NB} \left(S_{li} - S_{li}^{lim}\right)^2 \tag{11}$$

where k_v, k_s, k_m and k_i, are penalty factors, N_m is the number of existing meshes, N_i is the number of isolated loads.

In this study, the values of penalty factors have been considered 10.000.

3 Applied Approach

Genetic algorithms are global search techniques based on the mechanism of natural selection and genetics, elaborated by Charles Darwin in 1859. Their application field is very wide: economy, finance, function optimization, pacification, and more other domains. The reason of this huge number of applications is obviously the efficiency and the simplicity. The first works on genetic algorithms started in the fifteenth, when many American biologists simulated biologic structures on computers. Then between 1960 and 1970 John Holland, based on the previous works, developed fundamental principles of genetic algorithms within the context of mathematic optimization [16]. The single-objective optimization is based on minimizing and/or maximizing of just one objective function, which generally doesn't reflect the real physic of the system to be optimized.

4   Simulations and Results

In this section, the proposed algorithm is tested on 33 bus, 69 bus and 84 bus, then validated on Algerian distribution network (116 buses). IEEE networks are generally known, but the Algerian network consists of 116 bus, 124 lines containing 09 loop lines, this load is spread over 09 feeders. The nominal voltage of 116 buses network is 10 kV. The substation is connected to MV network via a 30/10 kV transformer. The upper and lower of voltages limits considered in this paper are 0.9 pu and 1.0 pu, respectively. The DG size considered is $P_{DG} = 0.3 * \left(\sum_{i=1}^{ND} P_{Di}\right)$ with $PF = 0.8$. This DG is placed in lowest bus voltage. This is to demonstrate the influence of wind based DG on the reconfiguration and the different parameters of distribution network. Table 1 presents the size and placement of wind for different distribution system.

Table 1. Size and placement of wind based DG

Distribution network	Wind based DG-unit		
	Bus location	Size	
		P (kW)	Q (kVAR)
33 bus	18	1115	836
69 bus	65	1141	856
84 bus	10	1200	900
116 bus	81	7166	5374

The optimal reconfiguration problem in presence of wind based DG using GA method based graphs theory has been tackled with the objective of minimizing active loss. It has been recall that for each reconfiguration requires a calculation of a load flow, by backward forward method. In this study, where four simulation cases are considered, case 1, presents initial reconfiguration (without reconfiguration and without DG installation), case 2 reconfiguration before DG installation, case 3 presents only DG installation and case 4 reconfiguration after DG installation. Following the different executions of the program under MATLAB software, the optimal parameters of GA method used in this simulation are, population size is 100, maximum iteration is 50, crossover probability is 0.9, mutation probability is 0.01 and one point crossover. Table 2, shows the switches state, active power losses and minimum bus voltage of each distribution network for different studied cases. In order to demonstrate the effectiveness of GA method comparisons were made with other works in literature (see Table 3). Figures 2, 3, 4 and 5 show voltage profile for same studied cases. From these figures it has been found that the improvement of the voltages is due to optimization of the reconfiguration and the presence of DG unit. It is the same reasoning with the total loss values. Case 4 (reconfiguration after DG installation) proved its effectiveness better than other cases by the minimization in addition to total losses with better voltage profile. From the results obtained, for the Algerian distribution network, the losses value in case 3 is greater than case 1, but after optimization of the reconfiguration

Table 2. Results of GA method

Distribution network	Before reconfiguration		After reconfiguration			Before reconfiguration		After reconfiguration		
	Before DG installation					After DG installation				
	Real power loss (kW)	Minimum bus voltage (pu)	Switches opened	Real power loss (kW)	Minimum bus voltage (pu)	Real power loss (kW)	Minimum bus voltage (pu)	Switches opened	Real power loss (kW)	Minimum bus voltage (pu)
33 bus	202.50	0.9131	9-14-7-32-37	139.51	0.937	132.33	0.940	9-12-7-34-37	53.712	0.976
69 bus	224.78	0.9092	14-70-69-56-61	99.58	0.942	66.65	0.966	12-70-69-57-21	42.10	0.973
84 bus	531.81	0.9285	62-7-86-72-13-89-90-83-92-39-34-42-55	469.80	0.950	450.38	0.947	96-7-86-72-13-89-90-83-92-39-34-42-84	419.3	0.953
116 bus	402.02	0.9696	99-75-79-105-19-121-68-60-107	367.65	0.975	684.90	0.974	103-75-79-101-29-121-27-70-25	242.71	0.986

Table 3. Comparisons with other works

Distribution network	Applied methods	Optimal reconfiguration (case 2)	Real power loss (kW)	Minimum bus voltage (pu)
33 bus	GA [17]	33-9-34-28-36	140.60	0.9371
	HBMO [18]	9-14-7-32-37	139.51	0.9378
	Proposed GA	**9-14-7-32-37**	**139.51**	**0.9378**
69 bus	F-GA [19]	12-55-61-69-70	99.62	0.9427
	Proposed GA	**14-70-69-58-61**	**99.58**	**0.9427**
84 bus	AIS-ACO [20]	7, 13, 34, 39, 42, 55, 62, 72,86,89, 90, 91, 92	469.88	0.9479
	HBMO [18]	7, 14, 34, 39, 42, 55, 62, 72, 83,86, 88, 90, 92	482.14	0.9529
	Proposed GA	**42-26-34-51-122-58-39-95-97-74-71-129-130-109-23**	**469.80**	**0.9532**

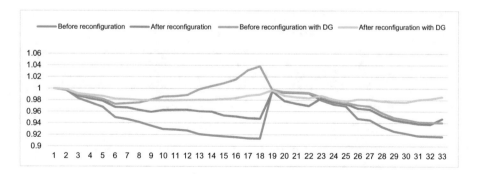

Fig. 2. Voltage profiles the different cases in 33 bus network.

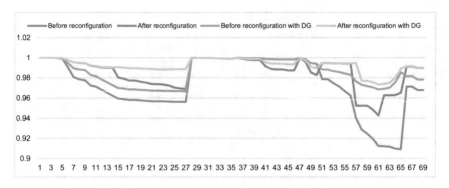

Fig. 3. Voltage profiles the different cases in 69 bus network.

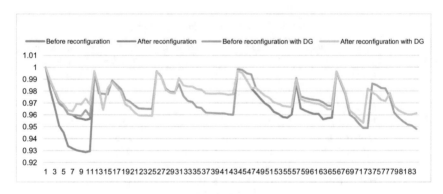

Fig. 4. Voltage profiles the different cases in 84 bus network.

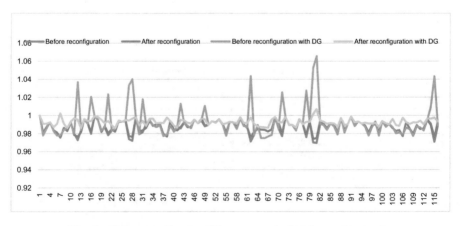

Fig. 5. Voltage profiles the different cases in 116 bus real network.

(case 4), the value of the demined losses decrease comparing to case 1, which confirms the importance of reconfiguring the network in presence of wind tubine.

5 Conclusion

In this paper, a genetic algorithm technique is proposed based on graph theory to optimize radial reconfiguration considering wind turbine based DG. The objective function considered is minimization of real power losses taking technical and topological constraints. The effectiveness of this method is shown in the quality of the results compared to the few works of literature, by validating the algorithm proposed on different IEEE distribution networks and a real distribution network.

References

1. Civanlar, S., Grainger, J., Yin, H., Lee, S.S.: Distribution feeder reconfiguration for loss reduction. IEEE Trans. Power Deliv. **3**(3), 1217–1223 (1988)
2. Shirmohammadi, D., Hong, H.W.: Reconfiguration of electric distribution networks for resistive line loss reduction. IEEE Trans. Power Deliv. **4**(1), 1492–1498 (1989)
3. Baran, M.E., Wu, F.F.: Network reconfiguration in distribution systems for loss reduction and load balancing. IEEE Trans. Power Deliv. **4**(2), 1401–1407 (1989)
4. Solo, A.M.G., Ramakrishna, G., Sarfi, R.J.: A knowledge-based approach for network radiality in distribution system reconfiguration. IEEE Trans. Power Eng. Soc. Gen. Meet. (2006)
5. Aoki, K., Nara, K., Satch, T., Kitagawa, M., Yamanaka, K.: New approximate optimization method for distribution system planning. IEEE Trans. Power Syst. **5**(1), 126–132 (2006)
6. Abnndams, R.N., Laughton, M.A.: Optimal planning of networks using mixed-integer programming. IEE Proc. **121**(2), 139–148 (1974)
7. Ahmadi, H., Marti, J.R.: Minimum-loss network reconfiguration: a minimum spanning tree problem. Sustain. Energy Grids Netw. **1**(xx), 1–9 (2015)
8. Jabr, R.A., Singh, R., Pal, B.C.: Minimum loss network reconfiguration using mixed-integer convex programming. IEEE Trans. Power Syst. **27**(2), 1106–1115 (2012)
9. Tomoiaga, B., et al.: Optimal reconfiguration of power distribution systems using a genetics algorithm based on NSGA-II. Energies **6**(3), 1439–1455 (2013)
10. Th Nguyen, T., et al.: Multi-objective electric distribution network reconfiguration solution using runner-root algorithm. Appl. Soft Comput. **52**, 93–108 (2017)
11. Chicco, G., Mazza, A.: Assessment of optimal distribution network reconfiguration results using stochastic dominance concepts. Sustain. Energy Grids Netw. **9**, 75–79 (2017)
12. Badran, O., et al.: Optimal reconfiguration of distribution system connected with distributed generations: a review of different methodologies. Renew. Sustain. Energy Rev. **73**, 854–867 (2017)
13. Mosbah, M., Mohammedi, R.D., Arif, S., Hellal, A.: Optimal of shunt capacitor placement and size in Algerian distribution network using particle swarm optimization. In: 8th International Conference on Modelling, Identification and Control (ICMIC), IEEE, Algiérs, January 2017

14. Mosbah, M., Hellal, A., Mohammedi, R.D., Arif, S.: Genetic algorithms based optimal load shedding with transient stability constraints. In: IEEE Proceedings of the International Electrical Sciences and Technologies in Maghreb, pp. 1–6 (2014)
15. Mosbah, M., et al.: Optimal sizing and placement of distributed generation in transmission systems. ICREGA-2016, Belfort, France, February, 8–10 (2016)
16. Holland, J.H.: Adaptation in Nature and Artificial Systems. The University of Michigan Press, Ann Arbor (1975)
17. Hong, Y.-Y., Ho, S.-Y.: Determination of network configuration considering multiobjective in distribution systems using genetic algorithms. IEEE Trans. Power Syst. **20**, 1062–1069 (2005)
18. Niknam, T.: An efficient multi-objective HBMO algorithm for distribution feeder reconfiguration. Exp. Syst. Appl. **38**(3), 2878–2887 (2011)
19. Liu, L., Chen, X.Y.: Reconfiguration of distribution networks based on fuzzy genetic algorithms. CSEE Proc. **20**, 66–69 (2000)
20. Ahuja, A., Das, S., Pahwa, A.: An AIS-ACO hybrid approach for multi-objective distribution system reconfiguration. IEEE Trans. Power Syst. **22**, 1101–1111 (2007)

Optimization of Irrigation with Photovoltaic System in the Agricultural Farms - Greenhouse: Case Study in Sahara (Adrar)

Zineb Mostefaoui[✉] and Sofiane Amara

Unité de Recherche Matériaux et Energies Renouvelables (URMER),
University of Tlemcen, Tlemcen, Algeria
zineb.mostefaoui@gmail.com, Sofiane.amara@yahoo.fr

Abstract. Renewable energy is an alternative solution for water pumping and irrigation for isolated and arid regions. This paper analyzes the irrigation requirements of greenhouse model representative and provides solution by photovoltaic system for irrigation.

According to the results of calculations, monthly water consumption varies between 1.02 m^3 and 9.70 m^3 for three products (Tomato, Muskmelon and Watermelon). The proposed technology can fully satisfy these water requirements.

Furthermore, this paper discusses the economic analyses for this technology, in comparison with the diesel system. The obtained results indicates that the Levelized cost of energy (LCOE) of PV system is acceptable compared to the diesel generator, where $LCOE_{PV} = 0.060 \text{ \$/kWh}$, $LCOE_{Diesel} = 0.260 \text{ \$/kWh}$, and the water price for PV system is 6.09 \$ and for diesel system is 10.43 \$, for daily consumption of 0.33 m^3. This proves the PV water pumping system is more economical than the diesel system.

Keywords: Greenhouse · Photovoltaic water pumping system · Irrigation Economic analyses

1 Introduction

The greenhouse cultivation has become a very good way to develop the agricultural sector. Nevertheless, these types of systems are generally provided to the burning of fossil fuels which increases the carbon-dioxide emissions (CO_2) or electrical appliances that consume more energy, since, uncertainty of fossil fuels availability and high costs their and interruption of electricity in those remote and scattered villages, limit the usability of irrigation systems [1]. Consequently, the paramount solution is to apply renewable energies, because they are natural and sustainable sources. In Algeria a study was done on photovoltaic water pumping system in the remote village of Ghardaia, to optimize the capacity of the different components of that system [2]. A study in China that uses a PV cells for generating electricity for powering a irrigation system which is composed of PV cells, water pump, irrigation wells and storage tank [3].

© Springer International Publishing AG 2018
M. Hatti (ed.), *Artificial Intelligence in Renewable Energetic Systems*, Lecture Notes in Networks and Systems 35, https://doi.org/10.1007/978-3-319-73192-6_42

In a study of the remote areas in Algeria presents an analysis of the economic feasibility of the PV pumping system in comparison with systems using diesel [4]. In rural areas in Oman, the optimum design of a PV system used to operate a water pumping system for irrigation, a comparison is made between the cost of the proposed system and the cost of diesel system; the comparison indicated that the system cost of energy is promising [5].

2 Estimation of Irrigation Water Needs

Adrar is located in south-west of Algeria, with latitude of $27.81°$ a longitude of $-0.18°$ and an altitude of 279 m. It is characterized very hot summers and cold winters. The greenhouse chosen has a total surface of $A_s = 200$ m^2. The determination of irrigation water needs for Tomato, Watermelon and Muskmelon, is based on the determination of ETP0 using the empirical formula [6]:

$$ETP0 = 0.0018 \cdot Rg - 0.02 \cdot M_P - 0.8$$

Where ETP0 the reference potential evapotranspiration (mm), Rg is the global radiation (J/cm^2) and M_P is the age of plastics in months (1, 2, 3…). The calculation of the corrected potential evapotranspiration, taking into to account sky condition [6]:

$$ETP = [(ETP0 + Tc) \cdot CEC] - Tc$$

Where *CEC* is the sky condition coefficient, Tc is the corrective term which is related to the value of the global radiation multiplied by the sky condition coefficient [6]. The actual evapotranspiration (ETA) was calculated by [6]:

$$ETA = ETP \cdot Kc$$

Where; Kc is the crop coefficient. The water requirements (in liters) are given by [6, 7]:

$$Bs = ETA \times A_S$$

3 Photovoltaic Water Pumping System

The PV water pumping included 6 solar PV panels (75 W/20 V), a centrifugal pump, a controller, and storage tank. The required hydraulic energy E_h in kWh/day is calculated from [8]:

$$E_h = \rho_{water}gHV/(3.6 \cdot 10^6)$$

The pumping head is supposed equal to H = 80 m, an average daily working time t_s equal to 5.5 h, the average electrical input power P_a is supposed to be equal to 300 W. The required electric energy E_a is determined by [4]:

$$E_a = P_a \times t_s$$

The energy produced by the PV array E_{PV} in kW is calculated by the following equation, where, F is the mismatch factor [4]:

$$E_{PV} = F * E_a$$

4   Economic Analyses

4.1   Life Cycle Calculation of Photovoltaic/Water Pumping System

The life cycle cost can be calculated using the following formula [9]:

$$LCC = C + M + R$$

Where C is the capital cost ($), M is the sum of all operation and maintenance costs ($), R is the sum of all equipment replacement costs ($). The initial capital cost of the PV pumping system can be calculated by [4]:

$$C = C_{pv} + C_{pump} + C_{aux}$$

Where C_{PV} is the initial capital cost of the photovoltaic modules, C_{pump} is the initial capital cost of the pump; C_{aux} is the initial auxiliary capital cost. The levelized cost of energy is calculated using [10]:

$$LCOE = \frac{\text{Lifecycle cost}}{\text{Lifetime energy production}}$$

The annualized life cycle costs are calculated using the following formulas [4]:

$$CRF = \frac{d}{1 - (1 + d)^{-T}}$$

$$PWF = \frac{1 + d}{1 + i}$$

$$C_y = C * CRF$$

$$C_k = R_k * PWF$$

$$R_y = \sum_k C_k$$

$$A_y = C_y + M_y + R_y$$

Where, PWF is the present worth factor, CRF is the capital recovery factor, d is the discount rate, T is the lifetime period, i is the interest rate, C_y is the annualized capital cost, C_k is the present worth of replacement at year k, R_k is the cost of replacement of a system component at year k, R_y is the present worth of all replacements incurred during the lifetime T, M_y is the yearly operating and maintenance cost of the initial capital cost, A_y is annualized life cycle cost.

4.2   Pumped Water Cost

The cost of m^3 of pumped water C_w by PV and diesel system pumping systems is calculated by [4]:

$$C_w = \frac{\text{Annualized life cycle of the system}}{\text{Total pumped water}}$$

5   Results and Discussion

Results indicated that there is a difference between the water consumption (Fig. 1). The water consumption of Tomato varies between a minimum of 1.28 m^3 in the beginning of cycle and a maximum of 9.70 m^3 in the fourth bouquet of flowering. The water consumption of Muskmelon varies between a minimum of 1.02 m^3 in the beginning of cycle and a maximum of 7.32 m^3 in the flowering period. Moreover, the minimal consumption of Watermelon is 1.53 m^3 in the beginning of cycle and the maximal consumption is 7.32 m^3 in the start of the harvest.

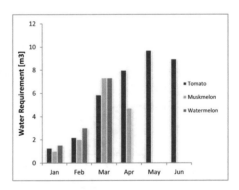

Fig. 1.   Water requirements of Tomato, Muskmelon and Watermelon.

The amount of water supplied by a PV/water pumping system throughout the growing cycle of Tomato is illustrated in Fig. 2 results show that the lowest amount of water pumped by system is in the beginning of cycle, while the large amounts is in the fourth bouquet of flowering.

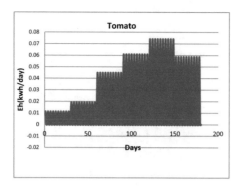

Fig. 2. Hydraulic energy necessary for Tomato.

The required hydraulic energy of Muskmelon is shown in Fig. 3 the maximum amount of water is in the beginning swelling of 0.055 kWh/day, while the minimum amount of water is in the Plantation of 0.008 kWh/day.

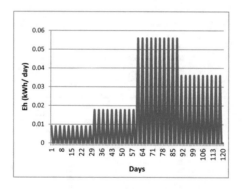

Fig. 3. Hydraulic energy necessary for Muskmelon.

For Watermelon the lowest amount of water pumped by system is in the beginning of cycle of 0.013 kWh/day, while the large amounts is in the harvest of 0.055 kWh/day, is shown in Fig. 4. The frequency of irrigation is only 2 days.

Table 1 below shows, the initial cost of the PV pumping system C = 3947.3 \$, moreover, the initial cost of diesel pumping system is C = 2850 \$. As shown in Fig. 5, the diesel pumping system has lower initial cost than the PV pumping system, but its other costs are higher if we compare with PV system.

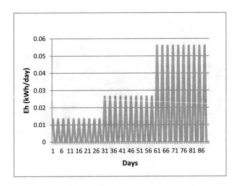

Fig. 4. Hydraulic energy necessary for Watermelon.

Table 1. Total initial investment costs of PV and diesel system ($) [4].

PV module	
Total cost of the modules	2056.5
Cost of motor-pump	1400
Support structure cost	172.82
Cables cost	6.91
Engineering and planning costs	311.08
Initial cost of PV system	3947.3
Diesel	
Unit cost	1450
Initial cost of diesel pumping system	2850

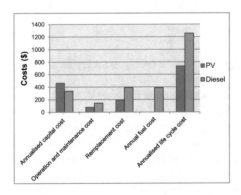

Fig. 5. Cost comparison of PV system and diesel system for pumping water.

Figure 6, shows a comparison of the Levelized cost of energy, the result shown that the LCOE of PV system is acceptable compared to the diesel generator, where $LCOE_{PV} = 0.060$ $/kWh, $LCOE_{Diesel} = 0.260$ $/kWh.

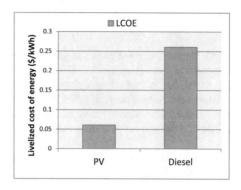

Fig. 6. Livelized cost of energy for Photovoltaic and diesel system.

In Fig. 7, the cost of water pumped by a PV system is 6.09 $ and 10.43 $ for the diesel generator, for daily consumption of 0.33 m^3. The cost of water pumped by a PV system lower than that pumped by a diesel generator.

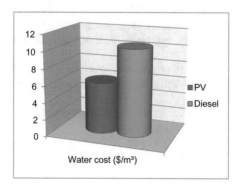

Fig. 7. Water cost for PV and diesel system.

6 Conclusion

This study considered photovoltaic energy applications in agricultural greenhouse, particularly irrigation system. A representative greenhouse model was presented to establish the irrigation water needs of greenhouse in Adrar region for an area about 200 m^2. Furthermore, irrigation water needs were determined for a PV water pumping system in rural area of Adrar, for three products (Tomato, Watermelon and Muskmelon). According to the obtained results the water consumption varies between 1.02 m^3 and a maximum of 9.70 m^3. The results shown that the LCOE of PV system is acceptable compared to the diesel generator, where $LCOE_{PV} = 0.060$ $/kWh, $LCOE_{Diesel} = 0.260$ $/ kWh, and the water price for PV system is 6.09 $ and for diesel system is 10.43 $, for daily consumption of 0.33 m^3. In conclusion, renewable energy can provide cheap and clean energy for agricultural greenhouse applications all over the world, especially in rural areas.

References

1. Chai, L., Ma, C., Ni, J.Q.: Performance evaluation of ground source heat pump system for greenhouse heating in northern China. Biosyst. Eng. **111**, 107–117 (2012)
2. Bakelli, Y., Arab, A.H., Azoui, B.: Optimal sizing of photovoltaic pumping system with water tank storage using LPSP concept. Solar Energy **85**, 288–294 (2011)
3. Yu, Y., Liu, J., Wang, H., Liu, M.: Assess the potential of solar irrigation systems for sustaining pasture lands in arid regions – a case study in Northwestern China. Appl. Energy **88**, 3176–3182 (2011)
4. Ould-Amrouche, S., Rekioua, D., Hamidat, A.: Modelling photovoltaic water pumping systems and evaluation of their CO2 emissions mitigation potential. Appl. Energy **87**, 3451–3459 (2010)
5. Kazem, H.A., Al-Waeli, A.H., Chaichan, M.T., Al-Mamari, A.S., Al-Kabi, A.H.: Design, measurement and evaluation of photovoltaic pumping system for rural areas in Oman. Environ. Dev. Sustain. 1–13 (2016). https://doi.org/10.1007/s10668-016-9773-z
6. Saouli Kessai, A., Souici, D.: Gestion De L'irrigation Localisee Sous Serre. Revue Des Regions Arides 2014 - Numero Special - No 35 (3/2014) - Actes Du 4eme Meeting International "Aridoculture Et Cultures Oasisennes: Gestion Des Ressources Et Applications Biotechnologiques En Aridoculture Et Cultures Sahariennes: Perspectives Pour Un Developpement Durable Des Zones Arides, December 2013
7. FAO: guidelines for computing crop water requirements bulletin no 56 de l'irrigation et du drainage (1989)
8. Al-Smairan, M.: Application of photovoltaic array for pumping water as an alternative to diesel engines in Jordan Badia, Tall Hassan Station: case study. Renew. Sustain. Energy Rev. **16**, 4500–4507 (2012)
9. Ghoneim, A.A.: Design optimization of photovoltaic powered water pumping system. Energy Convers. Manage. **47**, 1449–1463 (2006)
10. Darling, S.B., You, F., Veselka, T., Velosa, A.: Assumptions and the levelized cost of energy for photovoltaics. Energy Environ. Sci. **4**, 3133–3139 (2011)

Optimization and Characterization of Nanowires Semiconductor Based-Solar Cells

Fatiha Benbekhti[1(✉)], S. Tahiraoui[2], H. Khouani[3], and A. Baroudi[4]

[1] Research Unit of Materials and Renewable Energies (URMER), Electric and Electronic Engineering Department, Technology Faculty, Tlemcen University, Tlemcen, Algeria
F_benbekhti@yahoo.fr, Fatiha.benbekhti@mail.univ-tlemcen.dz
[2] Electric Engineering Department, Technology Faculty,
Hassiba Benbouali University of Chlef, Chlef, Algeria
Sd.tahraoui@univ.hb-chlef.dz
[3] Electric and Electronic Engineering Department, Technology Faculty,
Tlemcen University, Tlemcen, Algeria
hayet.kh20@gmail.com
[4] Research Unit of Materials and Renewable Energies (URMER), Technology Faculty,
Tlemcen University, Tlemcen, Algeria
Baroudi.eln@gmail.com

Abstract. The aim of our work is to be able to make significant improvements to the performance of silicon-based solar cells by the inclusion of silicon nanowires on the one hand, and on the other hand, by incorporating germanium into the substrate based on silicon. Our strategy consisted in particular of ordering the nanowires of silicon perpendicularly to the electrodes in order to improve significantly the collection of the photogenerated charges. The work we have presented in this paper is the modeling, simulation and optoelectronic characterization of photovoltaic cells based on silicon nanowires using a silicon-germanium substrate. The main objective is to improve the conversion efficiency of this type of solar cells by orientation of the silicon nanowires.

Keywords: Silicon · Germanium · Nanowires · Optimization · Efficiency Simulation

1 Introduction

The solar cells are promising systems to generate clean and renewable energy [1]. Many researches have made in order to replace fossil fuels by production a high-performance and low-cost solar cells as sustainable energy sources, compared with traditional photovoltaic devices based on silicon crystalline, and his high costs of manufacturing and installation. Semiconductor nanowires have recently emerged as a new class of materials with significant potential to reveal new fundamental physics and to propel new applications in quantum electronic and optoelectronic devices [2]. Semiconductor nanowires show exceptional promise as nanostructured materials for exploring physics in reduced dimensions and in complex geometries, as well as in one-dimensional nanowire devices.

© Springer International Publishing AG 2018
M. Hatti (ed.), *Artificial Intelligence in Renewable Energetic Systems*, Lecture Notes in Networks and Systems 35, https://doi.org/10.1007/978-3-319-73192-6_43

They are compatible with existing semiconductor technologies and can be tailored into unique axial and radial heterostructures. New generation of solar cells with high efficiency at economically viable costs is appeared as a cost-effective alternative to silicon-based photovoltaics, nanowires semiconductor-based solar cell, have attracted considerable attention recently and have shown promising developments for the next generation of solar cells [3]. The use of nanostructures represents a general approach to reducing both the cost and the size and to improve the photovoltaic efficiency. Solar cells must meet two criteria. First, they must absorb light, so they need a thick enough active material for maximum absorption. They also need to collect the electrons from the electron-hole pairs created by absorbed photons. The nanowires offer an interesting alternative. On the one hand, the nanowires can absorb on their length large amounts of light. On the other hand, the interest of nanowires is to decrease the distance between the n-p junction and the electron collector. P-n junction is achieved in two ways: the substrate is p-doped and nanowires n-doped, or the nanowires are p-type and covered by a conformal deposition of amorphous n-type silicon; which gives in both cases a p-n junction [1, 4].

Miniaturization has become the key to the optoelectronics development. The challenge is to continually increase the number of smaller components into infinitely small areas. The nanowires meet this requirement. Their use in devices requires their study for a good control of their dimensioning, their positioning, their distribution density, their shape. Their growth mechanisms should be well understood. Among the semiconductor nanowires manufactured, those based on silicon have many advantages. They are very promising in the manufacture of the new generation of photovoltaic cell. Indeed, the silicon industry is very advanced in manufacturing processes. This material benefits from several years of technology advance. It is the most used material in electronics because of its natural abundance and ease of machining. The silicon nanowires receive all the technical know-how acquired from the silicon processes customary in the industry. Our work falls within that framework [2, 4].

2   Materials and Methods

The simulation software Athena of Tcad-Silvaco provides general possibilities for the simulation of processes used in the semiconductor industry, such as diffusion, oxidation, ion implantation, etching, lithography, deposition processes. It allows fast and accurate simulations of all manufacturing steps used in optoelectronics and power components.

The solar cell under study is hetero-structure based on using of Germanium nanowires type p, on an oriented microcrystalline silicon substrate (111) and thin films of amorphous silicon (Fig. 1). A micromorph solar cell with the following structure was used in the simulation:

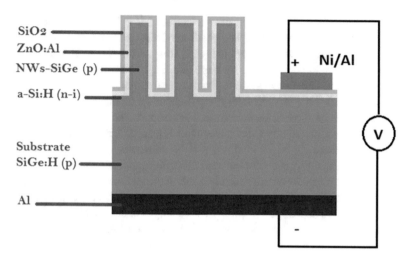

Fig. 1. Nanowires based-solar cell.

SiO2 (30 nm)/ZnO:Al (20 nm)/n-a-Si:H (10 nm)/i-a-Si:H(10 nm)/p-μc-NWs-SiGe:H(L = 2.5 μm-D = 100 nm)/p-μc-SiGe:H (2.5 μm)/Al (1 μm).

2.1 Substrate Preparation

The juxtaposition of materials with different electronic properties is an operating principle of current electronic components. The μc-SiGe:H hydrogenated microcrystalline silicon-germanium layers can be used as ultra-thin background absorbent layers in silicon-based multi-junction solar cells. Several studies have confirmed that a concentration of germanium in the layer greater than 20% leads to the formation of high defects, and therefore the variation in germanium is limited in this contribution to a maximum of about 22% [5].

The realization of silicon based components only requires the modulation of the conduction properties by introducing doping impurities in the silicon. Although the doping control is very technical, it is a routine activity of the microelectronics industry is well documented. The main dopants silicon thin-film is boron, p-type and arsenic n-type. Other dopants, phosphorus, gallium, Aluminum, etc., are used occasionally for very specific applications or even marginal. In this case, the oriented micro-crystalline silicon substrate (111) is p-doped with a low boron concentration ($Na = 10^{15}$ at/cm^3) and a logger 5 microns.

2.2 Nanowires Preparation

All the manufacture of semiconductor nanowires is made by several technological means. In classical microelectronics, two basic approaches, bottom-up and top-down approaches. The Bottom-up approach consists of assembling individual components such as atoms, molecules and aggregates to form more complex structures by synthesis (self-assembly). Upstream methods involve chemical synthesis of nanowires whose properties can be carefully controlled and adjusted during growth [6]. Downward approach is the method most used in microelectronic, it is based on the techniques of deposit, lithography and engraving. The position of the nanowires is well controlled on the substrate [8]. The nanowire procedure is as follows: A model is part of a resin (optical lithography, electron, near field, or by nano-printing) to form an etching mask. The pattern is then etched into a substrate to form nanostructures.

Lithographic is used as a preliminary step to define, after the deposition and removal of the resin from a hard mask formed from a set of pads. The etching of the substrate gives rise to the production of vertical nanowires [7]. In addition, the process is intrinsically costly and there are major challenges to be overcome before these technologies become high-throughput, cost-effective nanofil manufacturing [3, 7].

In our case, the top-down approach is used. After preparation of the substrate, p-doped microcrystalline silicon-germanium layer with a high concentration of boron ($Na = 10^{20}$ at/cm^3) is deposited by epitaxy with a length of 2.5 microns on the substrate. An etching is carried out to remove undesirable parts, so as to have very thin layers of 2.5 μm in length and a diameter of 100 nm on the substrate (Fig. 2).

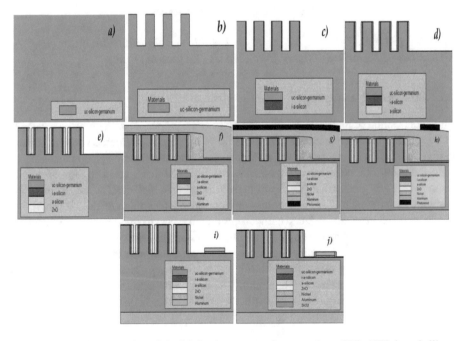

Fig. 2. Schematic illustration of the fabrication process for preparing a SiGe-NWs based silicon solar cell: (a) initial p-type silicon-germanium substrate, (b) etching Sige-NWs, (c) deposition and etching of an intrinsic amorphous silicon-germanium thin film, (d) deposition and etching of an intrinsic amorphous silicon-germanium thin film n-type emitter doped by phosphorus diffusion using solid dopant source, (e) deposition and etching of zinc oxide thin film, (f) deposition of Nickel and Aluminum layers, (g) deposition Photoresist layer, (h) layer anti-etchant pattern of front metal contacts, etching of Nickel and Aluminum layers, (i) strip of Photoresist layer, (j) making of metal contacts of NiAl and defined as a cathode in the front and making of metal contacts of Aluminum and defined as an anode on the back surface to make ohmic contacts.

2.3 Absorber Layer

Due to a high absorption coefficient of hydrogenated amorphous silicon (a-Si: H) in the visible range of the solar spectrum, a layer of one micrometer (1 μm) thick is sufficient to absorb 90% of the energy of usable sunlight [8]. After the etching of the nanowires on the substrate, a thin layer of the intrinsic amorphous silicon is deposited 100 nm thick. Then, the etching is done to remove an unwanted part. A deposition of a thin layer of amorphous silicon by n-type heavily doped epitaxy with arsenic (Nd = 10^{20} at/cm^3), with a thickness of 10 nm. Like previous steps, after engraving an epitaxy is done.

2.4 Electrodes

The Due to the low conductivity of the a-Si: H compared with c-Si, the addition of a conductive layer over its all surface is essential for current collection. In concern the illuminated face of the cell, the conductive transparent oxide (CTO) collects current, plays both the role of the conductive layer and the antireflection layer and consequently provides good contact with the metal electrodes (Fig. 1). In the absence of illumination of the back face of the cells, the conductive transparent oxide is replaced by a simple metal deposited on the amorphous layer [2, 8]. The Zinc oxide has very interesting electrical properties, its electronic structure has been widely studied, that the zinc oxide is a direct gap semiconductor at room temperature. The band gap of pure ZnO is about 3.37 eV, superior to conventional semiconductors, therefore ZnO has a gap about 367 nm which is therefore located in the near UV, ZnO is therefore transparent in the visible spectrum. In addition, the zinc oxide can be doped with a high concentration and it is possible to obtain high electrical conductivities of the order of 2×10^{-4} Ω cm^{-1}. These properties of transparency in visible light and good electrical conductivity are very interesting for applications requiring transparent electrodes [9, 10].

2.5 Antireflection Coating

The use of an antireflection coating (ARC) combined with silicon substrate textured, greatly reduces optical losses drastically affecting the power output of the solar cell by reducing the short-circuit current. Silicon oxide (SiO_2) and silicon nitride (Si_3N_4) have complementary properties. Silicon oxide with refractive index 1.46, has good transparency in the visible, good dielectric properties (Eg ~ 10 eV) and low mechanical stress but it is bad barrier to the diffusion of ions and dopants, the silicon nitride with 2.02 refractive index, is denser and therefore provides a better diffusion barrier. The final step is to deposit a thin layer of the oxide silicon with thickness of 100 nm and addition of electrodes. The structure is formed by the Silvaco simulator and shown in Fig. 3.

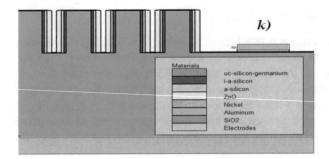

Fig. 3. Structure obtained after simulation

3   Results and Discussions

The amorphous nature of a-SiGe:H makes it easier to achieve conformal growth of thin films on rough or nanostructured surfaces than is the case with μc-Si:H. Combined with the relatively high deposition rate of hotwire chemical vapor deposition (HWCVD) as compared to PECVD, HWCVD a-SiGe:H is a feasible option for thin film silicon PV on textured light-scattering substrates. The recent developments in advanced light-scattering structures such as textures created with nano-imprint lithography and naturally grown ZnO nanorods, have motivated us to investigate whether a-SiGe:H HWCVD can be considered as active material for low cost thin film a comparison of a conventional and an unconventional method to improve the quality of forming electrodes on silicon solar cells [11].

The antireflection coating prevents the reflection of rays of sunshine and a loss of energy, as well creates a barrier into the connection zone, which increases the resistance between the electrode layer and the silicon substrate. The thickness of the deposited layer has an influence on the structure of the obtained electrode layer and resistance value of the resistance electrode [10].

In solar cells, the minority carrier lifetime, internal quantum efficiency and the solar cell efficiency are also affected by germanium despite although it is, electrically inactive in the silicon lattice [12]. Silicon-germanium nanowires were incorporated in single junction between hydrogenated amorphous Silicon thin films and microcrystalline Silicon-germanium substrate for forming heterostructure solar cell with an efficiency of 12.5%.

The simplest semiconductor junction that is used in solar cells for the separation of photogenerated charge carriers is the p-n junction, an interface between the p-type region and a n-type region of the semiconductor. Therefore, the basic semiconductor property of a material is the possibility of varying its conductivity by doping, must be demonstrated before the raw material can be considered a suitable candidate for solar cells, this was the case for amorphous silicon.

Indeed, the interest in germanium stems from the fact that it has a high mobility of carriers and a Bohr radius (11 nm) wider than that of silicon (by 4.3 nm). The quantum confinement effects are all the more important as the radius of the nanostructures is smaller than the Bohr radius. Thus, these effects are achieved for a nanowire diameter larger for germanium than for silicon (Fig. 4).

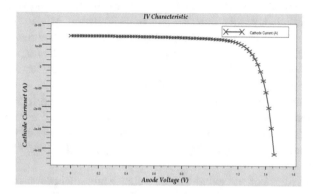

Fig. 4. IV Characteristic of solar cell obtained after simulation

The following figure (Fig. 5) presents the IV characteristic of the micromorph structure with and without germanium to see the effect of Ge incorporation in the structure.

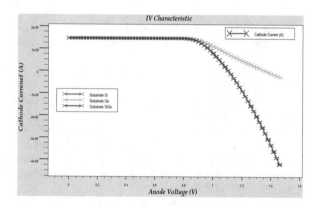

Fig. 5. IV Characteristic of solar cell obtained with and without Ge

From the figure (Fig. 5), we also see that the values of the short-circuit current and the efficiency are higher for a substrate of germanium compared with the substrate in Silicon and by using the substrate of SiGe the value of the short-circuit current is moved a little upwards which gives a slight increase in the efficiency value. We therefore come to the conclusion that the incorporation of germanium in a silicon substrate increases the efficiency of the cell, it reaches the 13.52%.

4   Conclusion

Our goal is to integrate nanowires in thin-film solar cells of silicon in order to improve the conversion efficiency and reduce the cost of manufacture. Miniaturization is one of the key process steps in the manufacture of high efficiency and low cost solar cells. The aim of the paper is the integration of Germanium on Silicon technology thin films solar cell. Based on the study results, the silicon substrate morphology has a huge influence on obtaining a minimal resistance value of electrodes.

However, the high defect density of low band gap a-Si:H films deposited by plasma enhanced chemical vapor deposition (PECVD), the occurrence of light induced degradation of devices with relatively thick absorber layers in particular and the high costs of Ge-containing source gasses, have thus far prevented a-SiGe:H in thin film silicon based solar cells to be viable.

Renewable sources of energy are the ideal choice and solar power is by far the most prominent energy source owing to its versatility, inexhaustible and environmental friendly features. The Si-NWs are an interesting alternative to conventional photovoltaic. The optimization of the cell parameters is therefore a compromise between optimization of the optical absorption and the optimization of the carriers' collection.

References

1. Peng, Q., Qin, Y.: ZnO Nanowires and Their Application for Solar Cells, China
2. Tian, J., Cao, G.: Semiconductor quantum dot-sensitized solar cells. Nano Rev. **4**, 22578 (2013)
3. Leprince-Wang, Y.: Élaboration et caractérisation des matériaux à l'échelle micro- et nanométrique, Ph.D. thesis, Marne La Vallée University (2003)
4. Tsakalakos, L., Balch, J., Fronheiser, J., Korevaar, B.A.: Silicon nanowire solar cells. Appl. Phys. Lett. **91**, 233117 (2007). USA
5. von Maydell, K., Grunewald, K., Kellermann, M., Sergeev, O., Klement, P., Reininghaus, N., Kilper, T.: Microcrystalline SiGe absorber layers in thin-film silicon solar cells. In: E-MRS Spring Meeting 2013 Symposium D - Advanced Inorganic Materials and Structures for Photovoltaics, 27–31 May 2013, Strasbourg, France, Energy Procedia, vol. 44, pp. 209–215 (2013)
6. Joyce, H.J., Gao, Q., Tan, H.H., Jagadish, C., Kim, Y., Zou, J., Smith, L.M., Jackson, H.E., Yarrison-Rice, J.M., Parkinson, P., Johnston, M.B.: III–V semiconductor nanowires for optoelectronic device applications. Prog. Quantum Electron. **35**, 23–75 (2011)
7. Boukhicha, R.: Croissance et caractérisation de nanofils de silicium et de germanium obtenus par dépôt chimique en phase vapeur sous ultravide. Ph.D. thesis, Paris-Sud 11 University, France (2011)
8. Laurent, K.: Etude expérimentale de la microstructure et des propriétés électriques et optiques de couches minces et de nanofils d'oxydes métalliques (d-Bi2O3 et ZnO) synthétisés par voie électrochimique. Université de Paris-Est (2008)
9. Dahou, F.Z.: Elaboration des couches minces originales pour réalisation de cellules solaires. Ph.D. thesis, Université d'Oran (2011)
10. Dobrzański, L.A., Musztyfaga, M., Drygała, A.: Final manufacturing process of front side metallisation on silicon solar cells using conventional and unconventional techniques. Stroj. Vestn. J. Mech. Eng. **59**(3), 175–182 (2013)

11. Veldhuizen, L.W., van der Werf, C.H.M., Kuang, Y., Bakker, N.J., Yun, S.J., Schropp, R.E.I.: Optimization of hydrogenated amorphous silicon germanium thin films and solar cells deposited by hot wire chemical vapor deposition. Thin Solid Films **595**, 226–230 (2015)
12. Emin, S., Singh, S.P., Han, L., Satoh, N., Islam, A.: Colloidal quantum dot solar cells. Elsevier Sol. Energy **85**, 1264–1282 (2011)

NPC Multilevel Inverters Advanced Conversion Technology in APF

Habiba Bellatreche[1(✉)] and Abdelhalim Tlemçani[2]

[1] Department of Electronic, University Saad Dahlab Blida, Blida, Algeria
bell-habiba@outlook.com
[2] Laboratory of Electrical Engineering and Automation,
University Yahia Fares, Medea, Algeria
h_tlemcani@yahoo.fr

Abstract. Multilevel inverters are most familiar with power converter applications. In this current paper, we discussed another possibility for DC/AC energy conversion. Converters based on current source inverters (CSI) configuration, take an important role in active power filtering in term of harmonics compensate and increasing system efficiency. To illustrate the design feasibility of the proposed structure of three and five level neutral point clamped (NPC) inverter, variable band hysteresis current regulation is used to determine the semiconductors switching signals. Association of shunt active power filter (APF) and multilevel inverter using type-2 fuzzy logic controller can seen in this analysis paper. All series of simulation results in MATLAB/Simulink environment are demonstrated and compared to illustrate the effectiveness of this scientific research.

Keywords: CS inverters · VS inverter · 3L-NPC · 5L-NPC · Shunt APF
Fuzzy logic controller · IT2 FLC · THD · HCC

1 Introduction

Power electronics is the technology behind the conversion of electrical energy from a source to the requirements of the end-user. Energy conversion equipment can be divided into four groups: AC/DC rectifiers, DC/DC converters, DC/AC inverters, and AC/AC transformers. Multilevel inverter configuration can be categorized into the Voltage Source Multilevel Inverter (VSI) and Current Source Multilevel Inverter (CSI) [6]. The difference between them is the type of power supply source or network.

A voltage source inverter is supplied by a DC voltage source to produces an AC output to the load; it is usually an AC/DC rectifier. VSI are widely used in industrial applications and renewable energy systems [3]. Their structure and control circuitry are simple. Whereas a current source inverter is supplied by a DC current source to delivers predetermine AC output; is usually an AC/DC rectifier with a large inductor to keep the current supply stable. The Multilevel (CSI) has the features of short circuit protection, lower voltage and current stress and less of total harmonic distortion (THD) in the output waveforms [7].

© Springer International Publishing AG 2018
M. Hatti (ed.), *Artificial Intelligence in Renewable Energetic Systems*, Lecture Notes
in Networks and Systems 35, https://doi.org/10.1007/978-3-319-73192-6_44

Nowadays, three commercial topologies of multilevel inverters exist: neutral point clamped (NPC), fling capacitors (FCs), and cascaded H-bridge (CHB). Neutral-point clamped inverter was presented in 1980 by Nabae. Because the NPC inverter effectively doubles the device voltage level without requiring precise voltage matching, the circuit topology prevailed in the 1980s [9]. Converters based on Multilevel CSI they are not well discussed in the literature and this research paper demonstrates how Multilevel (CSI) offers a better alternative solution in active power filtering. In these inverters, the dc-link capacitor voltage is kept constant using an intelligent controller which is type-2 fuzzy logic.

Variable band Hysteresis Current Control (HCC) has been a major research area in power electronics for many years. It still achieving a harmonic performance and offers several advantages in Multilevel (CSIs) compared with many variation of hysteresis proposed in literature [8].

In this paper, following three important points are discussed. In first three and five level CSI based on NPC topology is examined, then hysteresis approach is developed for generating the switching patterns. Special applications of these converters in active filtering of low voltage network using interval type-2 fuzzy controller is evaluated. All of the theoretical work has been confirmed and discussed in MATLAB/Simulink environment.

2   NPC Inverters

2.1   Three-Level Three-Phase NPC Inverter

Among the other multilevel topologies, 3L-NPC inverter has been widely used. The advantage is the lower current THD that reduces the filtering effort (less copper needed, lower losses in the filter). Using IGBTs and diodes with breakdown voltages (lower than the actual dc-link voltage) produce lower losses and the efficiency can be increased [1].

In Fig. 1 a simplified circuit diagram of 3L-NPC inverter. Phase (A) consists of (04) semiconductors IGBTs (T1, T2, and T3&T4), (04) antiparallel Free-Wheeling Diodes (D1, D2, D3&D4), (02) Clamping Diodes (D5&D6).

The dc-link capacitors divide the DC bus voltage into three levels namely $-Vdc/2$, 0 and $Vdc/2$. These voltage levels appear at the output of each leg K (phase A, B or C) of the inverter by appropriate switching of the power semiconductor devices. The middle point of the two capacitors is denoted as 'Z' which is the neutral point [2]. Table 1 represents the switching states for phase (A) of a 3L NPC. It can be obtained from (1).

$$\begin{cases} T_{K3} = \overline{T}_{K1} \\ T_{K4} = \overline{T}_{K2} \end{cases} \tag{1}$$

Table 1. Definition of switching states 3L-NPC inverter

Output level voltage	Device switching states (phase A)			
	T1	T2	T3	T4
$V_{dc}/2$	1	1	0	0
0	0	1	1	0
$-V_{dc}/2$	0	0	1	1

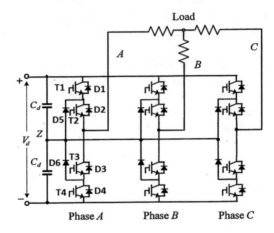

Fig. 1. Three level NPC inverter topology.

2.2   Five-Level Three-Phase NPC Inverter

The topology introduced in Fig. 2 is known as the five-level neutral-point-clamped (5L-NPC) converter [4]. The proposed 5L-NPC inverter consists of four (04) series connected capacitors C1, C2, C3 and C4.

The dc-link capacitors divide the DC bus voltage into five levels namely $+2V_{dc}$, $+V_{dc}$, 0, $-V_{dc}$ and $-2V_{dc}$. Each leg of the inverter has (08) bidirectional switches, (06) in series and (02) in parallel plus (02) diodes to get zero voltage. Each switch is composed of a transistor and a diode in antiparallel.

The Cd capacitors are shared by all three phases. The voltage across each capacitor is VDC/4 and each device voltage stress will be limited to one capacitor voltage level Vdc through clamping diodes. Switching states in Table 2 for leg (A) of 5L-NPC inverter can be obtained from (2).

$$\begin{cases} T_6 = \overline{T}_1 \\ T_4 = \overline{T}_2 \\ T_5 = \overline{T}_3 \\ T_7 = \overline{T}_1.T_2.T_3 \\ T_8 = \overline{T}_6.T_4.T_5 \end{cases} \tag{2}$$

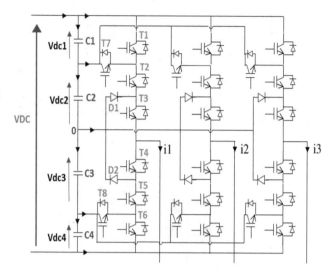

Fig. 2. Five level NPC inverter topology

Table 2. Definition of switching states 5L-NPC inverter

Output level voltage	Device switching states (phase A)		
	T1	T2	T3
$2V_{dc}$	1	1	1
V_{dc}	0	1	1
0	0	0	1
$-V_{dc}$	1	0	0
$-2V_{dc}$	0	0	0

3   Shunt Active Power Configuration

Shunt APF is used as prototype exposed in Fig. 3. It operates as an injecting current harmonic source, consequently it has the capability to insert equal and opposite harmonics current onto the power system.

It has advantages such as: small size, tuning is easy and accurate, elimination of harmonics currents, compensating reactive power, correction of power factor and rebalancing currents of nonlinear load [5].

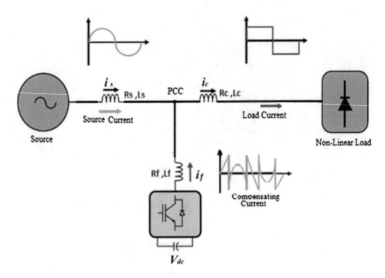

Fig. 3. Shunt APF Basic configuration

4 Harmonic Current Identification

The IEEE working groups of the Power Engineering Society and the Industrial Applications Society prepared recommended guidelines for power quality that the utility must supply and the industrial user can inject back onto the power distribution system. Table 3 of IEEE 519-1992 defines the voltage distortion limits that can be reflected back onto the utility distribution system [12].

Table 3. Voltage distortion limits.

Bus voltage at PCC	Total voltage distortion THD (%)
69 kV and below	5.0
69.001 kV through 161 kV	2.5
161.001 kV and above	1.5

Several harmonic current identification methods have been investigated by many researchers such as synchronous detection, average mode, instantaneous active and reactive powers method [10]. The main task of harmonic currents compensation is to determine the harmonic current references to be generated by the active filter, after it controlled to generate the identified harmonic currents and inject them in the main electrical power grid.

Synchronous reference frame is one of popular methods harmonic component extraction. The transformation from 3 axis to 2 axis is principally used, the PLL circuit is providing sin (wt) and cos (wt) for synchronization. To obtain the reference harmonic current, first the load current is measured then high pass filter separated the harmonic components ($\tilde{I}_{Ld}$, $\tilde{I}_{Lq}$) from the dq load currents (I_{Ld}, I_{Lq}). The load currents are compared to the fundamental component and the error is the reference harmonics signals [11]. Figure 4 demonstrates the procedure of SRF.

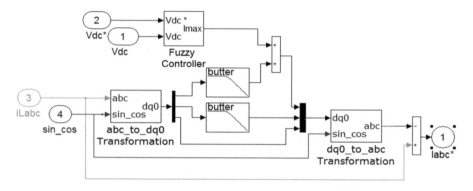

Fig. 4. Synchronous reference frame block diagram.

5 Switching Control Strategy

For reason that it is characterized by unconditioned stability, very fast response and high quality of precision, the new variable hysteresis band significantly improves the output voltage harmonic performance compared to a fixed-band HCC.

In Fig. 5, the desired currents and real compensating currents are compared to create the errors values.

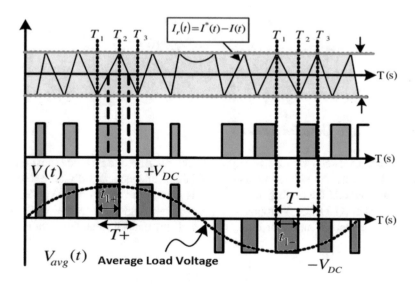

Fig. 5. Variable band HCC.

Current errors signals become the inputs to the hysteresis block control in order to drive the inverter in manner to minimize the error which in turn controls all the system. For 3L-NPC inverter, HCC algorithm is given in (3).

5.1   HCC for 3L-NPC Inverter

$$\begin{cases} \begin{cases} \begin{cases} [(e_k \geq \Delta I)] \wedge [(e_k \leq 2\Delta I)] \\ \vee \\ [(e_k \leq -\Delta I)] \wedge [(e_k \geq -\Delta I)] \end{cases} \Rightarrow T_{K1} = 1 \wedge T_{K2} = 0 \\ [(e_k > 2\Delta I)] \Rightarrow T_{K1} = 1 \wedge T_{K2} = 0 \\ [(e_k < -2\Delta I)] \Rightarrow T_{K1} = 1 \wedge T_{K2} = 1 \end{cases} \tag{3}$$

5.2   HCC for 5L-NPC Inverter

$$\begin{cases} e_k > 2\Delta I \Rightarrow T_1 = 0, T_2 = 0, T_3 = 0 \\ \Delta I < e_k < 2\Delta I \Rightarrow T_1 = 1, T_2 = 0, T_3 = 0 \\ -\Delta I < e_k < \Delta I \Rightarrow T_1 = 0, T_2 = 0, T_3 = 1 \\ -2\Delta I < e_k < -\Delta I \Rightarrow T_1 = 0, T_2 = 1, T_3 = 1 \\ e_k < -2\Delta I \Rightarrow T_1 = 0, T_2 = 0, T_3 = 1 \end{cases} \tag{4}$$

Where: ΔI is the width of hysteresis tolerance band.
And

$$\begin{cases} e_{k=(a,b,c)} = I_{k\_ref} - I_k \\ \begin{cases} I_{k\_upper} = I_{k\_ref} + \Delta I \\ I_{k\_lower} = I_{k\_ref} - \Delta I \end{cases} \end{cases} \tag{5}$$

6   DC Voltage Control Loop

Capacitor voltage is considered as a supply source for APF and its value must be kept constant. To ensure the voltage fluctuations of the semiconductors do not exceed the limits prescribed and the performance of filter is maintained an Interval Type-2 Fuzzy Logic Controller (IT2 FLC) is proposed. It is based on a linguistic description, does not require a mathematical model of the system, used membership functions with values varying between 0 and 1 [13] and overcomes the problem of uncertainty and non-linearity of Multi level NPC based shunt APF. IT2 FLC is defined as [14].

$$\tilde{A} = \int_{x \in X} \int_{u \in J_x} 1/(x, u), J_x \subseteq [0, 1] \tag{6}$$

Note that J_x is an interval set:

$$J_x = \left[\underline{\mu_{\tilde{A}}}(x), \overline{\mu}_{\tilde{A}}(x) \right] \tag{7}$$

Where:

$$\underline{\mu_{\tilde{A}}}(x) = \min(J_x), \forall x \in X \tag{8}$$

$$\overline{\mu}_{\tilde{A}}(x) = \max(J_x), \forall x \in X \tag{9}$$

The FOU ($\tilde{A}$) can also be expressed as

$$\text{FOU}(\tilde{A}) = \bigcup_{\forall x \in X} J_x \tag{10}$$

IT2 FLC is implemented as exposed in Fig. 6. It needs 2 inputs error **e(k)** and variation in error **Δe(k)**. The output after a limit is considered as the amplitude of the reference current $\mathbf{I_{max}(k)}$ takes care of the active power demand of load and the losses in the system.

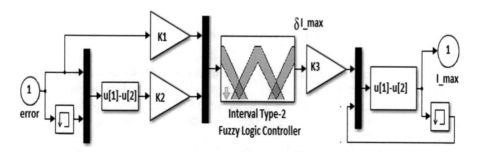

Fig. 6. IT2 FLC block diagram.

IT2 FLC is defined as (11)

$$\begin{cases} e(k) = V_{dc}^*(k) - V_{dc}(k) \\ \Delta e(k) = e(k) - e(k-1) \\ I_{\max}(k) = I_{\max}(k-1) + \delta I_{\max}(k) \end{cases} \tag{11}$$

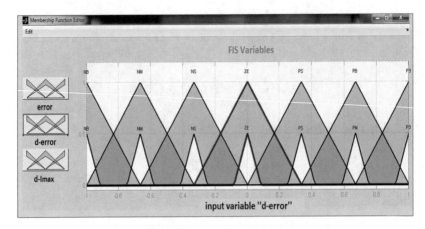

Fig. 7. Membership functions for e, Δe & δImax

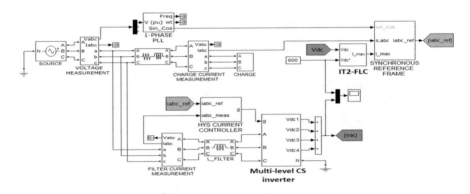

Fig. 8. Control strategy circuit diagram

Seven triangular membership functions are used for the implemented IT2 FLC and seven fuzzy levels or sets are chosen as: NB (negative big), NM (negative medium), NS (negative small), ZE (zero), PS (positive small), PM (positive medium), and PB (positive big) [15] to convert these numerical variables into linguistic variables as in Fig. 7.

The Control strategy circuit is clearly defined in Fig. 8. The dc-link capacitor voltages whether in 3L-NPC or 5L-NPC are equals and utilized for the generation of I_{max}; it is estimated by regulating the dc-bus capacitor voltage of multilevel inverter.

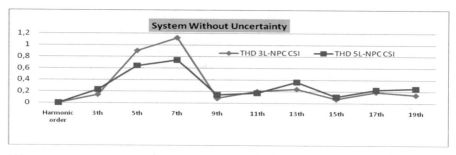

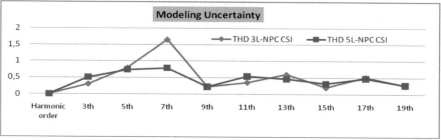

Fig. 9. Comparative between THD Graphs of current source for 3L-NPC & 5L-NPC inverter

7   Simulation Results

This part presents the details of the simulation in MATLAB/SIMULINK and Excel environment. Following parameters are considered for the study.

- Network (Vs = 220 V, Fs = 50 Hz)
- Non Linear Load ($R_L = 5\Omega$, $L_L = 8.10^{-3}H$)
- Shunt APF (Cf = 2200.10^{-6F} F, Lf = $1.10^{-3}H$)
- IT2 FLC & HCC (Vdc* = 600 V, Triangular MFs, 49 rules base, $\Delta I = 0.1A$).

In order to evaluate the feasibilities of the used multilevel inverters (3L-NPC 15L-NPC) and its performance in active power filtering, we analyzed for comparison the THD of current source for each inverter, in addition we investigated in proposed IT2-FLC performance to maintain and keep the dc-link capacitor constant.

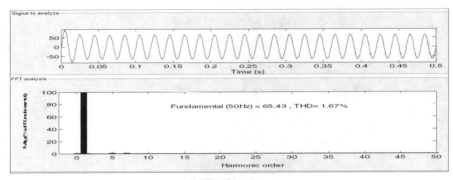

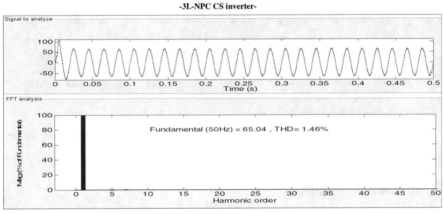

Fig. 10. Current source phase (A) and it harmonics spectrum

- Using 3L-NPC or 5L-NPC CS inverter, the THD can be greatly reduced in Fig. 9 lowers than 5%; limit imposed by the IEEE-519 standards.
- Current source waveform is well shaped and approximately sinusoidal after filtering.
- The dc-link capacitor voltage is well regulated and maintained at a constant value of 600 V in shorter time with a very limited fluctuation (Figs. 10 and 11).

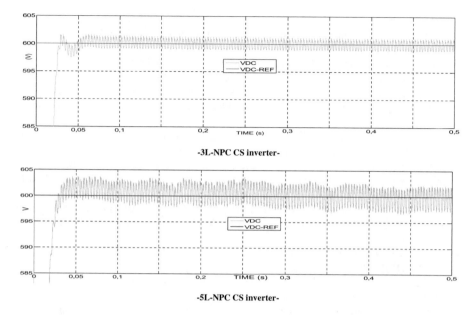

Fig. 11. Capacitor voltage and it reference

8 Conclusion

In this evaluate study, an important assessment of current source multilevel inverter has been presented and tested in presence of uncertainties. Multilevel inverters have advantages because of their poor total harmonic distortion and the quality of the current source is improved as the number of levels increase. Thereafter we focus on the robustness analysis of IT2 FLC and it concluded that it is more robust because of the presence additional degree of freedom provides by footprint of uncertainty, plus it is able to cope with non-linearity and uncertainty.

References

1. Ingo, S.: Note application, SEMIKRON: Innovation and service (2015)
2. Lin, B.R., Yang, T.Y.: Three-level voltage-source inverter for shunt active filter. IEE Proc. Electr. Power Appl. **151**(6), 744–751 (2004)
3. Luo, F.L., Ye, H.: Advanced Conversion Technologies, Power Electronics. Taylor & Francis, Boca Raton (2010)
4. Abdelkrim, T., El Madjid, B., Karima, B.: Feedback control of three-level PWM rectifier: application to the stabilization of DC voltages of five-level NPC APF. Arch. Control Sci. **20**(3), 317–339 (2010)
5. Mikkili, S., Panda, A.K.: PI and fuzzy logic controller based 3-phase 4-wire shunt active filter for mitigation of current harmonics with Id-Iq control strategy. J. Power Electron. **11**(6), 914–921 (2011)

6. Barbosa, P.G., Braga, H.A.C., Teixeira, E.C.: Boost current multilevel inverter and its application on single phase grid connected photovoltaic system. IEEE Trans. Power Electron. **21**(4), 1116–1124 (2006)
7. Suroso, T.N.: A single-phase multilevel current source converter using H-bridge and DC current modules. Int. J. Power Electron. Drive Syst. **4**(2), 165–172 (2014)
8. Davoodnezhad, R., Holmes, D.G.: Three-level hysteresis current regulation for a three phase neutral point clamped inverter. In: Power Electronics and Motion Control Conference, pp. 47–52 (2012)
9. Luo, F.L., Ye, H.: Multilevel DC/AC inverters in advanced DC/AC inverters: applications in renewable energy. Taylor & Francis Group, LLC (2013)
10. Benyettou, L., Benslimane, T., Bentata, K., Abdelkhalek, O.: Open transistor faults characterization novel method for cascaded H-Bridge five-level three-phase shunt APF. AMSE J. Ser. Modell. A **88**(1), 53–70 (2015)
11. Pigazo, A.: A recursive park to improve the performance of synchronous reference frame controllers in shunt active power filters. IEEE Trans. Power Electron. **24**(9), 2065–2075 (2009)
12. Shah, N.: Harmonics in Power Systems: Causes, Effects and Control. Siemens Industry, Inc, DRWP-DRIVE-0613, June 2013
13. Rathika, P., Devaraj, D.: Fuzzy logic-based approach for adaptive hysteresis band DC voltage control in shunt active. Int. J. Comput. Electr. Eng. **2**(3), 404 (2010)
14. Mendel, J.M.: Type-2 fuzzy sets and systems: an overview. IEEE Comput. Intell. Mag. **2**, 20–29 (2007)
15. Mikkili, S., Panda, A.K.: Simulation and real-time implementation of shunt active filter Id–Iq control strategy for mitigation of harmonics with different Fuzzy membership functions. IET Power Electron. **5**(9), 1856–1872 (2012)

Optimization Study of Hybrid Renewable Energy System in Autonomous Site

Samia Saib[1(✉)], Ahmed Gherbi[1], and Ramazan Bayindir[2]

[1] Laboratory of Automatics, (LAS), Department of Electrical Engineering,
Faculty of Technology, Setif 1 University, Setif, Algeria
saib_soumia@yahoo.fr, gherbi_a@gmail.com
[2] Department of Electrical and Electronics Engineering, Faculty of Technology,
Gazi University, Ankara, Turkey
bayindir@gazi.edu.tr

Abstract. This study aims to investigate the optimal sizing and the reliability of the (PV/wind/battery) hybrid energy system to supply a stand-alone site during their 20 years of lifetime. In this case, an Equivalent Loss Factor (ELF) as reliability index is used and applied in this work. A metaheuristic method such as particle swarm optimization (PSO) is used to solve the optimization problem and compared with other new and improved PSO Algorithms as Fast Convergence and Time Varying Acceleration. A comparative analysis is performed between this study and previous works for a standalone system with a hybrid energy system considering the economic cost using PSO method. Simulation results using Matlab software prove the reliability of the hybrid PV/wind/battery system during its lifetime and the improved PSO method by FC-TVAC algorithm has shown its performance in the optimisation study.

Keywords: Hybrid renewable energy system · Storage energy
Optimization · Reliability · Improved PSO

1 Introduction

Actually, renewable energy is used more than the conventional energy since its environmental and economic benefits. It has gained more applications in several areas, predominantly photovoltaic (PV) and wind systems connected to the grid or in a standalone mode. The use of the battery energy system is very important mostly for the autonomous system with these hybrid systems to store energy and to feed the load when the demand will not be sufficient. In this work, a study has been carried about the (PV/wind) hybrid renewable energy systems (HRES) with energy storage for supplying a load in a remote area, with the objective to obtain an optimal sizing with a minimum cost, taking into account the reliability of the system during its lifetime. Several works available in the literature have presented various studies about the optimization of the hybrid energy system with renewable and conventional sources, in autonomous or an island site using various metaheuristic methods and simulated through different softwares considering the reliability of the system as presented in [1–5]. In reference [6], the authors presented an optimization study with a sensitivity analysis of PV/wind/

© Springer International Publishing AG 2018
M. Hatti (ed.), *Artificial Intelligence in Renewable Energetic Systems*, Lecture Notes
in Networks and Systems 35, https://doi.org/10.1007/978-3-319-73192-6_45

diesel/battery hybrid energy system supplying a load in the site of Iran. They have used Homer software to assure the reliability and the cost of the studied system. The authors in [7, 8] offered a hybrid energy system as photovoltaic panels, wind turbine, diesel generator and battery storage energy to feed an autonomous system. The results of the proposed works, have determined that the PV/diesel hybrid system gets a good solution regarding the economic efficiency and pollution. And have shown that the hybrid system is more economical than the PV system, wind turbine and storage system or diesel alone. After this presentation of some previous works concerning the optimization of hybrid energy system in autonomous site about the sizing, cost and reliability. A new study has been developed by implementing an improved PSO technique in the optimization problem for the hybrid energy system supply a stand-alone system. A comparative study has been achieved with other previous works regarding the economic cost. In this present work, a PSO approach is applied in the optimization process and compared with other improved PSO algorithm in order to determine the optimal size of a solar panel, wind turbine and battery system with a minimum cost in considering the system reliability. Simulation results have been performed under Matlab environment. Data of solar irradiation and wind speed are extracted from the site of the renewable energy development center of Bouzaréah Algiers (CDER) located at (36°48′N, 3°1′E, 345 m), north of Algeria during one year.

2   Modeling of the Hybrid System

The hybrid system studied consists of a PV panel, wind turbine and battery energy storage connected to a DC bus. It supplies an AC load via a DC/AC inverter as illustrated in Fig. 1.

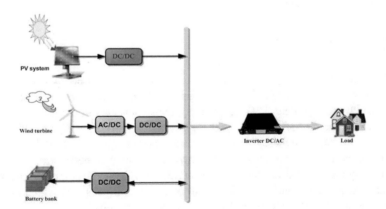

Fig. 1. The hybrid energy system connected to the autonomous site.

2.1   Photovoltaic System

The output power of the PV system can be determined by the following equation [1]:

$$P_{pv} = \frac{G}{G_0} P_{pv} \eta_{pv} \tag{1}$$

Where G is the solar radiation (W/m^2), G_0 is the solar radiation at standard test condition equal to 1000 W/m^2, P_{pv} is the rated power of the PV panel in (kW), η_{pv} is efficiency of photovoltaic DC/DC converter.

2.2   Wind System

The appropriate model of wind turbine is related to the type and the generated power. The output power of the wind system is given by the following expression [9]:

$$P_w = \begin{cases} 0 & \\ P_n \left(\frac{v - v_{ci}}{v_n - v_{ci}} \right) & v_{ci} \leq v \leq v_n \\ P_n & v_n \leq v \leq v_{max} \end{cases} \tag{2}$$

With P_n is the nominal power, v_{ci}, v_{max}, v are respectively lower-cut, upper-cut and predestined wind speeds.

2.3   Battery System

A battery energy storage system (BESS) is included in this study for managing the energy flow between the hybrid system and the load. For supplying the electrical energy to the load by the hybrid renewable energy source, an appropriate power management is required and can be summarized in the following steps:

1. When the power generated by the (PV/Wind) energy system is more than the power required, the excess energy will be stored in the batteries.
2. When there is any shortage of the power produced by the (PV/Wind) hybrid system, the stored energy in the batteries will be discharged and used to supply the load.

The stored energy in the batteries should be greater than the minimum energy $E_{bat.min}$ and less than the maximum energy $E_{bat.max}$.

3   Economic Criterion

The present economic cost (PEC) of the hybrid energy system consists of the capital, replacement and maintenance costs for the 20 years of the project time. The system cost is given by the following equation [10]:

$$PEC = \sum_{j}^{n} Q_j \times \left(PP_j + Y \times RO_j + W \times MA_j \right) \tag{3}$$

Where n is the number or capacity of the renewable systems, Q is the number of units, PP, RO, MA are the purchased, operation and maintenance costs of the components. Y, W considered to be constant to convert the replacing and maintenance cost to present cost which are in the next formulas [11, 12]:

$$Y_i = \sum_{k=1}^{m_1} \frac{1}{(1+K_r)^{l \times m_2}} \tag{4}$$

$$W = \frac{(1+K_r)^R - 1}{K_r(1+K_r)^R} \tag{5}$$

Where: m_1, m_2 are respectively the numbers of lifespan and replacement of HRES for life span of project, K_r is real interest rate and is equal to 6%. In this study, an Equivalent Loss Factor (ELF) index is taken into account with the purpose to ensure the reliability of the system. ELF is estimated by [13]:

$$ELF = \frac{1}{h} \sum_{i=1}^{h} \frac{L(i)}{d(i)} \tag{6}$$

Where h is the hourly time during one year, $L(i)$ and $d(i)$ are respectively the total loss and demand load. $ELF \leq 0.01$ is takes into account in the optimization formulation. The optimization problem should be satisfied the following constraints:

$$E_{bat,min} \leq E_{bat}(t) \leq E_{bat,max}, \; N_i > \; = 0, \; ELF < \; = 0.01.$$

4   Optimization Technique

A particle swarm optimization is performed in the optimization study. In the PSO metaheuristic method, each solution of the optimization problem is needed a particle which is moving in the search area. All the Particles contain memory and two random factors. The PSO algorithm uses a number of particles, which move through the solution space, and are evaluated according to some fitness criterion after each step to get an optimum. Each particle is presented by its position x_i and velocity v_i which achieved using the following equations:

$$v_i^{k+1} = \omega.v_i^k + c_1.rand_1\left(P_{best} - x_i^k\right) + c_2.rand_2\left(g_{best} - x_i^k\right) \tag{7}$$

$$x_i^{k+1} = x_i^k + v_i^{k+1} \tag{8}$$

Where v_i^k and x_i^k are the particle's velocity and position, c_1 and c_1 are weight values, p_{best} is the best solution attained but far from the particle, g_{best} is the accepted solution which the position defined by the neighbors of particle, r_1 and r_2 are random values in the range 0–1, w is inertia weight. The formula of this inertia is given by the next expression:

$$\omega = \omega_{max} - \left(\frac{\omega_{max} - \omega_{min}}{iter_{max}}\right) iter \tag{9}$$

With $iter$ is the current iteration and $iter_{max}$ is the maximum number of iterations.

4.1 Fast Convergence Based PSO

This technique is founded on the balance of the diversity of position of the particle by developing a newly parameter, which is the mean particle size (Pms). A study has been made with the next improvements. The new third parameter of ith particles expression and the velocity are demonstrated through the following expressions [14]:

$$Pms_i = (a_{i_1} + a_{i_2} + \ldots\ldots\ldots + a_{i_D})/d \tag{10}$$

With d is the size of the swarm in the search space.

$$V_i^{k+1} = \omega.V_i^k + c_1.rand_1\left(P_{best,i} - X_i^k\right) + c_2.rand_2\left(G_{best} - X_i^k\right) + c_3.rand_3\left(Pmd_i - X_i^k\right) \tag{11}$$

Where c_3 is the average factor, and $(c_1 + c_2 + c_3) \geq 4$.

4.2 Time Varying Acceleration Coefficients Based PSO

The concept of Time Varying Acceleration Coefficients (TVAC) based PSO is to improve and converge towards the global optima in the optimization formulation in the swarm. This is reached by varying the acceleration coefficients with time, in the case of the cognitive component is decreased and the social component is rising in the space of the swarm. The accelerations coefficients are given by:

$$\begin{cases} c_1 = (c_{1_s} - c_{1e})\dfrac{iter}{iter_{max}} + c_{1e} \\ c_2 = (c_{2s} - c_{2e})\dfrac{iter}{iter_{max}} + c_{2e} \end{cases} \tag{12}$$

Where: $c_{1_s}, c_{1e}, c_{2s}, c_{2e}$ are social acceleration coefficients, with starting and end cognitive values. The inertial weight is formulated as follows [15]:

$$\omega = (\omega_{max} - \omega_{min})\left(\frac{iter_{max} - iter}{iter_{max}}\right) + \omega_{min} \tag{13}$$

5 Simulation Results and Discussion

Simulation results, are carried out under Matlab environment and the load energy is required 500 kW during this project. Profiles of wind speed and solar irradiation used in this study during one year are illustrated in Figs. 2 and 3, and it's appeared that are

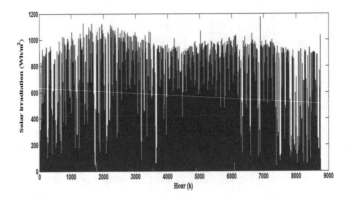

Fig. 2. Hourly solar radiation of one year.

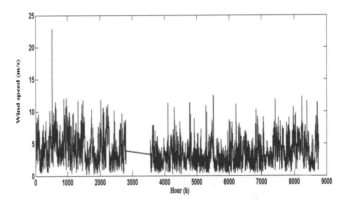

Fig. 3. Hourly wind speed of one year.

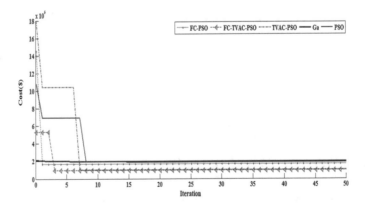

Fig. 4. Convergence of metaheuristic methods.

Table 1. Optimal sizing of the hybrid energy system.

Methods	PV system	Wind system	Battery system	Cost (MUS$)
PSO	105	549	677	0.186
FCPSO	364	315	209	0.164
TVACPSO	318	395	315	0.101
FC-TVAC-PSO	294	573	632	0.0965
GA	267	584	682	0.205

very satisfied. The result obtained shown in the Table 1, The sizing of the hybrid energy system with the system cost under various suggested metaheuristic methods. As can be noted, the improved FC-TVAC_PSO algorithm given an optimal solution compared the other methods. Figure 4 illustrates the convergence of the metaheuristic methods applied in the optimization process, where the convergence has been proven by the improved PSO algorithms and it's faster than the PSO and genetic algorithms. The reliability of the (PV/wind/battery) configuration system is considered by estimating the ELF index. In this study, this factor reaches 0.0001 during one year and it's lower than the maximum ELF index (equals to 0.01).

A comparative study between previous works and this study about the economic cost of the standalone system related to the (PV/wind/battery) hybrid energy system is achieved through the PSO algorithm, taking into account the same parameters and the number of population, performed in different region, site and the system prices. The results obtained in the Table 2 illustrate the difference between both studies and it can be noted that the system cost is lower and it is considered better compared with other works.

Table 2. Comparison of the economic costs.

References	Previous work Cost (MUS$)	Simulation results Cost (MUS$)
[1]	1.29769	0.186
[9]	17.744	0.186
[16]	2.47	0.186
[17]	3.1370	0.186

6 Conclusion

In this paper, an optimization study has been dealt for a hybrid energy system including photovoltaic system, wind turbine and storage energy system with battery bank to feed a load in the remote area. Taking into account the system reliability, the ELF index has been evaluated and has reached 0.0001, which is less than the maximum ELF required in this work. Particle Swarm Optimization metaheuristic method is applied and compared with other improved PSO algorithm, in the objective to obtain an optimum solution about the cost and sizing of the system. The best economic cost and the convergence are selected by the FC-TVAC-PSO method and are better compared with the other methods. Also, the studied system has proved its efficiency concerning the economic cost than a previous work. It can be concluded that the (PV/wind/battery) hybrid energy system is more economical and reliable during its lifetime.

References

1. Bashir, M., Sadeh, J.: Size optimization of new hybrid stand-alone renewable energy system considering a reliability index. In: Environment and Electrical Engineering International Conference EEEIC 2011, Venice, Italy, 18–25 May 2012 (2012)
2. Wu, Y.K., Shih, M.C.: Review of the optimal design on a hybrid renewable energy system. In: MATEC Web of Conferences, ACPEE 2016, vol. 55, article no. 06001 (2016)
3. Semaoui, S., Hadj Arab, A., Bacha, S., Azoui, B.: Optimal sizing of a stand-alone photovoltaic system with energy management in isolated areas. Energy Procedia 36, 358–368 (2013)
4. Ganesan, E., Dash, S.S., Samanta, C.: Modeling, control, and power management for a grid-integrated photovoltaic, fuel cell, and wind hybrid system. Turk. J. Elec. Eng. Comp. Sci. 24, 4804–4823 (2016)
5. Salehin, S., Rahman, M.M., Islam, A.K.M.S.: Techno-economic feasibility study of a solar PV-Diesel system for applications in northern part of Bangladesh. Int. J. Renew. Energy. Res. 5(4), 1221–1229 (2015)
6. Hatam, A., Hossein, K.K.: Optimization and sensitivity analysis of a hybrid system for a reliable load supply in Kish_IRAN. I. J. Adv. Renew. Energy Res. 1(4), 33–41 (2012)
7. Hassiba, Z., Cherif, L., Ali, M.: Optimal operational strategy of hybrid renewable energy system for rural electrification of a remote Algeria. Energy Procedia 36, 1060–1069 (2013)
8. Abdelhamid, K., Rachid, I.: Techno-economic optimization of hybrid photovoltaic/wind/diesel/battery generation in a stand-alone power system. Sol. Energy 103, 171–182 (2014)
9. Fatemeh, J., Gholam, H.R.: Optimum design of a hybrid renewable energy system. In: Renewable Energy - Trends and Applications, 9 November 2011, pp. 232–250. InTech, Croatia (2011)
10. Saber, A.N., Iraj, F., Mohamad, J.H., Peyman, N.: Optimization of hybrid PV/Wind/FC system using hunting search algorithm. In: 6th Iranian Conference on Electrical and Electronics Engineering ICEEE 2014, 19–21 August 2014, Iran, pp. 1–11 (2014)
11. Saeid Lotfi, T., Farid Lotfi, T., Mohammad, G.: Optimal design of a hybrid solar - wind-diesel power system for rural electrification using imperialist competitive algorithm. I. J. Renew. Energy Res. 3(1), 404–411 (2013)
12. Gangwar, S., Bhanja, D., Biswas, A.: Cost reliability and sensitivity of a stand-alone hybrid renewable energy system–a case study on a lecture building with low load factor. J. Renew. Sustain. Energy 7 (2015)
13. Garcia, R.S., Weisser, D.: A wind–diesel system with hydrogen storage joint optimization of design and dispatch. J. Renew. Energy 31, 2296–2320 (2006)
14. Amaresh, S., Sushanta, K.P., Sabyasachi, P.: Fast convergence particle swarm optimization for functions optimization. J. Procedia Technol. 4, 319–324 (2012)
15. Krishna, T.C., Manjaree, P., Laxmi, S.: Particle swarm optimization with time varying acceleration coefficients for non-convex economic power dispatch. Electr. Power Energy Syst. 31, 249–257 (2009)
16. Saber, A.N., Mahdi, H.: Economic designing of PV/FC/Wind hybrid system considering components availability. I. J. Modern Educ. Comput. Sci. 7, 69–77 (2013)
17. Saber, A.N., Abbas, R.G., Saheb, K.: PV/FC/Wind hybrid system optimal sizing using PSO modified algorithm. Bul. Inst. Polit. Iasi, t. LVIII (LXII) 4, 51–66 (2012)

An Improved Technique Based on PSO to Estimate the Parameters of the Solar Cell and Photovoltaic Module

Z. Amokrane[1(✉)], M. Haddadi[1], and N. Ould Cherchali[2]

[1] Department Electronique, Laboratoire des Dispositifs de Communication et de Conversion Photovoltaïque, Ecole Nationale Polytechnique, El Harrach, Algeria
amokrane_zouhir@yahoo.fr
[2] Laboratoires de Recherche en Électrotechnique et en Automatique (LREA), Université Yahia-Fares de Médéa, Médéa, Algeria
nocherchali@yahoo.fr

Abstract. Solar cell/module modeling involves the formulation of the non-linear current versus voltage (I-V) curve. Determination of parameters plays an important role in solar cell/module modeling. This paper presents an application of the improved PSO search method and Particle Swarm Optimization technique for identifying the unknown parameters of solar cell and photovoltaic module models, namely, the series resistance, shunt resistance, generated photocurrent, saturation current, and ideality factor that govern the current-voltage relationship of a solar cell/module. For the confirmation of accuracy of the proposed method, a measurement data of 57 mm diameter commercial (R.T.C. France) silicon solar cell and a module consisting of 36 polycrystalline silicon cells (Photowatt-PWP 201) has been selected and the best optimal value of each parameter has been obtained using Improved PSO. Comparative study among different parameter estimation techniques is presented to demonstrate to verify the accurateness and the effectiveness of the proposed approach.

Keywords: Photovoltaic · Artificial intelligence · Solar cell
Parameter estimation · Improved PSO

1 Introduction

Renewable energy sources are proving to be economical and clean due to efficient cells and new, advanced technology. Solar energy is obviously advantageous compared to any other source of energy for the environment and, therefore, it becomes popular. This growing demand urges simulator developers to model photovoltaic cells, so they hurt design engineers to simulate and check their solar energy systems. The modeling of the PV cell/module involves the identification of equivalent circuit parameters. Based on the type of the photovoltaic model circuit, circuit parameters may vary. To extract the exact parameters of the photovoltaic model, there are different methods: analytical methods, optimization method and iterative method. Among these methods, the optimization methods are very efficient to extract the parameters of the photovoltaic model.

© Springer International Publishing AG 2018
M. Hatti (ed.), *Artificial Intelligence in Renewable Energetic Systems*, Lecture Notes in Networks and Systems 35, https://doi.org/10.1007/978-3-319-73192-6_46

Today's researchers use the PV model with a single diode for simulation purposes. Particle Swarm Optimization (PSO) and genetic algorithms (GA) have proven to be very effective in solving a variety of complex problems. However, they have strengths and weaknesses. The extracting of the parameters is performed for the single diode model by Newton Raphson's iterative method in [4] and by GA in [1–3]. The modeling of (PV) is performed by MATLAB/SIMULINK [5–11]. For solar array model, the evaluation of the characteristics and the parameters extracted by experimental analysis were performed in [3]. The extraction of the parameters of the single diode PV model is performed by PSO and GA in [12]. However, the above approaches do not make it possible to propose a reliable single diode model as a function of consistency, accuracy and computation time. This paper presents a new PSO algorithm improved by a constraint factor.

2   Solar Cell and PV Module Modelling

Although various models have been developed in the literature to describe the behavior of solar cells and PV module, only one model is used practically, i.e. the single diode model and the dual diode model. This model is simple, easy to solve and suitable for electrical engineering applications. In sub-sections A and B, this model is presented briefly.

2.1   Solar Cell (Single-Diode Model)

Single diode model can be represented using equivalent electrical circuit as shown in Fig. 1. It consists of a photo-generated current source Iph, a series resistance Rs, a shunt resistance Rsh, and an anti-parallel diode. This model is known as single diode model and is widely used to represent the solar cell behavior [16]. The output current I, is given as:

$$I = I_{ph} - I_D - I_{sh} \tag{1}$$

Considering Shockley equation for the diode current, Id, and substituting the current of the shunt resistor, Ish, Eq. (1) is rewritten as:

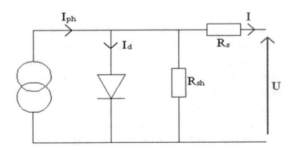

Fig. 1. Single-diode model equivalent circuit.

$$I = I_{ph} - I_s \left(\exp\left(\frac{V + R_s.I}{n.V_t}\right) - 1 \right) - \frac{V + R_s.I}{R_{sh}} \tag{2}$$

Where I_{ph} is the photocurrent, I_s is the saturation current, n *is* the diode ideality factor, Rs the series resistance and Rsh is the shunt resistance, so there are 5 unknown parameters. V is the solar cell output voltage, Vt (= kT/q) represents the thermal voltage, T is the temperature at Standard Test Conditions (STC = 25 °C, A.M 1.5 and 1000 W/m^2), q is the electronic charge ($1.60217646 \times 10^{-19}$ C) and K is the Boltzmann's constant ($1.3806503 \times 10^{-23}$ J/K).

2.2   Photovoltaic (PV) Module Model (Single-Diode Model)

The PV module consists of series and parallel solar cell combinations, that is, series strings are connected in parallel with each other. A typical model configuration of a PV module (using single diode model) is shown in Fig. 1, and the terminal equation that relates the currents and voltages of a PV module arranged in Np parallel strings and Ns series cells is mathematically expressed in the following equation:

$$I = I_{ph}.N_P - I_s.N_P \left(\exp\left(\frac{\frac{V}{N_P} + \frac{R_s}{N_P}I}{n.V_t}\right) - 1 \right) - \frac{V\frac{N_P}{N_S} + R_s.I}{R_{sh}} \tag{3}$$

Where N_S and N_P are the number of cells connected in series and in parallel, respectively [17].

3   Problem Formulation

Equations (2)–(3) are implicit nonlinear functions that involve the output current produced by the PV module or solar cell in both sides of the equation. Furthermore, the parameters Iph, Is, Rs, Rsh and n vary as a function of temperature and irradiance. Such functions have no explicit analytical solutions. Numerical methods, curve fitting techniques, and different optimization methods are often utilized to solve such functions. In this paper, the estimation problem is formulated as a non-linear optimization. The improved optimization technique PSO is used to estimate the parameters of the solar cell and the PV module by minimizing the objective function [13].

In order to minimize the summation of individual absolute errors (IAE), a new objective function has been proposed, and to form it, the I−V data relation of the unique diode model of Eq. (5) is written as follows:

$$f\left(I, V, I_{ph}, I_s, R_s, R_{sh}, n\right) = 0 \tag{4}$$

The objective function of the AEIs for all data measurements is defined as follows:

$$f = \sum_{i=1}^{N} \left| f\left(I, V, I_{ph}, I_s, R_s, R_{sh}, n\right) \right| \tag{5}$$

Where Vi and Ii are ith measured voltage and current pair values, respectively and N is the number of data points.

During the improved PSO optimization process, the objective function is minimized with respect to the parameter set. Theoretically, the objective function must have a zero value when the exact values of the parameters are obtained. However, it is expected to achieve a very low non-zero value due to the presence of measurement noise errors. Consequently, the solution obtained is better when the objective function is small [19].

4   Particle Swarm Optimization (PSO)

Particle swarm optimization is a heuristic global optimization method put forward by Kennedy and Eberhart in 1995; It is developed from swarm intelligence and is based on the research of bird and fish flock movement behavior. This algorithm is used to solve complex problems of optimism. Due to its many advantages including its easy implementation, and its simplicity, the algorithm can be used widely in several areas such as function optimization. It uses a number of particles (candidate solutions) which fly around in the search space to find the best solution. Meanwhile, the particles all look at the best particle (best solution) in their paths. In other words, particles consider their own best solutions as well as the best solution found so far. Each particle in PSO should consider the current position, the current velocity, the distance to pbest, and the distance to gbest in order to modify its position. PSO was mathematically modeled as follows:

$$v_i^{t+1} = w.v_i^t + c_1.rand.\left(pbest_i - x_i^t\right) + c_2.rand.\left(gbest_i - x_i^t\right) \tag{6}$$

$$x_i^{t+1} = x_i^t + v_i^{t+1} \tag{7}$$

where v_i^t is the velocity of particle i at iteration t, w is a weighting function, c_j is an acceleration coefficient, rand is a random number between 0 and 1, x_i^t is the current position of particle i at iteration t, pbesti is the pbest of agent i at iteration t, and gbest is the best solution so far [18].

The first part of Eq. (6) provides exploration ability for PSO. The second and third parts represent private thinking and collaboration of particles respectively. At the beginning, PSO randomly places the particles in a problematic space. In each iteration, Eq. (6) calculates the particle velocities. Equation (7) calculates the particle positions after defining the velocities. The process of changing particles' positions will continue until an end criterion is met.

5 Proposed Improved Particle Swarm Optimization Algorithm

In the literature [6], Clerc presented a sort of particle swarm optimization with constriction factor and the definition of a simple particle swarm optimization with constriction factor:

$$v_i^{t+1} = \chi \left(v_i^t + c_1.rand.\left(pbest_i - x_i^t\right) + c_2.rand.\left(gbest_i - x_i^t\right)\right). \tag{8}$$

$$\chi = \frac{2}{\left|2 - l - \sqrt{l^2 - 4l}\right|}, \quad l = c_1 + c_2, \quad l > 4 \tag{9}$$

In the proposed algorithm, the parameters are set to: l = 4.2, c1 = 1.40, c2 = 2.80, which gives a contraction factor F equal to 0.6417 so the two coefficients are 0.6417 * 1.40 * rand = 0.8984 * rand   and   0.6417 * 2.80 * rand = 1.7968 * rand, equivalent to All elements of the new formula for updating the speed multiplied by the weighting factor, taking into account the distinct function of various elements; In other words the algorithm is improved in terms of convergence.

The pseudo code of the procedure is as follows:

For each particle

Initialize particle

END

Do

For each particle

Calculate fitness value

If the fitness value is better than the best fitness value (pBest) in history

       Set current value as the new pBest

END

Choose the particle with the best fitness value of all the particles as the gBest

For each particle

Calculate particle velocity according equation (a)

Update particle position according equation (b)

End

While maximum iterations or minimum error criteria is not attained.

6   Optimization Results

The performance of proposed improved PSO algorithm (ICFPSO) in estimating solar cell/PV module model parameters is tested on single-diode model. Measurement data of 57 mm diameter commercial (R.T.C. France) silicon solar cell and PV module in which 36 polycrystalline silicon cells (Photowatt-PWP 201) are used for validation. The data have been taken from the system under 1 sun (1000 W/m^2) at 33 °C for solar cell and at 45 °C for the PV module.

6.1   Case Study 1: Solar Cell Model Using Single-Diode Model

The experimental I-V data from a silicon solar cell (57 mm diameter commercial RTC France), has been used for parameter estimation using the single diode models. The experimental data has been extracted at 33 °C and irradiation of 1000 W/m^2 with 26 pairs of I-V data used [15].

Solar cell model case is tested in this section to validate of the proposed estimation method. And this by minimizing the objective function as illustrated in Eq. (5) to achieve an optimal set of parameters that reflects the characteristics of this solar cell. Thus, a value of zero for the objective function would produce an optimal solution. Table 1 shows the extracted parameters from experimental data (Voltage and current measurements) of a 57 mm diameter commercial (R.T.C France) silicon solar cell at 33 °C and at 1000 W/m^2. After comparing the results obtained by the proposed method with the results obtained from the references mentioned in Table 1, and it has been shown that most of the parameters extracted using to the proposed method are very similar to those reported in the other references.

Table 1. Estimated parameters of solar cell model using different methods.

Case	Item	ICFPSO	PS	Easwarakhanthan et al. (1986)	Bouzidi et al. (2007)	GA
Solar cell	I_{ph}	0.7609	0.7617	0.7608	0.7607	0.7619
	$I_{SD}(lA)$	0.9877	0.9980	0.3223	0.3267	0.8087
	$R_s(X)$	0.0365	0.0313	0.0364	0.0364	0.0299
	$G_{sh}(S)$	0.0189	0.0156	0.0186	0.0166	0.0236
	n	1.4754	1.6000	1.4837	1.4816	1.5751

Table 2 show a curve fitting is performed next an evaluation the goodness of fit of the obtained solution.

Table 2. Curve fitting of the estimated solar cell parameters.

Measurement	$V_a(V)$	$I_a(A)$	IAE based on ICFPSO	IAE based on PS	IAE based on Easwarakhanthan et al. (1986)	IAE based on Bouzidi et al. (2007)	IAE based on GA
1	−0.2057	0.7640	0.000854	0.000537	0.000109	0.000347	0.000936
2	−0.1291	0.7620	0.001575	0.001343	0.000686	0.000383	0.001666
3	−0.0588	0.7605	0.001223	0.001747	0.000879	0.000717	0.001999
4	0.0057	0.7605	0.000874	0.000739	0.000321	0.000355	0.000927
5	0.0646	0.7600	0.000531	0.000314	0.000919	0.000835	0.000445
6	0.1185	0.7590	0.000425	0.000453	0.000931	0.000739	0.000533
7	0.1678	0.7570	0.000964	0.001622	0.000120	0.000410	0.001662
8	0.2132	0.7570	0.000487	0.000737	0.000826	0.000448	0.000755
9	0.2545	0.7555	0.000684	0.001151	0.000369	0.000085	0.001178
10	0.2924	0.7540	0.000247	0.001032	0.000261	0.000251	0.001127
11	0.3269	0.7505	0.000328	0.001817	0.001044	0.001581	0.002074
12	0.3585	0.7465	0.001547	0.001005	0.001182	0.001688	0.001571
13	0.3873	0.7385	0.000651	0.000628	0.002309	0.002682	0.001707
14	0.4137	0.7280	0.002135	0.003040	0.000775	0.000843	0.001185
15	0.4373	0.7065	0.000087	0.003405	0.003065	0.002581	0.000530
16	0.4590	0.6755	0.002418	0.005220	0.004330	0.002930	0.001094
17	0.4784	0.6320	0.002790	0.006581	0.006168	0.003427	0.001150
18	0.4960	0.5730	0.002494	0.005747	0.010241	0.005676	0.000829
19	0.5119	0.4990	0.001452	0.002477	0.016846	0.009993	0.004796
20	0.5265	0.4130	0.001028	0.000112	0.022874	0.013259	0.007163
21	0.5398	0.3165	0.001684	0.002691	0.030060	0.017316	0.009007
22	0.5521	0.2120	0.000435	0.003910	0.036806	0.020599	0.008115
23	0.5633	0.1035	0.004720	0.003590	0.043444	0.023594	0.004458
24	0.5736	−0.0100	0.003463	0.005423	0.054194	0.030641	0.001648
25	0.5833	−0.1230	0.000235	0.000334	0.059145	0.031672	0.010085
26	0.5900	−0.2100	0.000284	0.000339	0.069445	0.039175	0.014679
Total IAE			0.033581	0.055993	0.367349	0.212223	0.081320

The parameters extracted using the proposed method are substituted in Eq. (6) to evaluate the fitness. A similar procedure is carried out with the results obtained from other references mentioned. The optimal value is zero for each of the 26 equations. It should be noted that the IAE associated with most measures is lower in the case of improved PSO results. In addition, the sum of the AEAs is 0.033581, which is much lower than those obtained using a methods mentioned.

Figure 2 shows the pattern of IAE for all methods mentioned. It is obvious that the parameters extracted using proposed method generated the best IAE profile. From this figure it can be seen that a mentioned methods have diverged above measurement 15 or so.

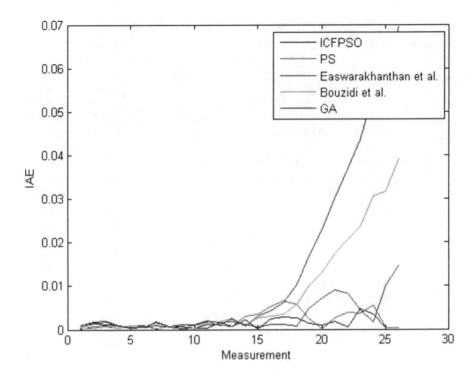

Fig. 2. IAE profiles for different estimation algorithms (solar cell case).

6.2   Case Study 2: PV Module Model Using Single-Diode Model

In this section, the measurements were taken using the solar module mentioned above to characterize an I-V data set. It is interesting to mention that the lowering of irradiation levels will not affect the results of the estimate in general. However, by reducing the following, 0.25 sun will reduce the unsaturated region of the I-V curve, which will result in the collection of many data points in the saturation region. From the point of view of estimation, this will lead to poor estimation results. In a similar way to the solar cell model, the parameters are estimated using the proposed improved PSO technique.

The estimated parameters as well as the curve adjustment values are given in Tables 3 and 4, respectively. The sum of the IAEs is equal to 0.048532. In this case, the maximum absolute error recorded was 0.005614.

Table 3. Estimated parameters of PV module model using different methods.

Case	Item	ICFPSO	PS	Easwarakhanthan et al. (1986)	Bouzidi et al. (2007)	GA
PV module	I_{ph}	1.0313	1.0313	1.0318	1.0339	1.0441
	$I_{SD}(lA)$	2.6386	3.1756	3.2875	3.0760	3.4360
	$R_s(X)$	1.2357	1.2053	1.2057	1.2030	1.1968
	$G_{sh}(S)$	0.0012	0.0014	0.0018	0.0018	0.0018
	N	48.2889	48.2889	48.4500	48.1862	48.5862

Table 4. Curve fitting of the estimated PV module parameters.

Measurement	$V_a(V)$	$I_a(A)$	IAE based on presented ICFPSO	IAE based on PS	IAE based on Easwarakhanthan et al. (1986)	IAE based on Bouzidi et al. (2007)	IAE based on GA
1	0.1248	1.0315	0,004351	0.002135	0.002197	7.74758E-05	0.010193793
2	1.8093	1.0300	0,001354	0.003030	0.003783	0.001645391	0.008698485
3	3.3511	1.0260	0,000521	0.001267	0.002651	0.000495255	0.009911526
4	4.7622	1.0220	0,005145	0.000558	0.001406	0.000771843	0.011228492
5	6.0538	1.0180	0,002524	0.002262	0.000236	0.001971045	0.012457628
6	7.2364	1.0155	0,000419	0.001986	0.001009	0.001243466	0.011728394
7	8.3189	1.0140	0,004151	0.000419	0.003879	0.001553998	0.008880327
8	9.3097	1.0100	0,001986	0.002528	0.006421	0.003985862	0.006327239
9	10.2163	1.0035	0,002282	0.006023	0.010319	0.007721693	0.00237697
10	11.0449	0.9880	0,005478	0.006603	0.011258	0.008442889	0.001333667
11	11.8018	0.9630	0,005614	0.006499	0.011449	0.008368167	0.000977468
12	12.4929	0.9255	0,001965	0.005437	0.010586	0.007217233	0.001607955
13	13.1231	0.8725	0,004784	0.002350	0.007565	0.003931467	0.004322128
14	13.6983	0.8075	0,001874	0.002308	0.007422	0.003598334	0.004074905
15	14.2221	0.7265	0,000227	0.000119	0.004707	0.000824159	0.006303696
16	14.6995	0.6345	0,000723	0.001255	0.003093	0.000681443	0.007322338
17	15.1346	0.5345	0,001384	0.000617	0.003074	0.000404116	0.00662565
18	15.5311	0.4275	0,000884	0.001154	0.001730	0.001261056	0.007121865
19	15.8929	0.3185	0,000254	0.000390	0.002341	1.36166E-05	0.005535372
20	16.2229	0.2085	0,000845	0.001615	0.002547	0.001034466	0.00423231
21	16.5241	0.1010	0,000914	0.005205	0.005052	0.004482183	0.000525321
22	16.7987	−0.0080	0,000257	0.000561	0.000669	0.000225555	0.004951548
23	17.0499	−0.1110	0,000039	0.000051	0.002283	0.000751118	0.0052044
24	17.2793	−0.2090	0,000087	0.000244	0.003185	0.000524973	0.004701146
25	17.4885	−0.3030	0,000478	0.002267	0.006750	0.002956545	0.006836252
Total IAE			0.048532	0.056883	0.115612	0.064183	0.153479

The IAE profiles for different estimation methods are shown in Fig. 3 again, the parameters estimated using improved PSO yielded the best profile of the IAE. the reductions in IAE as well as the sum of the IAEs are very perceptible when the results of improved pso are compared with other competing methods.

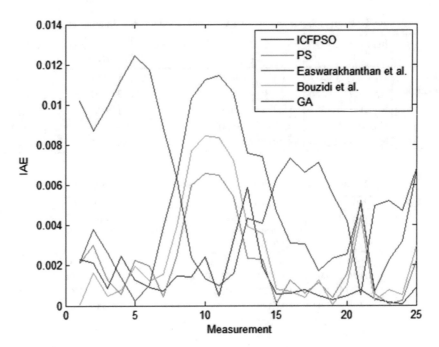

Fig. 3. IAE profiles for different estimation algorithms (PV module case).

7 Conclusions

This paper presents a modified PSO variant for the estimation of parameters of solar cells and photovoltaic modules. The proposed algorithm is implemented and tested using measurement data of a solar cell and a photovoltaic module, the results are compared with others benchmark algorithms to validate the accuracy and performances of the proposed techniques. ICFPSO shows better performance in terms of convergence rate and avoidance of local minima. It is observed that ICFPSO gives the highest accuracy that mentioned algorithms due to the fact that lower error is produced. Therefore, it can be concluded that the proposed ICFPSO improves the problem of trapping in local minima with very good convergence speed compared to the others existing meta-heuristics algorithms.

References

1. AlRashidi, M.R., AlHajri, M.F., El-Naggar, K.M., Al-Othman, A.K.: A new estimation approach for determining the I-V characteristics of solar cells. Sol. Energy **85**(7), 1543–1550 (2011)
2. Clerc, M.: The swarm and the queen: towards a deterministic and adaptive particle swarm optimization. In: Proceedings of the Congress of Evolutionary Computation, Washington, pp. 1951–1957 (1999)

3. Ye, M., Wang, X., Xu, Y.: Parameter extraction of solar cells using particle swarm optimization. J. Appl. Phys. **105**(9), 094502–094508 (2009)
4. Bouzidi, K., Chegaar, M., Nehaoua, N.: New method to extract the parameters of solar cells from their illuminated I–V curve. In: 4th International Conference on Computer Integrated Manufacturing, Setif, Algeria, November 2007
5. Safari, S., Ardehali, M.M., Sirizi, M.J.: Particle swarm optimization based fuzzy logic controller for autonomous green power energy system with hydrogen storage. In: 2011 Global Conference on Renewable Energy and Energy Efficiency for Desert Regions "GCREEDER 2011". Energy Conversion and Management, vol. 65, pp. 41–49, January 2013
6. Orioli, A., Di Gangi, A.: A procedure to calculate the five-parameter model of crystalline silicon photovoltaic modules on the basis of the tabular performance data. Appl. Energy **102**, 1160–1177 (2013)
7. Lineykin, S.: Five-parameter model of photovoltaic cell based on STC data and dimensionless. In: Electrical Engineering and Electronics, pp. 1–5 (2012)
8. Lo Brano, V., Orioli, A., Ciulla, G., Di Gangi, A.: An improved five-parameter model for photovoltaic modules. Sol. Energy Mater. Sol. Cells **94**(8), 1358–1370 (2010)
9. Askarzadeh, A., Rezazadeh, A.: Parameter identification for solar cell models using harmony search-based algorithms. Sol. Energy **86**(11), 3241–3249 (2012)
10. Chegaar, M., Ouennough, Z., Guechi, F., Langueur, H.: Determination of solar cells parameters under illuminated conditions. J. Electron. Devices **2**, 17–21 (2003)
11. Easwarakhanthan, T., Bottin, J., Bouhouch, I., Boutrit, C.: Nonlinear minimization algorithm for determining the solar cell parameters with microcomputers. Int. J. Sol. Energy **4**, 1–12 (1986)
12. Askarzadeh, A., Rezazadeh, A.: Artificial bee swarm optimization algorithm for parameters identification of solar cell models. Appl. Energy **102**, 943–949 (2013)
13. Ishaque, K., Salam, Z., Mekhilef, S., Shamsudin, A.: Parameter extraction of solar photovoltaic modules using penalty-based differential evolution. Appl. Energy **99**, 297–308 (2012)
14. Gong, W., Cai, Z.: Parameter extraction of solar cell models using repaired adaptive differential evolution. Sol. Energy **94**, 209–220 (2013)
15. Alam, D., Yousri, D., Eteiba, M.: Flower pollination algorithm based solar PV parameter estimation. Energy Convers. Manage. **101**, 410–422 (2015)
16. Bai, J., Liu, S., Hao, Y., Zhang, Z., Jiang, M., Zhang, Y.: Development of a new compound method to extract the five parameters of PV modules. Energy Convers. Manage. **79**, 294–303 (2014)
17. Ismail, M., Moghavvemi, M., Mahlia, T.: Characterization of PV panel and global optimization of its model parameters using genetic algorithm. Energy Convers. Manage. **73**, 10–25 (2013)
18. Hamid, N.F.A., Rahim, N.A., Selvaraj, J.: Solar cell parameters identification using hybrid Nelder-Mead and modified particle swarm optimization. J. Renew. Sustain. Energy **8**, 015502 (2016)
19. Sirjani, R., Shareef, H.: Parameter extraction of solar cell models using the lightning search algorithm in different weather conditions. J. Sol. Energy Eng. **138**, 041007 (2016)

Materials in Renewable Energy

Robust and Efficient Control of Wind Generator Based on a Wound Field Synchronous Generator

T. Khalfallah[1(✉)], B. Cheikh[1], A. Tayeb[1], M. Denai[2],
and D. M'Hamed[3]

[1] L2GEGI Laboratory, Department of Electrical Engineering,
Ibn Khaldoun University, Tiaret, Algeria
tahir.commande@gmail.com, bochradz@yahoo.com,
allaoui_tb@yahoo.fr
[2] School of Engineering Technology, University of Hertfordshire,
Hatfield AL10 9AB, UK
m.denai@herts.ac.uk
[3] CAOSEE Laboratory, Department of Electrical Engineering,
Béchar University, Béchar, Algeria
doumicanada@gmail.com

Abstract. This paper presents a new contribution for the control of Wind-turbine energy driven wound field synchronous generator (WFSG) connected to the grid via a back to back converter. The goal is to track the maximum power point via a dedicated control. For this purpose, a robust backstepping controller with integral actions is applied to the WFSG. This control is based on both feedback laws and Lyapunov technique. In order to verify the validity of the proposed method, this control is applied to an accurate dynamic model of the whole generator. The simulation results show the performances.

Keywords: Wind turbine · Wound field synchronous generator
Back to back converter · Maximum power point tracking · Backstepping control

1 Introduction

Wind energy is one of the attractive renewable resources and grows up rapidly [1]. In this field, the variable generation systems are becoming the most important and fastest growing application. The wind turbines generators operating at variable speeds are more efficient and less sensible to the grid disturbances [2, 3].

In this aspect, the WFSG is considered to be promising because of its high efficiency as the whole stator current is used to produce the electromagnetic torque [4]. In addition, it permits to avoid using permanent magnets.

The main benefit of the WFSG with salient pole is that it allows the direct control of the power factor of the machine, consequently the stator current may be reduced under these circumstances [5]. The power converter machine side is called "Stator Side Converter" (SSC) and the converter Grid-side power is called "Grid Side Converter" (GSC).

© Springer International Publishing AG 2018
M. Hatti (ed.), *Artificial Intelligence in Renewable Energetic Systems*, Lecture Notes
in Networks and Systems 35, https://doi.org/10.1007/978-3-319-73192-6_47

The SSC converter controls the active power and reactive power produced by the machine. As the GSC converter, it controls the DC bus voltage and power factor on the grid side. Among the existing control design techniques, the backstepping approach is attractive control design technique due to its robustness; it is based on a recursive algorithm for control designing for a class of nonlinear systems. It uses the Lyapunov stability principle which can be applied to a large set of nonlinear systems [6–9].

In this paper, a technique to control two power converters is presented; this technic is based on the integral backstepping control. The dynamic performances of the global system are analysed by simulations in Matlab/Simulink environment. The wind generator is modeled first, then maximum power point tracking (MPPT) algorithm will be presented. Thereafter, we present a model of the WFSG in the d-q reference, and the general principle of control of both power converters which is based on integral backstepping technique; finally, some remarks will conclude the work.

2   System Description and Modeling

The topology of the wind energy conversion system (WECS) presented in this study is depicted in Fig. 1. It consists of a wind turbine, a gearbox, a WFSG, and back-to back converters. The rotor winding of the WFSG is connected to DC bus by DC/DC converter, whereas the stator winding is fed by the back-to-back bidirectional, which is the Pulse Width Modulation Voltage Source Converter (PWM-VSC). The back-to-back converter contains a Stator Side Converter (SSC) and a Grid Side Converter (GSC), which are connected by a dc bus. The control system of the WECS consists of the generator side control sub-system and the grid side control sub-system. The Maximum Power Point Tracking (MPPT) algorithm is based on the tip speed ratio (TSR) method. The GSC controller maintains the dc-link voltage at the desired value by exporting

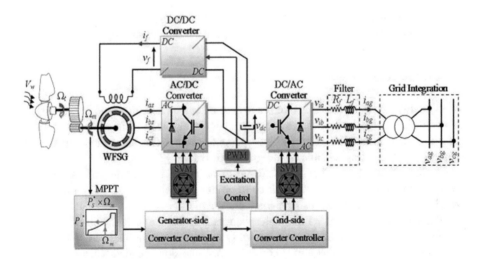

Fig. 1. Wind energy conversion system structure.

active power to the grid. It can control the injected reactive power to the grid. The generator-side converter controls the power flow from the WFSG to the DC bus via stator currents. The quadrature component controls the active power, whereas the direct component controls the reactive power.

2.1 Modeling of the Wind Turbine and Gearbox

The aerodynamic power that the wind turbine extracts from the wind is expressed by the following equation [10]:

$$P_a = \frac{1}{2}\rho\pi R^2 V_w^3 C_p(\lambda, \beta) \tag{1}$$

Where ρ is the air density, R is the wind turbine blade radius, V_w is the wind velocity (m/s), and C_p is called the power coefficient, which is a function of both the blade pitch angle B and the TSR, λ, the tip speed ratio is defined as [11]

$$\lambda = \frac{R\Omega_t}{V_w} \tag{2}$$

Where Ω_t is the turbine speed.

The mechanical equation of the shaft, including both the turbine and the generator masses, is given by

$$J\frac{d\Omega_m}{dt} = T_m - T_{em} - f\Omega_m \tag{3}$$

J and f are the total moment of inertia and the viscous friction coefficient appearing at the generator side, T_m is the gearbox torque, T_{em} is the generator torque, and Ω_m is the mechanical generator speed.

$$T_m = \frac{T_a}{G}\Omega_t = \frac{\Omega_m}{G}T_a = \frac{P_a}{\Omega_t} \tag{4}$$

G is the gear ratio and T_a is the aerodynamic torque.

2.2 WFSG Modeling

The WFSG is selected to use in the wind turbine because of their characteristics which already mentioned above. The modeling of the WFSG in d–q frame is given by [10]

$$v_{ds} = -r_s i_{ds} + \omega_e L_q i_{qs} - \omega_e M_{sQ} i_Q - L_d \frac{di_{ds}}{dt} + M_{sf}\frac{di_f}{dt} + M_{sD}\frac{di_D}{dt} \tag{5}$$

$$v_{qs} = -r_s i_{qs} - \omega_e L_d i_{ds} + \omega_e M_{sf} i_f + \omega_e M_{sD} i_D - L_q \frac{di_{qs}}{dt} + M_{sQ}\frac{di_Q}{dt} \tag{6}$$

$$v_f = -r_f i_f + L_f \frac{di_f}{dt} - M_{sf} \frac{di_d}{dt} + M_{fD} \frac{di_D}{dt} \tag{7}$$

$$0 = r_D i_D + M_{fD} \frac{di_f}{dt} - M_{sD} \frac{di_d}{dt} + L_D \frac{di_D}{dt} \tag{8}$$

$$0 = r_Q i_Q + L_Q \frac{di_Q}{dt} - M_{sQ} \frac{di_{qs}}{dt} \tag{9}$$

With v_{ds}, v_{qs}, i_{ds}, and i_{qs} are voltages and currents in d-q frame; v_f and i_f are voltage and current of the main field winding; i_D and i_Q are direct and quadrature damper currents; r_s is stator resistance; r_f is main field resistance; r_D, r_Q are dampers resistances; L_d and L_q are direct and quadrature stator main inductances; L_D and L_Q are direct and quadrature dampers inductances; L_f is main field inductance; M_{sf} is mutual inductance between direct stator winding and main field one; M_{fD} is mutual inductance between main field winding and direct damper one; M_{sQ} is mutual inductance between stator and quadrature damper; M_{sD} is mutual inductance between stator and direct damper; ω_e is the electrical angular speed of the WFSG.

3   WFSG-Based Wind Turbine Control

3.1   Modeling of the Wind Turbine and Gearbox

The TSR is a method of MPPT used to extract maximum power from wind. According to (2), the optimal turbine speed is achieved when TSR has optimal value. If TSR remains at optimal value, it is ensured that the extracted energy is maximum.

Therefore, the turbine speed reference is [13, 14]

$$\Omega_{t\_ref} = \frac{\lambda_{opt} V_w}{R} \tag{10}$$

3.2   Robust Integral Backstepping Control Design of WFSG

(1)   Speed Control

To solve the speed tracking problem, the following tracking error variable is defined as

$$z_\Omega = \Omega_m^* - \Omega_m + K_\Omega' \int_0^t \left(\Omega_m^* - \Omega_m \right) dt \tag{11}$$

With $K_\Omega' \int_0^t \left(\Omega_m^* - \Omega_m \right) dt$ is the integral term added to the rotor speed tracking error.

Next, in order to design the speed control v_{qs} which is designed to force i_{qs} to track i_{qs}^*, we take the time derivative of (11) and by replacing i_{qs} by i_{qs}^*, which can be considered as a new input, it follows that

$$\dot{z}_\Omega = \dot{\Omega}_m^* + \frac{P}{J}\left(L_q - L_d\right)i_{ds}i_{qs} + \frac{P}{J}M_{sf}i_f i_{qs} - \frac{P}{J}M_{sQ}i_Q i_{ds} + \frac{f}{J}\Omega_m - \frac{T_m}{J} + K_\Omega'\left(\Omega_m^* - \Omega_m\right)$$

$$(12)$$

Choosing the following candidate Lyapunov function $V_\Omega = \frac{1}{2}z_\Omega^2$ and taking the time derivative along the trajectories of (12), we get

$$\dot{V}_\Omega = z_\Omega\left\{\dot{\Omega}_m^* + \frac{P}{J}\left(L_q - L_d\right)i_{ds}i_{qs} + \frac{P}{J}M_{sf}i_f i_{qs} - \frac{P}{J}M_{sQ}i_Q i_{ds} + \frac{f}{J}\Omega_m - \frac{T_m}{J} + K_\Omega'\left(\Omega_m^* - \Omega_m\right)\right\}$$

$$(13)$$

Following the backstepping methodology, the virtual control input i_{qs}^* is chosen as

$$i_{qs}^* = \frac{J}{pM_{sf}i_f + p\left(L_q - L_d\right)i_{ds}}\left[K_{\Omega m}z_{\Omega m} - \dot{\Omega}_m^* - \frac{f}{J}\Omega_m - K_\Omega'\left(\Omega_m^* - \Omega_m\right) + \frac{T_m}{J}\right] \quad (14)$$

Then $\dot{V}_\Omega = -k_\Omega z_\Omega^2$, with $k_\Omega > 0$.

(2) **Quadratic Current Control**

Since the virtual input i_{qs}^* is designed to stabilize the dynamics (12), now to design the control input v_{qs}, we introduce the following tracking error:

$$z_q = i_{qs}^* - i_{qs} + z_q' \quad (15)$$

Where $z_q' = k_q' \int_0^t \left(i_{qs}^* - i_{qs}\right)dt$ is an integral action. Consider the following Lyapunov candidate function:

$$V_q = \frac{1}{2}z_q^2 + \frac{1}{2}z_q'^2 \quad (16)$$

By taking the time derivative of V_q, and by replacing the suitable terms, it follows that

$$\dot{V}_q = z_q\left\{\frac{di_{qs}^*}{dt} - \frac{di_{qs}}{dt} + K_q'\left(i_{qs}^* - i_{qs}\right)\right\} + z_q' K_q'\left(i_{qs}^* - i_{qs}\right)$$

$$= z_q\left\{\frac{di_{qs}^*}{dt} + \frac{r_s}{L_q}i_{qs} + \omega_e\frac{L_d}{L_q}i_{ds} - \omega_e\frac{M_{sf}}{L_q}i_f - \omega_e\frac{M_{sD}}{L_q}i_D - \frac{M_{sQ}}{L_q}\frac{di_Q}{dt} + \frac{1}{L_q}v_{qs} + K_q'\left(i_{qs}^* - i_{qs}\right)\right\}$$

$$+ z_q' K_q'\left(i_{qs}^* - i_{qs}\right)$$

$$(17)$$

By choosing the control input v_{qs} as

$$v_{qs} = -K_q L_q z_q - L_q \frac{di^*_{qs}}{dt} - r_s i_{qs} - \omega_e L_d i_{ds} + \omega_e M_{sf} i_f + \omega_e M_{sD} i_D + M_{sQ} \frac{di_Q}{dt} \quad (18)$$

It follows that

$$\dot{V}_q = -K_q z_q^2 + \left(z_q + z'_q\right) K'_q \left(i^*_{qs} - i_{qs}\right) \quad (19)$$

Since $i^*_{qs} - i_{qs} = z_q - z'_q$, then (19) becomes

$$\dot{V}_q = -\left(K_q - K'_q\right) z_q^2 - K'_q z'_q{}^2 \quad (20)$$

Therefore, the condition to make sure that the derivative is still negative, is $K_q > 0$ and $K_q > K'_q$.

Where K_q is parameter introduced by the backstepping method.

Then, under the control action v_{qs}, the current i_{qs} tracks the desired current reference i^*_{qs}, i.e. $i_{qs} \rightarrow i^*_{qs}$; and since i^*_{qs} is designed such that the rotor speed tracks the desired reference, i.e. $\left(\Omega_m \rightarrow \Omega^*_m\right)$.

(3) Direct Current Control

Now, to eliminate the reluctance torque, the current reference is fixed to zero, i.e. $\left(i^*_{ds} = 0\right)$. To apply the vector control, let us define the following tracking error:

$$z_d = i^*_{ds} - i_{ds} + k'_d \int_0^t \left(i^*_{ds} - i_{ds}\right) dt \quad (21)$$

Defining the following candidate Lyapunov function:

$$V_d = \frac{1}{2} z_d^2 + \frac{1}{2} z'_d{}^2 \quad (22)$$

The time derivative is computed as

$$\dot{V}_d = z_d \left[\frac{di^*_{ds}}{dt} - \frac{di_{ds}}{dt} + K'_d \left(i^*_{ds} - i_{ds}\right)\right] + z'_d K'_d \left(i^*_{ds} - i_{ds}\right) \quad (23)$$

Since $i^*_{ds} = 0$, then replacing $i^*_{ds} - i_{ds} = z_d - z'_d$ in the above equation, we obtain

$$\dot{V}_d = z_d \left[\frac{r_s}{L_d} i_{ds} - \omega_e \frac{L_q}{L_d} i_{qs} - \frac{M_{sf}}{L_d} \frac{di_f}{dt} + \omega_e \frac{M_{sQ}}{L_d} i_Q - \frac{M_{sD}}{L_d} \frac{di_D}{dt} + \frac{v_{ds}}{L_d}\right] + \left(z_d + z'_d\right) K'_d \left(z_d - z'_d\right) \quad (24)$$

Then, by choosing the control input v_{ds} as follows:

$$v_{ds} = -K_d L_d z_d - r_s i_{ds} + \omega_e L_q i_{qs} + M_{sf}\frac{di_f}{dt} - \omega_e M_{sQ} i_Q + M_{sD}\frac{di_D}{dt} \tag{25}$$

And by substituting in (24), we obtain

$$\dot{V}_d = -\left(K_d - K_d'\right)z_d^2 - K_d' z_d'^2 \tag{26}$$

Therefore, the condition to ensure that the derivative is still negative is: $K_d > 0$ and $K_d > K_d'$.

Where K_d is parameter introduced by the backstepping method.

Then, this implies that under the action of the control v_d, the current ids tracks the desired reference, i.e. $i_{ds} \rightarrow i_{ds}^* = 0$.

Finally, combining the action of the control inputs v_{qs} with i_{qs}^* and v_{ds}, the control objectives are achieved.

After obtaining the v_{ds} and v_{qs} control signals; they are turned into three phases referential by means of the inverse Park transformation and are given as a reference to the PWM block (pulse width modulation) in order to generate the Converter signals pulse as shown in Fig. 2.

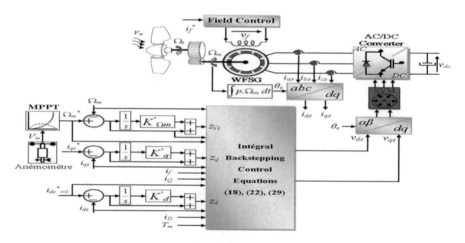

Fig. 2. Simulation scheme of integral backstepping control applied for WFSG-wind turbine.

4 Simulation Results and Discussion

To evaluate the performance of the proposed control strategy of the WFSG wind system from Fig. 1, the implementation in the software MATLAB/SIMULINK environment has been considered. The parameters of the turbine and WFSG used are given in Appendix.

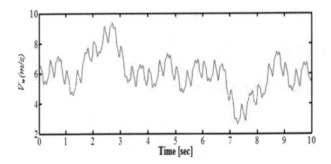

Fig. 3. Random of the wind.

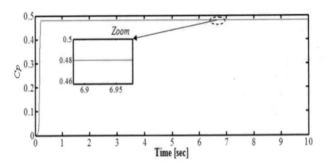

Fig. 4. Power coefficient.

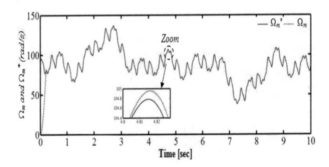

Fig. 5. WFSG speed and its reference.

A more practical wind speed profile (Fig. 3) is now used to demonstrate further the dynamic performance of the WFSG with the proposed control. During the simulation, the d-axis command current of the WFSG side converter control system, i_{ds}, is set to zero; whereas, for the grid side inverter, the q-axis command current, i_{qs}, is set to zero. Figures 3, 4, 5, 6, 7, 8 and 9 show the dynamic response of the proposed controller under random wind. It can be seen that the wind speed increases, the rotor angular velocity increases proportionally too, the power coefficient will drop to maintain the rated output power. The WECS operate under MPPT control. The initial pitch angle B keeps the value of $0°$, the tip speed ratio maintains the optimal value 8.1, and the power coefficient C_p is the maximum around 0.48 as shown in Fig. 4.

Figure 5 shows the speed tracking results of the WFSG. By applying the proposed control scheme, the optimal speed command is accurately tracked to extract the maximum power from the wind energy at any moment.

In Fig. 6 the generated torque reference follows the optimum mechanical torque of the turbine quite well. The decoupling effect of the between the direct and quadratic stator current of the WFSG is illustrated in Fig. 7.

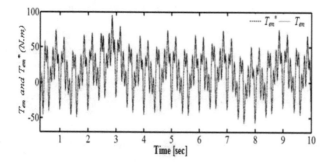

Fig. 6. Electromagnetic torque and its reference.

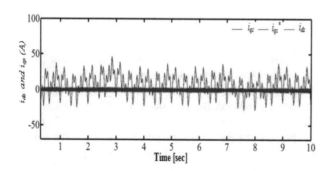

Fig. 7. Direct and quadratic stator current.

Figure 8 shows the simulation result of DC-link voltage that remains a constant value. Thus, proves the effectiveness of the established regulators. As can be seen in Fig. 9, the reactive power is kept at zero, which indicates that, the unity power factor control is achieved. Also, the active power is proportional to the wind speed.

The simulation results demonstrate that the controller works very well and shows very good dynamic and steady-state performance. The control algorithm can be used to extract maximum power from the variable-speed wind turbine under fluctuating wind.

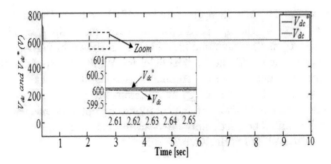

Fig. 8. DC-link voltage and these zoom.

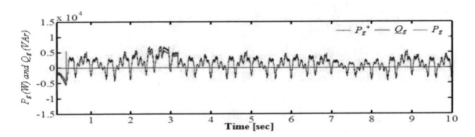

Fig. 9. Grid active and reactive powers.

5   Conclusion

In this paper, a new approach has been presented for the variable speed wind turbines with wound field synchronous generators to improve the reliability and efficiency of the wind energy conversion system. The proposed method is a nonlinear backstepping with integral action which guaranteed by Lyapunov stability analysis. Simulation results prove the effectiveness of our strategy as regards robust stability and transient performances and ability to track the maximum power from the wind power when exciting the system with a variable wind speed. As well as, the results show that this technique can be successfully applied to design the nonlinear control algorithms of wind turbine model.

Appendix

The wind turbine generator system used for the experimental has the following parameters:

(1) Wind turbine parameters:

$P_t = 10\,\text{KW}; \rho = 1.225\,\text{kg.m}^{-2}; R = 3\,\text{m}; G = 5.4; f = 0.017\,\text{N.m.s}^{-1}.$

(2) Generator parameters:

$$S_n = 7.5\,\text{KVA}; \quad r_s = 1.19\,\Omega; \quad r_f = 3.01\,\Omega; \quad U_{rms} = 400\,\text{V}; \quad x_d = 1.4\,\text{p.u};$$
$$x_q = 0.7\,\text{p.u}; T''_{d0} = 522\,\text{ms}; x'_d = 0.099\,\text{p.u}; x''_d = 0.049\,\text{p.u}; T'_d = 40\,\text{p.u}; T''_d = 3.7\,\text{ms};$$
$$T_a = 6\,\text{ms}.$$

References

1. Tseng, H.C., Cheng, S.: Robust sensorless control of PMSG with MRAS in variable speed wind energy conversion system. In: Proceeding of ICMA, pp. 1917–1922 (2011)
2. Xu, D., Luo, Z.: A novel AC-DC converter for PMSG variable speed wind energy conversion systems. In: IEEE 6th International Power Electronics and Motion Control Conference, IPEMC 2009, pp. 1117–1122 (2009)
3. Laverdure, N., Roye, D., Bacha, S., Belhomme, R.: Techniques des systèmes éoliens: intégration au réseau électrique. Revue 3EI, décembre 2004, pp. 14–25 (2004). ISSN 1252 - 770 X
4. Ames, R.L.: AC Generators: Design and Application. Wiley, New York (1990)
5. Topal, D.E., Ergene, L.T.: Designing a wind turbine with permanent magnet synchronous machine. Istanb. Univ. J. Electr. Electron. Eng. **11**, 1311–1317 (2011)
6. Loucif, M., Boumédiène, A.: NonLinear integral backstepping control for DFIG under wind speed variation using MATLAB/PSB. In: Third International Conference on Power Electronics and Electrical Drives ICPEED 2014, Oran, 10–11 December 2014
7. Tahri, A., Hassaine, S., Moreau, S.: A robust control for permanent magnet synchronous generator associated with variable speed wind turbine. J. Electr. Eng. **14**(2), 17–23 (2015)
8. Hamida, M.A., Glumineau, A., Leon, J.: Robust integral backstepping control for sensorless IPM synchronous motor controller. J. Franklin Inst. **349**, 1734–1757 (2012)
9. Bossoufi, B., Karim, M., Lagrioui, A., Taoussi, M., Derouich, A.: Observer backstepping control of DFIG-Generators for wind turbines variable-speed: FPGA-based implementation. Renew. Energy **81**, 903–917 (2015)
10. Bianchi, F.D., Battista, H.D., Mantz, R.J.: Wind Turbine Control Systems: Principles, Modelling and Gain Scheduling Design. Springer, Berlin (2007)
11. Qiao, W., Qu, L., Harley, R.G.: Control of IPM synchronous generator for maximum wind power generation considering magnetic saturation. IEEE Trans. Ind. Appl. **45**, 1095–1105 (2009)
12. Barakat, A., Tnani, S., Champenois, G., Mouni, E.: Analysis of synchronous machine modeling for simulation and industrial applications. Simul. Modell. Pract. Theory **18**, 1382–1396 (2010)
13. Nasiri, M., Milimonfared, J., Fathi, S.H.: Modeling, analysis and comparison of TSR and OTC methods for MPPT and power smoothing in permanent magnet synchronous generator-based wind turbines. Energy Convers. Manage. **86**, 892–900 (2014)
14. Ganjefar, S., Ghassemi, A.A., Ahmadi, M.M.: Improving efficiency of two-type maximum power point tracking methods of tip-speed ratio and optimum torque in wind turbine system using a quantum neural network. Energy **67**, 444–453 (2014)

Using Phase Change Materials (PCMs) to Reduce Energy Consumption in Buildings

I. Bekkouche[1(✉)], A. Benmansour[1], and R. Bhandari[2]

[1] Materials and Renewable Energy Research Unit (URMER),
University of Tlemcen, Tlemcen, Algeria
ismailbekkouche5@gmail.com,
halim.benmansour13@gmail.com
[2] Institute for Technology and Resources Management in the Tropics
and Subtropics (ITT), Cologne, Germany
ramchandra.bhandari@th-koeln.de

Abstract. Algeria's energy needs are actually satisfied mainly through the use of hydrocarbons. Demand for energy is growing exponentially because the population is growing massively and the building sector is in full expansion in order to address and solve the problems of housing, health and education. In 2011, Algeria initiated a dynamic green energy program by launching an ambitious development project aims at developing a renewable energy power source of 22000 MW and to save more than 60 Million TOE in energy efficiency by the year 2030. Our research topic is linked to the energy efficiency program initiated in Algeria. The objective of this article in to present a study carried out on the thermal performance of Phase Change Materials (PCMs) and their integration in buildings in order to achieve thermal comfort and to reduce the energy consumption. Then, an investigation on the capacity of using (PCMs) in Algeria is presented for a sustainable development and rational energy consumption.

Keywords: Energy transition · Energy efficiency · Phase change materials
Sustainable development · Building · Algeria

1 Introduction

Algerian building sector responsible on more than 40% of electricity consumption, because the growing population by more than 41 million and growing construction especially Algeria is an emerging country (Housing construction, Schools and administrations, development of industry and agriculture, etc.). Moreover, Algerians are heavy consumers of energy, because it is subsidized by the state and the cost is very low, it is one of the lowest in the world. This huge consumption has a big impact on greenhouse emissions with take into consideration Algeria is a signatory of the Kyoto Protocol (GHG reduction). For reduce this consumption need a serious attention to improve thermal insulation in buildings. Therefore, what is the technologies and applications, which have ability to reduce this consumption? different types of PCMs have ability to reduce energy consumption for achieve thermal comfort in building.

© Springer International Publishing AG 2018
M. Hatti (ed.), *Artificial Intelligence in Renewable Energetic Systems*, Lecture Notes
in Networks and Systems 35, https://doi.org/10.1007/978-3-319-73192-6_48

Many studies have investigated the effect of PCM on energy efficiency of buildings in Mediterranean climate. Among these studies, Tenorio et al. conducted a study on Energy efficiency Indicators for assessing construction systems storing Renewable Energy by application to Phase Change Material for bearing façades in Mediterranean climate and the results show that reduce buildings energy consumption effective by Ventilated facades which contain PCMs, and contributing to energy savings, PCMs adapt well with Spain's climate particularly, it reduce or eliminate energy consumption for air conditioning by deploying the cooling power stored overnight [2]. Guarino et al. have undertaken a parametric analysis to Phase change materials applications to optimize cooling performance of buildings in the Mediterranean area (Palermo) with five scenarios investigated and the results show that using PCMs reached at least 40% of cooling consumption reduction and air Temperature during peak hours in summer can be reduced by more than 7–8 °C, also conclude that integration PCMs in building reduce temperature fluctuations in an indoor environment aiming to obtain a more inert space to enhance the indoor comfort conditions [20]. In this paper, reviews on thermal performance of PCMs in different elements of buildings, and discussion on possibility of using it in Algerian buildings.

2   Phase Change Materials (PCM)

Every material could undergo phase change depending on the temperature it is subjected to. Additionally, phase change also depends on other properties of a material such as latent heat, specific heat capacity, thermal conductivity etc. [3]. PCM are applicable to energy storage through temperature regulation within some specific ranges of application in building (heating and cooling) [4]. This is because increasing the temperature in a certain point breaks up the chemical bonds in the material and heat absorbtion starts the endothermic process where the material changes state from solid to liquid, as temperature decreases. The material releases energy and returns to its original state (i.e. solid) and this phenomenon gives thermal comfort to room. This leads to a more stable and comfortable indoor climate, and decreases load on cooling and heating [5]. Therefore, desirable properties for PCMs is the latent thermal energy storage especially in building with low thermal mass [3, 5].

3   Classification of Phase Change Materials

Phase Change Materials (PCMs) for practical purposes can be classified into four categories: solid–solid, solid–liquid, gas–solid and gas–liquid. Only the solid–liquid category can be used for building sector (for cooling and heating) because the other varieties have technical limitations. there is a wide variety of PCMs on the market with different melting point ranges. The most common classifications of PCMs are organic, inorganic and eutectic, as presented in Fig. 1 [6].

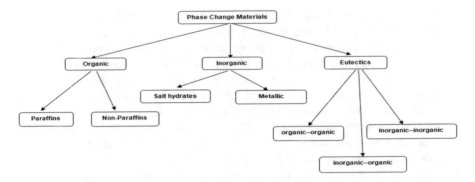

Fig. 1. Classification of phase change materials (PCMs) [3, 4, 17, 19]

4 Methods of PCMs Application in Building Components

There are many methods to integrate PCM into conventional construction materials and elements, which improve thermal comfort and reduce energy consumption.

- ***Direct incorporation***
  This method is simple by mix liquid or powder of PCMs directly in building materials such a gypsum, cement paste, mortar during production, it is practicable and economical method.
- ***Immersion***
  This technology incorporates PCMs directly in building structure components e.g. concrete, brick blocks, wallboards.
- ***Encapsulation***
  It comes in two forms viz.: macro-encapsulation and micro-encapsulation. Their integration before being used into construction elements need meeting requirements of strength, durability, stability and reliability protect [17, 19].

5 Literature Review

See Tables 1 and 2.

Table 1. List of several recent research studies on PCMs integration in different building components and summarizes the major results, almost all conclude that integration PCMs in buildings reduce energy consumption and control indoor temperature for get on thermal comfort.

Building components	PCM name	Melting temperature	Location	Study method	Main results	Reference
Wall	M91	29 °C	Cyprus	Simulation (TRNSYS)	This study showed that energy savings by combined case (PCM with insulation) is 66.2% compared to base case, combined case has the highest energy and money savings by 20,567 kWh/yr, 3003 Euro/yr respectively. also the payback period is reduced to 7½ years from 14½ just by PCM	[7]
	PCM29	29 °C	Greece	Simulation (DesignBuilder 4.0)	This study showed that the climate conditions, location and the surface area and the thickness of the PCM layer with position in the building is major factors which strengthens the performance of PCM, the conventional building with insulation and PCM layer are better compared with no insulation and lightweight building	[8]
	PCM24 and PCM26	24 °C and 26 °C respectively	Greece	Simulation (MATLAB and TRNSYS)	This numerical study revealed that the PCM24 reduce 29% of annual cooling loads while PCM26 reduce 16% and the cooling is low 25.7% compared by no-PCM	[9]
Floor	PCM GR27 Paraffin	27 °C	Spain	Experimental and simulation	This experimental study showed that PCM has a notable effect on the thermal response to the temperature by delaying it and reduce thermal diffusivity by increase quantity. But the organic PCM paraffin need a sufficient protection from fire behaviour for could be used in radiant floor	[10]
	PCM38 and PCM18	38 °C and 18 °C	China	Simulation	Analyse thermal performance of double layer from different kind of PCM in floor show that optimal melting temperatures exist, the fluctuations and the heat fluxes reduced on the floor surface, also notable increase energy released in peak period by 41.1% and 37.9% during heating and cooling	[11]
Roof	PCM29	29 °C	Greece	Simulation (DesigBuilder 4.0)	This study shows that the integration of PCMs on the ceiling achieve an important energy savings than external or internal wall	[8]
	Paraffin	21,7 °C	Morocco	Experimental	This experimental study shows that the integration of PCM in the roof of building reduce temperature by 1.5 °C inside, it also decreases the amplitude of thermal oscillations of south and west walls, plus it the PCM has a good insulation and storage power and increase thermal inertia, finally the ceiling with PCM reaches 11.2 W/m²	[12]
Window	PCM RT28HC Paraffin	27 °C to 29 °C	Portugal	Experimental	This experiment study reveals the PCM potential as thermal regulation of indoor spaces of buildings, 18% to 22% temperature reduced on indoor compartment by PCM in window shutter and decreased also temperature on peaks about 6% and 11% (maximum and minimum)	[13]
	Paraffin MG29	27 °C to 29 °C	China	Experimental and simulation	This study shows the performance of PCMW to reduce the gain of solar heat on double glazed window and shifts the peak cooling load when solar radiation is high, plus it the heat which enter building reduced by 18.3% through PCMW in sunny summer days	[14]
Heating and cooling systems	PCM Paraffin Wax	22 °C	Czech Republic	Experimental	This experimental study show possibility of optimizing the performance and capacity of thermal energy storage in buildings by PCM, improve the thermal comfort inside, and possibility to use PCMs in active systems for heating and cooling by combine it with the PV system	[15]

Table 2. List of review articles which study classification of PCMs and their integration in buildings to improve thermal performance to buildings.

The summary of review articles	Reference
This paper is an investigation of using PCMs in cooling applications in buildings in different climate, presents factors which affecting on performance of PCM, finally reviewed proposed solutions as some authors determine that effectiveness of melting temperature show more in warm climate	[4]
Study categorization of PCM and techniques of how can be integrated into buildings in different building elements (windows, shutters, walls, floor, roof, concrete) and also in different applications (cooling, active, passive, thermal comfort), with an environmental assessment	[5]
Reviews recent studies on PCMs applications and techniques to how integrate in active and passive cooling buildings envelope which incorporating method in walls is dominate and organic type has a reasonable price	[6]
Survey literatures of recent research and development activities of PCM technology in building applications, description of PCM and study different models used in building components with collecting data	[16]
Study the different desirable properties for selection PCMs and their methods of incorporating in building, classification of PCMs and analyse different measurement of thermal properties and conductivity	[17]
Investigate on different researches about using PCMs in free cooling systems of buildings, makes economical and environmental analysis, also discuss that PCMs able to reduce CO_2 emission	[18]
Summaries to previous works experimental and numerical of selection criteria to PCM and incorporating methods in buildings, also definitions to different measurements of thermal properties	[19]

6 Possibility to Integrate PCMs in Algerian Buildings

6.1 Climate and Buildings Situation in Algeria

North of Algeria experience mild Mediterranean climate cold in winter and hot in summer. However, because of the humidity as a result of the region's closeness to the sea, the weather can be uncomfortable. In contrast, the south has a desert climate, warm and dry summer.

Buildings in Algeria are poorly insulated. This is because of no energy savings mechanism to improve thermal performance, this is the result of absence awareness to importance of insulation in buildings and the government subsidy that make energy cost very low.

This climate conditions and absence of insulation are two strong reasons for increasing demand on energy from heating and cooling. To solve this problem and keep our resources a need to take a serious decision for improve insulation in buildings necessary.

6.2   Capacity of Using PCMs in Algerian Houses

Almost of all literature reviews in this paper are research studies on integration of PCMs in buildings envelope in Mediterranean climate which known by cold winter and hot summer. The results achieved are very effective for energy savings, decrease indoor temperature, increase thermal inertia and improve thermal performance of buildings.

Is observed in this study that melting temperature range is very important factor for chose right PCM in building Applications. Temperature range is between 20 °C to 30 °C [7], also PCM designed for 29 °C melting temperature used in many studies indicate is the best choice [7, 8, 13, 14], second factor is the climate conditions from external temperature and solar radiation. Orientation of the building has a significant role on the performance of PCM especially in the west which access to the best result in studies because it is more prone to the sun.

There are many methods for integrating PCM in different elements of buildings (wall, roof, floor and windows) or materials (brick, concrete and cement) by encapsulation or direct incorporation which is more cheap and easy to apply.

Position of PCM in building play a big role on the thermal performance of building. When PCM is applied on external side toward outside factors is more effective on energy consumption by shifting the ventilation systems load from the peak period to the off-peak period.

The high performance of PCM is achieved when is used in roof. This arrangement achieved the greatest energy saving compared with other elements of building [8]. This is because it receives more solar radiation, also reduce the temperature fluctuation, and has ability to reduce temperature on different walls of building [12].

Currently it is observed that there is increase in cooling and heating demand which increase energy demand. For this PCM has the ability to optimize thermal energy storage in building to achieving acceptable thermal comfort and control indoor temperature.

Economically PCM is also able to reduce energy consumption which decrease cost especially when it is combined by insulation. When a comparison made between insulation combined by PCM and insulation without PCM, Energy and money achieved are 20,567 kWh/yr, 3003 Euro/yr respectively which are very high, also payback period (7½ years) is shorter compared by payback period (14½ years) of PCM case [12].

All these points give Algeria ability to improve thermal performance in buildings by PCM in north and about hot climate in the south, Bouguerra et al. have undertaken a study on performance of PCM for cooling of buildings in Mild climates (Djelfa-Algeria) focusing on hot season. The results show that PCMs has ability to reduce energy approximately 20% for cooling energy and when integrate PCM in ceiling the widest temperature variation give the best compromise gain/cost and is relatively easy to realize even in existing houses [1]. All these positive results indicate that PCM has a huge potential of used for Algerian climate conditions.

7 Conclusion

The result of this study prove that phase change Materials (PCMs) have ability to reduce energy consumption in building sector (cooling in summer and heating in winter), this mean that PCMs reduce also Co_2 emissions in houses and give an acceptable thermal comfort. There are different types of PCM which give important to how select suitable one for any application of building, PCM available in the markets and by low cost. We gave even studying on previous literature reviews which done in Mediterranean climate and achieved important results in every cases, these results show the need to include this technologies in Algeria. Finally need a strong thinking to how improve energy efficiency in our buildings by using these materials by integrating them with renewable energy. This will conserve our energy resources and save our environment from greenhouse gas emission. Finally Algerian policy need to support this field of development for it is fully successful.

References

1. Bouguerra, E.H., Retiel, N.: Performance of phase change materials for cooling of buildings in mild climates. In: 30th International Plea Conference, CEPT University, Ahmedabad, December 2014
2. Tenorio, J.A., Sánchez-Ramos, J., Ruiz-Pardo, Á., Álvarez, S., Cabeza, L.F.: Energy efficiency indicators for assessing construction systems storing renewable energy: application to phase change material-bearing façades. Energies 8, 8630–8649 (2015)
3. Salam, M.R.H.A.A.: Simulation and optimization of solar thermal system integrated with PCM thermal energy storage for seawater desalination. Faculty of Engineering, Kassel University Kassel, Germany (175), February 2011
4. Souayfane, F., Fardoun, F., Biwole, P.H.: Phase Change Materials (PCM) for cooling applications in buildings: a review. Energy Build. (2016). http://dx.doi.org/10.1016/J. Enbuild.2016.04.006
5. Kalnæs, S.E., Jelle, B.P.: Phase change materials for building applications: a state-of-the-art review and future research opportunities. Energy Build. (2015). http://www.sciencedirect.com/science/article/pii/S0378778815001188
6. Akeiber, H., Nejat, P., Majid, M.Z.A., Wahid, M.A., Jomehzadeh, F., Famileh, I.Z., Calautit, J.K., Hughes, B.R., Zaki, S.A.: A review on phase change material (PCM) for sustainable passive cooling in building envelopes. Renew. Sustain. Energy Rev. 60, 1470–1497 (2016)
7. Panayiotou, G.P., Kalogirou, S.A., Tassou, S.A.: Evaluation of the application of phase change materials (PCM) on the envelope of a typical dwelling in The Mediterranean region. Renew. Energy 97, 24–32 (2016)
8. Karaoulis, A.: Investigation of energy performance in conventional and lightweight building components with the use of phase change materials (PCMs): energy savings in summer season. Procedia Environ. Sci. 38, 796–803 (2017)
9. Stamatiadou, M.E., Katsourinis, D., Founti, M.: Computational assessment of a full-scale mediterranean building incorporating wallboards with phase change materials. Indoor Built Environ. 26(10), 1429–1443 (2016)
10. Laia, H., Mazo, J., Delgado, M., Zalba, B.: Fire behaviour of a mortar with different mass fractions of phase change material for use in radiant floor systems. Energy Build. 84, 86–93 (2014)

11. Jin, X., Zhang, X.: Thermal analysis of a double layer phase change material floor. Appl. Therm. Eng. **31**, 1576–1581 (2011)
12. Mourid, A., El Alami, M., Najam, M.: Passive study of energy efficiency of a building with PCM on the roof during summer in Casablanca. Power Energy Eng. **4**, 26–37 (2016)
13. Silva, S., Vicente, R., Rodrigues, F., Samagaio, A., Cardoso, C.: Performance of a window shutter with phase change material under summer Mediterranean climate conditions. Appl. Therm. Eng. **84**, 246–256 (2015)
14. Zhong, K., Li, S., Sun, G., Li, S., Zhang, X.: Simulation study on dynamic heat transfer performance of PCM-filled glass window with different thermophysical parameters of phase change material. Energy Build. **84**, 246–256 (2015)
15. Skovajsa, J., Koláček, M., Zálešák, M.: Phase change material based accumulation panels in combination with renewable energy sources and thermoelectric cooling. Energies **84**, 246–256 (2016)
16. Akeiber, H.J., Wahid, M.A., Hussen, H.M., Mohammad, A.T.: Review of development survey of phase change material models in building applications. Sci. World J. (2014). http://dx.doi.org/10.1155/2014/391690
17. Memon, S.A.: Phase change materials integrated in building walls: a state of the art review. Renew. Sustain. Energy Rev. **31**, 870–906 (2014)
18. Kamali, S.: Review of free cooling system using phase change material for building. Energy Build. **80**, 131–136 (2014)
19. Zhou, D., Zhao, C.Y., Tian, Y.: Review on thermal energy storage with phase change materials (PCMs) in building applications. Appl. Energy **92**, 593–605 (2012)
20. Guarino, F., Longo, S., Cellura, M., Mistretta, M., La Rocca, V.: Phase change materials applications to optimize cooling performance of buildings in the Mediterranean area: a parametric analysis. Energy Procedia **78**, 1708–1713 (2015)

Design and Modeling of Miniature On-Chip Spiral Inductor for DC-DC Converter

Abdelhadi Namoune[1], Rachid Taleb[2(✉)], and Fayçal Chabni[2]

[1] Laboratoire d'Electronique de Puissance Appliquée (LEPA),
Electrical Engineering Department, USTO-MB University, Oran, Algeria
n2abdelhadi@yahoo.fr
[2] Laboratoire Génie Electrique et Energies Renouvelables (LGEER),
Electrical Engineering Department, Hassiba Benbouali University, Chlef, Algeria
rac.taleb@yahoo.fr, chabni.fay@gmail.com

Abstract. In this paper, we study the effect of the technological parameters of on chip inductor, the characteristics of on chip inductor is simulated using the MATLAB simulation. All of inductor is fabricated in the standard silicon process with metal layers. The simulated results are accurate only when technological parameters used in ensemble, such as the metal conductivity, metal thickness, oxide thickness between the top metal and the substrate, and substrate conductivity on inductance, quality factor and self-resonant frequency.

Keywords: Integrated inductor · Technological parameter · Spiral inductor

1 Introduction

One may that has assisted designers in satisfying demands for small size and low cost is the possibility of using on chip inductive devices in silicon radio frequency integrated circuits. On chip inductor play important roles in impedance matching, tuning and filtering. On aspect that limits their use is the amount of consumed silicon area. Overall die size is a direct driver in production cost. However, the continuing advance of process technologies, such as providing thicker and higher conductivity metal [1], thicker and lower permittivity dielectric layers, and lower conductivity substrate [2], makes high quality on chip passive devices more readily available [3]. Many design and analysis techniques of chip inductors have been investigated to correlate performances with technological parameters.

This paper contains of as follows Sect. 2 discuss about design and modeling of the spiral inductor, Sect. 3 about results and discussion of different technological parameters and Sect. 4 about conclusion.

2 Design and Modeling of Square Inductor

The lateral parameters of a spiral are shown in Fig. 1. The main parameters are the number of turns (n), the metal width (w), the spacing between adjacent turns (s), the inner and outer diameters (d_{in} and d_{out}), and the average diameter (d_{avg}). The fill ratio (ρ) is given by either of the following expression [4] (Table 1):

© Springer International Publishing AG 2018
M. Hatti (ed.), *Artificial Intelligence in Renewable Energetic Systems*, Lecture Notes in Networks and Systems 35, https://doi.org/10.1007/978-3-319-73192-6_49

$$\rho = \frac{d_{out} - d_{in}}{d_{out} + d_{in}} = \frac{n(w + s) - s}{d_{avg}} \tag{1}$$

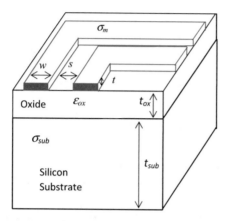

Fig. 1. Transverse section of a spiral planar inductor

Table 1. Technological parameters of the integrated inductor

Symbol	Technological parameters
t_{ox}	Thickness of oxide layer (distance between the substrate and coil)
t	Metal thickness
σ_m	Metal conductivity
σ_{si}	Substrate conductivity
ρ_{Si}	Substrate resistivity
ε_{ox}	Permittivity of the dielectric between substrate and coil
t_{sub}	Substrate thickness
ε_{Si}	Permittivity of the substrate

The model for a micro inductor is shown in Fig. 2 [5]. The spiral π model includes the series inductance (Ls), the series resistance (Rs), the feed-forward capacitance (Cs), the inductor to substrate capacitance (Cox), and the substrate resistance (Rsi) and capacitance (Csi).

The inductance (Ls) calculation of spiral inductor is determined by approximate expressions derived in [6]. The expressions include electromagnetic principles using current sheet approximations obtained for discrete inductors. The equation for the series inductance:

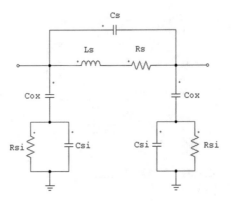

Fig. 2. Equivalent electrical model

$$L_s = \frac{2\mu.n^2.d_{avg}}{\pi}\left[\ln\left(\frac{2,067}{\rho}\right) + 0,178.\rho + 0,125.\rho^2\right] \tag{2}$$

Where μ is the magnetic permeability of free space ($\mu = 4\pi \cdot 10^{-7}$ H/m).
The series resistance is calculated by the following equation [7]:

$$R_s = \frac{1}{\sigma.\delta.w.\left(1 - e^{\frac{-t}{\delta}}\right)} \tag{3}$$

Where l and w are the length and width of the spiral, σ is the metal conductivity, t is the metal thickness, and δ is the skin length given by [8]:

$$\delta = \sqrt{\frac{2}{2.\pi.f.\mu.\sigma}} \tag{4}$$

Where f is the frequency.

The series resistance expression models the increase of resistance at higher frequencies due to the skin effect [9].

The series capacitance Cs represents the parasitic capacitive coupling between input and output of the inductor, can be estimated using the following formula [10]:

$$C_s = \frac{t.\varepsilon_0.l^2}{s} \tag{5}$$

Where, ε_0 is the permittivity of free space.

Csi and *Rsi* are the capacitance and resistance of the silicon substrate and *Cox* is the oxide capacitance between the spiral and the silicon substrate calculated [11] as:

$$C_{ox} = \frac{1}{2} l.w \frac{\varepsilon_{ox}}{t_{ox}} \tag{6}$$

$$R_{si} = \frac{2}{l.w.G_{sub}} \tag{7}$$

$$C_{si} = \frac{1}{2} l.w.C_{sub} \tag{8}$$

Where *Gsub* and *Csub* are the conductance and capacitance per unit area of the silicon substrate and *tox* is thickness of the oxide layer separating the spiral and the substrate.

The efficiency of integrated micro coil is calculated [12] according by relation:

$$Q = 2\pi . \frac{\text{Stocked energie}}{\text{dissipated energie}} \tag{9}$$

3   Results and Discussion

The simulation procedure pursued for the analysis of inductor is meshing and porting of the device. After porting S-parameters are obtained to calculate inductance and Quality factor. Now from S-parameters, Y-Parameters are obtained. Using Y-Parameters the inductance and quality factor values are intended using the expressions revealed in Eqs. (10) and (11) respectively.

$$L = \frac{\text{Img}\left(\frac{1}{Y_{11}}\right)}{2.\pi.f} \tag{10}$$

$$Q = \frac{\text{Img}\left(\frac{1}{Y_{11}}\right)}{\text{Real}\left(\frac{1}{Y_{11}}\right)} \tag{11}$$

A MATLAB simulation was achieved to study the influence of different technological parameters on the value of the inductance and quality factor.

3.1   Metal Conductivity

Figure (3a) show how the metal conductivity influences on inductance value. Changing the metal conductivity nearly cannot modify inductance value. Figure (3b) show how the metal conductivity influence on Quality factor value. It can be seen that increasing the metal conductivity increases the quality factor value. The self resonant frequency will not be changed by the metal conductivity. The inductor with higher metal conductivity has higher Q because it has lower series resistance.

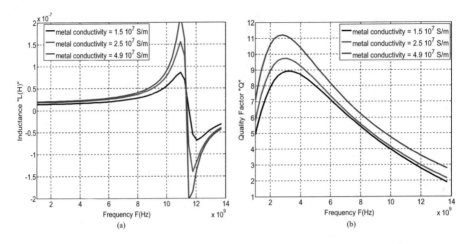

Fig. 3. Effect of the metal conductivity (a) on inductance, (b) on quality factor

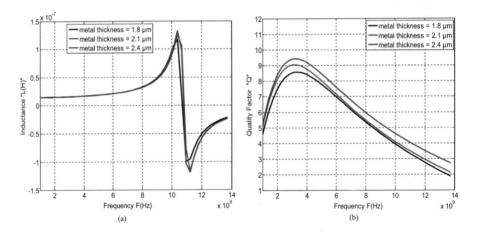

Fig. 4. Effect of the metal thickness (a) on inductance, (b) on quality factor

3.2   Metal Thickness

Figure (4a) illustrates how the metal thickness influences on inductance value. Varying the metal thickness approximately cannot change inductance value. Figure (4b) show how the metal thickness influences on quality factor value. The maximum quality factor of the inductor could be enhanced by escalating the metal thickness. The self resonant frequency will not be changed by the metal thickness.

The inductor with higher metal thickness has higher quality factor because it has lower series resistance.

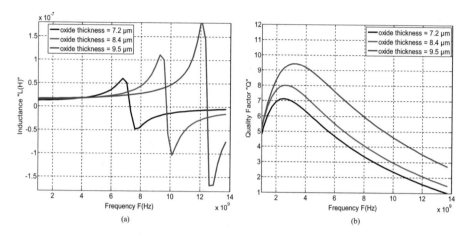

Fig. 5. Effect of the oxide thickness (a) on inductance, (b) on quality factor

3.3   Oxide Thickness

Figure (5a) demonstrates how the oxide thickness influences on inductance value. Changing the oxide thickness almost cannot vary inductance value. Figure (5b) illustrates how the oxide thickness influences on quality factor value. A wider oxide layer diminishes the capacitance of the silicon substrate, which advances the self resonant frequency. Mounting the oxide thickness will elevate the quality factor and the self resonant frequency.

3.4   Substrate Conductivity

Figure (6a) illustrates how the substrate conductivity influences on inductance value. Changing the substrate conductivity almost cannot alter inductance value. Figure (6b)

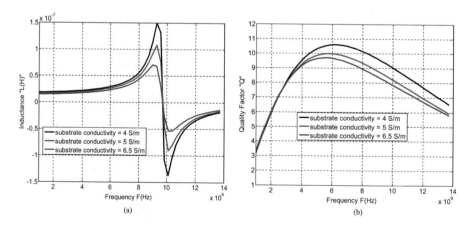

Fig. 6. Effect of the substrate conductivity (a) on inductance, (b) on quality factor

illustrates how the substrate conductivity influences on quality factor value. As the substrate conductivity reduces the quality factor augment. It is clear that substrate conductivity is the key bounding factor for the quality factor in a certain frequency band.

4 Conclusion

The square spiral inductor is designed for RF circuits in this paper. The parametric characteristics of this design was presented and the design procedure outline its performance in terms Quality factor and inductance. The effective quality factor and inductance was also deliberate experimentally by considering dynamic variation of the technological parameters within the frequency range of (1–14) GHz. Better performance of the design is obtained with the optimized parameters from three values. Maximum value of quality factor obtained in the analysis is 2 at 4 GHz and the self resonant frequency of 11 GHz.

References

1. Fukuda, Y., Inoue, T., Mizoguchi, T., Yatabe, S., Tachi, Y.: Planar inductor with ferrite layers for DC-DC converter. IEEE Trans. Magn. **39**(4), 2057–2061 (2003)
2. Sia, C.B., Lim, W.M., Ong, B.H., Tong, A.F., Yeo, K.S.: Modeling and layout optimization techniques for silicon-based symmetrical spiral inductors. Prog. Electromagn. Res. **143**, 1–18 (2013)
3. Wang, C., Kim, N.-Y.: Analytical optimization of high-performance and high-yield spiral inductor in integrated passive device technology. Elsevier Microelectron. J. **43**, 176–181 (2012)
4. Sandrolini, L., Reggiani, U., Puccetti, G.: Analytical calculation of the inductance of planar ZIG-ZAG spiral inductors. Prog. Electromagn. Res. **142**, 207–220 (2013)
5. Piruouznia, P., Ganji, B.A.: Analytical optimization of high performance and high quality factor MEMS spiral inductor. Prog. Electromagn. Res. M **34**, 17–179 (2014)
6. Dai, C.-L., Hong, J.-Y., Liu, M.-C.: High Q factor CMOS-MEMS inductor. In: Symposium on Design, Test, Integration and Packaging of MEMS/MOEMS, pp. 138–141 (2008)
7. Tseng, S.-H., Hung, Y.-J., Juang, Y.-Z., Lu, M.S.-C.: A 5.8-GHz VCO with CMOS-compatible MEMS inductors. Elsevier Sens. Actuators A **139**, 187–193 (2007)
8. Zhao, P., Wang, H.G.: Resistance and inductance extraction using surface integral equation with the acceleration of multilevel green function interpolation method. Prog. Electromagn. Res. **83**, 43–54 (2008)
9. Pan, S.J., Li, L.W., Yin, W.Y.: Performance trends of on-chip spiral inductors for RFICs. Prog. Electromagn. Res. **45**, 123–151 (2004)
10. Huang, F., Lu, J., Jiang, N., Zhang, X., Wu, W., Wang, Y.: Frequency-independent asymmetric double-π equivalent circuit for on-chip spiral inductors: physics-based modeling and parameter extraction. IEEE J. Solid-State Circuits **41**(10), 2272–2283 (2006)
11. Lai, I.C.H., Fujishima, M.: A new on-chip substrate-coupled inductor model implemented with scalable expressions. IEEE J. Solid-State Circuits **41**(11), 2491–2499 (2006)
12. Kang, M., Gil, J., Shin, H.: A simple parameter extraction method of spiral on-chip inductors. IEEE Trans. Electron Devices **52**(9), 1976–1981 (2005)

Optimization of Copper Indium Gallium Diselenide Thin Film Solar Cell (CIGS)

A. Aissat[1,2(✉)], A. Bahi Azzououm[1], F. Benyettou[2],
and A. Laidouci[1,2]

[1] LATSI Laboratory, Faculty of Technology, University Saad Dahlab,
BP270, 09000 Blida, Algeria
sakre23@yahoo.fr
[2] FUNDAPL Laboratory, Faculty of Sciences, University Saad Dahlab,
BP270, 09000 Blida, Algeria

Abstract. We performed modeling and simulation of copper indium gallium diselenide (CIGS) thin film solar cell. CIGS absorbers today have a typical thickness of about 1–2 μm. However, on the way toward mass production, it will be necessary to reduce the thickness even further. We investigated the influence of the alloy compositions $x = [Ga]/\{[In] + [Ga]\}$, the CIGS absorbers thickness and the temperature. Optimal results are obtained with a thickness of about 2 μm and a temperature of 318 K. It was also shown that, the short-circuit current density (J_{sc}) decreases when the x composition increases. Very high J_{sc} of 40.72 mA/cm^2 was obtained, when x = 0.2. In contrary, the open-circuit voltage (Voc), the fill factor (FF) and the efficiency (η) of the solar cell are increasing with the increase of the x composition. An optimal efficiency of about 30.34% was obtained with x = 0.9. Moreover, a comparison with published data for the Cu(In,Ga)Se$_2$ cells have shown an excellent agreement.

Keywords: Semiconductor · CIGS · Thin film · Solar cell · Photovoltaic

1 Introduction

Thin film solar cells have the potential for low-cost and large-scale terrestrial photovoltaic applications. A number of semiconductor materials including polycrystalline CdTe, CIGS and amorphous silicon (a-Si) materials have been developed for thin-film photovoltaic solar cells [1]. CuIn$_{1-x}$Ga$_x$Se$_2$ (CIGS) has attracted great interest as an absorber layer in thin film photovoltaic (PV) devices because of the high power conversion efficiency that it provided [2–6]. The polycrystalline copper indium gallium diselenide (CIGS) thin film is an element of the I–III–VI2 group of chalcopyrite semiconductors, compared to other Cu-chalcopyrite thin film solar cells as well as CdTe and amorphous Si thin film solar cells [1], CIGS gives the highest conversion efficiency.

Recently, the conversion efficiency of CIGS thin film solar cells has been improved to 20.3% by ZSW (Centre for Solar Energy and Hydrogen Research) [4]. The optical and electrical properties of CIGS are in function of the alloy composition x measured as the atomic ratio $x = [Ga]/\{[In] + [Ga]\}$. The efficiency of PV devices exhibits a

© Springer International Publishing AG 2018
M. Hatti (ed.), *Artificial Intelligence in Renewable Energetic Systems*, Lecture Notes in Networks and Systems 35, https://doi.org/10.1007/978-3-319-73192-6_50

dependence on CIGS composition, and the maximum is found to occur at the value of x ~ 0.3, when considered as an average through the depth of the absorber layer [7].

This paper addresses a simulation study to optimize the CIGS based thin film solar cells through the investigation of the influence of different alloy compositions [Ga]/{[In] + [Ga]}, also the influences of CIGS absorbers thickness and the temperature.

2 Structure of CIGS Solar Cells and Simulation

The structure of CIGS solar cells include five layers, from bottom to up: Substrate: CIGS solar cells can use glass, foil, and more flexible material such as Polyimide as substrates. Back contact: molybdenum is the back contact of choice because it can tolerate the harsh reactive ambient of the selenization processes at high temperature and form an Ohmic contact with CIGS. Absorption layer: p type CIGS thin film is formed with a little Cu-poor in the total composition. It should be noted that the ratio Ga/In + Ga was controlled. Buffer layer: CdS is usually used as the buffer material.

Front contact: Transparent and contacting oxide (TCO) window bilayers include intrinsic ZnO and Al doped ZnO (AZO) (Fig. 1).

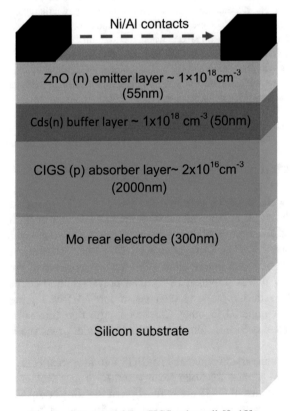

Fig. 1. Structure of the CIGS solar cell [8–10].

The device characteristics of the CIGS solar cells are studied numerically using the Silvaco-Atlas software. To investigate the effect of the variation of the $Ga/(In + Ga)$ ratio, x, in a CIGS solar cell, physical parameters are required, and are obtained from measurements and previous literatures to make the accurate model [8–10]. Table 1 summarizes the physical parameters used in the simulation.

Table 1. Base parameters for CIGS solar cells.

	ZnO	CdS	CIGS
Thickness (μm)	0.055	0.050	2
$E_g(eV)$	3.3	2.4	1.2
ε_r	9	10	13.4
$\chi_e(eV)$	4	3.75	3.89
$\mu_n(cm^2/Vs)$	50	10	300
$\mu_p(cm^2/Vs)$	5	1	10
$N_A(1/cm^3)$	0	0	8.10^{16}
$N_D(1/cm^3)$	5.10^{17}	5.10^{17}	5.10^{17}
$NC(1/cm^3)$	$2.2.10^{18}$	$2.2.10^{18}$	$2.2.10^{18}$
$NV(1/cm^3)$	$1.8.10^{19}$	$1.8.10^{19}$	$1.8.10^{19}$

2.1 Effects of Ga/(In + Ga) on CIGS Thin Films Solar Cell

To investigate the effect of the variation of the $Ga/(In + Ga)$ ratio, x, in a CIGS solar cell, firstly, we will report the modeling and simulation results of CIGS solar cells with x = 0.3, in comparison with the previous reported experimental results as shown in Table 2. Secondly, we will show the simulation results of CIGS solar cell with absorber layer of 2 μm for variable x ratio. The I–V curve obtained using ATLAS^{TM} is shown in Figs. 2 and 3.

Table 2. CIGS cell parameters and characteristics.

	Simulation	Experimental [11, 12]
x = $Ga/(In + Ga)$	0.3	0.3
$E_g(eV)$	1.2726	1.27
$J_{SC}(mA/cm^2)$	39.8959	31.2
$V_{OC}(V)$	0.737769	0.752
FF	78.2406	77.73
Efficiency (%)	23.0293	18.3

The difference between the simulation and the experimental results with absorber layer of 2 μm for x = 0.3 were due to different values of the basic parameters used and especially to the thickness of the absorber. The large difference in the short-circuit current is because ATLAS^{TM} uses the entire surface of the superior layer as a contact,

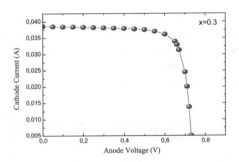

Fig. 2. I–V curve for a CIGS solar cell with x = 0.3.

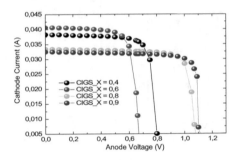

Fig. 3. I–V curve for a CIGS solar cell with variable x composition.

in the other hand the actual cells presents a grid of contacts at the top are used as a contact of the cell. The difference in the efficiencies was due to large difference in short-circuit current.

According to the results of Table 3, we observe that the short-circuit current density (J_{sc}) decreases when the compositions x increase, a very high J_{sc} of 40.72 mA/cm^2 was obtained when x = 0.2. In contrary, the open-circuit voltage (V_{oc}), the fill factor (FF) and the efficiency (η) of the solar cell increase when the composition x increase. An optimal efficiency of about 30.34% was obtained with x = 0.9.

Table 3. Characteristics of CIGS cells with absorber layer of 2 μm for variables compositions x.

	CIGS absorber = 2 μm				
x = Ga/(In + Ga)	0.2	0.4	0.6	0.8	0.9
J_{SC}(mA/cm^2)	40.7226	38.3248	35.279	33.4837	32.5563
V_{OC}(V)	0.67045	0.804901	0.940859	1.07906	1.14862
FF	77.1967	78.9545	80.0911	80.893	81.1564
Efficiency (%)	21.0766	24.3553	26.5843	29.2274	30.3483

2.2   Effects of CIGS Absorber Thickness on CIGS Thin Films Solar Cell

To reduce production time and reduce cost. The thickness of a solar cell based on a thin layer is a very important parameter, so the choice of an optimum thickness is the goal of several researchers. The standard thickness of the CIGS absorber layer in CIGS thin-film solar cells is presently 1.5–2 µm. If this thickness could be reduced with no loss in performance, it would lead to even more effective solar cells and in boosting efficiencies to new record levels. To investigate the effect of CIGS absorbers thickness, firstly, we report the modeling and simulation of CIGS solar cells with absorber layer of 1.2, 1.6 and 1.8 µm at x = 0.3, and secondly at x = 0.9 as shown in Table 4.

Table 4. Characteristics of CIGS cells for variables absorbers thickness.

	Ga/(In + Ga) = 0.3			Ga/(In + Ga) = 0.9		
CIGS absorber thickness (µm)	1.2	1.5	1.8	1.2	1.5	1.8
$J_{SC}(\text{mA/cm}^2)$	38.6222	39.3271	39.7289	30.8465	31.7628	32.3169
$V_{OC}(\text{V})$	0.7345	0.7364	0.7374	1.1445	1.1469	1.1481
FF	78.0754	78.1867	78.2281	80.982	81.0866	81.1409
Efficiency (%)	22.1479	22.6441	22.918	28.592	29.5391	30.1068

The I–V curve obtained is shown in Fig. 4.

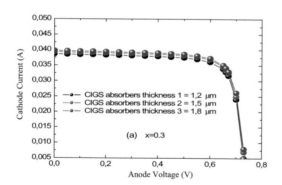

Fig. 4. I–V curve for a CIGS solar cell with variable CIGS absorbers thickness and for fixed x compositions.

According to the results in Table 4, we observe that, whatever the ratio x, the short-circuit current density increase when the thickness increase, a very high J_{sc} of 39.72 mA/cm² was obtained at x = 0.3 when the thickness is about 1.8 µm. Similarly, the V_{oc}, FF and the η of the solar cell increase when the thickness increase. An optimal efficiency of about 30.10% was obtained at x = 0.9 when the thickness is about 1.8 µm.

2.3 Effects of Temperature on CIGS Thin Films Solar Cell

Temperature is an important parameter in the behavior of the solar cells. According to the results of Table 5, we observe that, whatever the ratio x, the increase in temperature results in a net decrease in the open circuit voltage, as well as a decrease in the maximum power (a variation of 15 K results in a decrease of 10 mW of the maximum power). The optimum operating temperature used in our calculations is 318 K.

Table 5. Characteristics of CIGS cells for variables temperature.

	$Ga/(In+Ga) = 0.3$			$Ga/(In+Ga) = 0.9$		
Temperature (K)	288	303	318	288	303	318
$J_{SC}(mA/cm^2)$	40.1625	40.0283	39.8959	32.835	32.6914	32.5563
$V_{OC}(V)$	0.7952	0.7665	0.737769	1.2066	1.1776	1.14862
FF	80.556	79.4144	78.2406	82.9127	82.045	81.1564
Efficiency (%)	25.7284	24.3675	23.0293	32.8492	31.5894	30.3483

The I–V curve obtained using ATLAS^{TM} is shown in Fig. 5. At higher temperature, the band gap energy has been slightly reduced, which may accelerate recombination between the valence band and the conduction band. Although more free electrons are produced in the conduction band, the high temperature band gap energy is unstable which can lead to the recombination of electrons and holes while traversing across the regions.

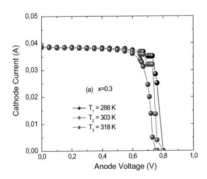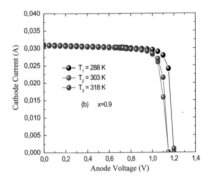

Fig. 5. I–V curve for a CIGS solar cell with variable temperature and for fixed x compositions: x = 0.3 (b) x = 0.9

3   Conclusion

Based on the Silvaco-Atlas software, we presented numerical simulations of CIGS thin film solar cells under AM1.5. An optimal results with conversion efficiency around 23.02% were obtained at x = 0.3 with a thickness of about 2 μm and a temperature of

318 K, it's in agreement with the high record conversion efficiency found experimentally in the CIGS solar cell. From this study, we found that the open-circuit voltage (V_{oc}), the fill factor (FF) and the efficiency (η) of the solar cell are increasing with the increase of the x composition. In contrary, we observe that, whatever the ratio x, the increase in temperature results in a net decrease in the open circuit voltage, as well as a decrease in the maximum power (a variation of 15 K results in a decrease of 10 mW of the maximum power).

References

1. Elbar, M., Tobbeche, S.: Numerical simulation of CGS/CIGS single and tandem thin film solar cells using the Silvaco-Atlas software. Energy Procedia **74**, 1220–1227 (2015)
2. Repins, I., Contreras, M.A., Egaas, B., DeHart, C., Scharf, J., Perkins, C.L., To, B., Noufi, R.: 19.9%-efficient ZnO/CdS/CuInGaSe$_2$ solar cell with 81.2% fill factor. Prog. Photo Volta. Res. Appl. **16**, 235–239 (2008)
3. Chirilă, A., Reinhard, P., Pianezzi, F., Bloesch, P., Uhl, A.R., Kranz, L., Keller, D., Gretener, C., Tiwari, A.N.: Potassium-induced surface modification of Cu(In,Ga)Se$_2$ thin films for high-efficiency solar cells. Nat. Mater. **12**, 1107–1111 (2013)
4. Jackson, P., Hariskos, D., Lotter, E., Paetel, S., Wuerz, R., Menner, R., Wischmann, W., Powalla, M.: New world record efficiency for Cu(In,Ga)Se$_2$ thin-film solar cells beyond 20%. Prog. Photovolt. Res. Appl. **19**, 894–897 (2011)
5. Jackson, P., Hariskos, D., Wuerz, R., Kiowski, O., Bauer, A., Friedlmeier, T.M., Powalla, M.: Properties of Cu(In,Ga)Se$_2$ solar cells with new record efficiencies up to 21.7%. Physica Status Solidi Rapid Res. Lett. **9**, 28–31 (2015)
6. Jackson, P., Hariskos, D., Wuerz, R., Powalla, M.: Compositional investigation of potassium doped Cu(In,Ga)Se$_2$ solar cells with efficiencies up to 20.8%. Physica Status Solidi Rapid Res. Lett. **8**, 219–222 (2014)
7. Aryal, P., Ibdah, A.-R.: Parameterized complex dielectric functions of CuIn$_{1-x}$Ga$_x$Se$_2$: applications in optical characterisation on of compositional non-uniformities and depth profiles in materials and solar cells. Prog. Photovolt. Res. Appl. **10**, 1002/p ip.2774 (2016)
8. Kuo, S.Y., Hsieh, M.Y., Lai, F.I., Liao, Y.K., Kuo, H.C.: Modeling and optimization of sub-wavelength grating nanostructures on Cu(In,Ga)Se$_2$ solar cell. Jpn. J. Appl. Phys. **51** (2012). Article ID 10NC14
9. Kuo, S.-Y., Hsieh, M.-Y., Han, H.-V., et al.: Dandelion-shaped nanostructures for enhancing omnidirectional photovoltaic performance. Nanoscale **5**(10), 4270–4276 (2013)
10. Hsieh, M.-Y., Kuo, S.-Y., Han, H.-V., et al.: Enhanced broadband and omnidirectional performance of Cu(In,Ga)Se$_2$ solar cells with ZnO functional nanotreearrays. Nanoscale **5** (9), 3841–3846 (2013)
11. Fotis, K.: Modeling and simulation of a dual-junction CIGS solar cell using Silvaco ATLAS. Naval Postgraduate School, Monterey, California, December 2012
12. Efficiency and band gap energy. http://www.grc.nasa.gov/WWW/RT/RT1999/images/5410hepp-f3.jpg

Nano-sensor Based on Ionic Liquid Functionalized Graphene Modified Electrode for Sensitive Detection of Tetrahydrofuran

Mohamed Kadari[1(✉)], Mostefa Belarbi[2], El Habib Belarbi[1], and Yassine Chaker[1]

[1] LSCT Research Laboratory, University of Tiaret, Tiaret, Algeria
kadarimohamed@yahoo.fr, belarbi.Hb@gmail.com,
chakeryacine@hotmail.com
[2] LIM Research Laboratory, University of Tiaret, Tiaret, Algeria
belarbimostefa@yahoo.fr

Abstract. This paper develops characterization of graphene and ionic liquids in order to develop material, which allows several applications like sensors and energy storage. Several steps were applied in order to synthesise Nano-sensor. First, we synthesised Oxide Graphene. We mix ionic liquids with (GO) in order to maximise conductivity. We add new elements like the toxic substance tetrahydrofuran, which that it was adsorbed by GO-IL. This new material constitutes needs to product the Sensor based on ionic liquid functionalized graphene modified electrode for sensitive detection of Tetrahydrofuran.

Keywords: Nano sensor · Graphene · Ionic liquids · Toxic substances
Tetrahydofuran

1   Introduction

Graphene is a two-dimensional molecule thick of an atom, exclusively made up of sp^2 hybrid carbon atoms. First obtained in 2004 [1], graphene can be formally regarded as the fundamental component of all sp^2 carbon allotropes, including graphite, carbon nanotubes and fullerenes. In particular, graphite is a set of graphene sheets stacked one on top of the other and held together by weak van der Waals forces, while fullerenes and carbon nanotubes can be considered wrapped and rolled graphene, respectively.

The structure of the graphene is free of defects, with atoms all of the same type, bound by strong and flexible bonds: this is the origin of the extraordinary properties of this material. Moreover, electrons can move across the graphene network without encountering obstacles due to imperfections in the structure or presence of heteroatoms. As a result, electrons can move much faster than they do in metals or semiconductors.

The outstanding properties of graphene include high values of Young's modulus (~1100 GPa), breaking strength (125 GPa), thermal conductivity (5×10^3 $Wm^{-1}K^{-1}$) [2], mobility of charge carriers (2×10^5 $cm^2V^{-1}s^{-1}$) and a specific surface area (2630 m^2g^{-1}). Graphene and its derivatives are promising candidates as components in applications such

© Springer International Publishing AG 2018
M. Hatti (ed.), *Artificial Intelligence in Renewable Energetic Systems*, Lecture Notes in Networks and Systems 35, https://doi.org/10.1007/978-3-319-73192-6_51

as energy storage materials, [3] paper-shaped materials, [4] polymer composites, liquid crystal devices and mechanical resonators.

Graphene represents a high energy density, power density and specific capacitance in relation of its highly power structure based supercapacitors [3]. The porous nature of the graphene and highly reduced graphene enhances the accessibility for ion diffusion and high conduction. In order to enhance this characteristic, we used ionic liquids. Section 2 presents synthesis of (GO) with ionic liquids and Sect. 3 present DRX and UV-vis Characterization.

Ionic liquids are liquids that consist almost exclusively of cations such as imidazolium, pyridinium, quaternary phosphonium and quaternary ammonium, and anions such as halogen, triflate, tetrafluoroborate and hexafluorophosphate, which exist in the liquid state at relatively low temperatures. Their characteristic features include almost no vapor pressure, non-flammability, non-combustibility, high thermal stability, relatively low viscosity, wide temperature ranges for being liquids, and high ionic conductivity generally of the order of 10^{-1} $S.m^{-1}$ [5]. Their electrochemical window is wide, which has been exploited in the field of fuel cells.

It is most interesting to note that their physicochemical properties are adjustable by changing the nature of the anion and/or the cation. The possible combinations of cations/anions are very numerous and continuously evolving, such that in general the properties of ionic liquids are governed by Coulomb forces, hydrogen bonding and Van der Waals interactions [6].

Chemical and physical methods have been proposed for the preparation of graphene. They include production from colloidal suspensions, [7, 8] electrochemical methods, micromechanical exfoliation, chemical vapor depositionand epitaxial growth. In our work, we have synthesized the graphene oxide associated with the ionic liquid, and also the graphene oxide alone as a control sample, followed by a characterization by DRX and UV spectroscopy.

The purpose of this part of the search work is to test adsorption of (GO) associated with ionic liquids in order to adsorb toxic substances, [9] for the chosen (THF) by (GO-IL) and a basis to energy storage which can be exploited. UV characterization is exploited before in order to prove that our (GO-IL) material can adsorb (THF).

2 Experiments

2.1 Preparing Graphene Oxide Using Hummers Updated Method

Graphene oxide (GO) was synthesized from graphite powder using the Hummers method (modified) [10–12]. Briefly, the first step, a preparation of a mixture of graphite with 0.5 g of sodium nitrate, then with constant stirring, 23 ml of sulfuric acid are added. After one hour, 3 g of $KMnO_4$ was added gradually to the solution while maintaining the temperature below 20 °C to prevent overheating and explosion. The mixture is stirred at 35 °C. for 12 h. This gives rise to a first oxidation, which is produced by dimanganese heptoxide (Mn_2O_7) and permanganyl cation (MnO_3^+) formed by permanganate in contact with the solvent and $NaNO_3$, In view of the oxidizing nature of the latter [13]. The functional groups created on the graphite comprise epoxides, tertiary alcohols, ketones, carboxylic acids and lactones. This first stage also begins the exfoliation of graphite in graphene. The resulting solution was diluted by adding 500 ml of deionized water with vigorous stirring. The slow addition of demineralised water to this stage causes an exotherm and the now aqueous medium transforms the previous oxidizing species (Mn_2O_7 and MnO_3^+) into the permanganate ion (MnO_4^-), also capable of oxidizing graphene. Under these conditions, the alkenes are capable of undergoing oxidative cleavage leading to two carbonyls (Fig. 1a) [13]. The epoxides can be hydrolysed to hydroxyls (Fig. 1d) and the ketones on the (GO) rims may be prone to cleavage leading to a ketone and a carboxylic acid (Fig. 1c), [13] this second Oxidation thus alters the sp2 domains still intact and increases the relative number of carbonyl, hydroxyl, lactone and carboxylic acid functions with respect to the number of epoxide functions [13]. To complete the reaction with $KMnO_4$, the suspension was treated with 5 ml H_2O_2 (30%) hydrogen peroxide. The mixture was then washed with HCl and H_2O respectively, followed by filtration and drying, thus graphene oxide sheets were obtained (Sample 1).

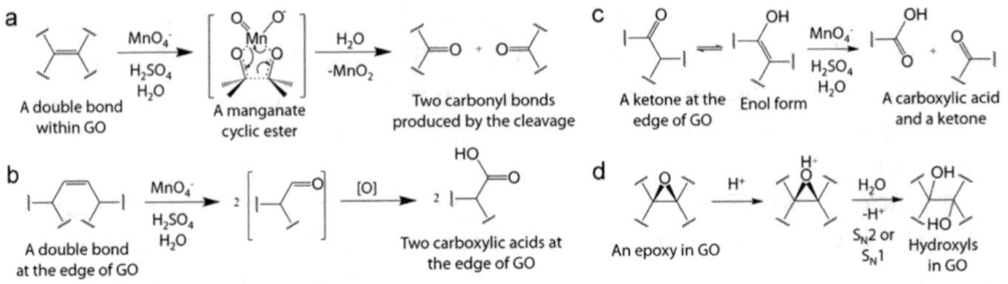

Fig. 1. Possible mechanisms during the second oxidation step of the Hummers method: (a) oxidative cleavage of a C=C double bond via a manganese cyclic ester intermediately resulting in two carbonyl bonds, (b) oxidative cleavage of a C=C double bond producing two carboxylic acids, (c) oxidative cleavage of a ketone forming one carboxylic acid and one ketone, and (d) acid-catalyzed hydrolysis of an epoxy producing two hydroxyl bonds [13].

2.2   Preparing Graphene Oxide Associated to Ionic Liquid Using Ultrasons (First Methode)

This time a simple method was chosen, a mixture of 0.5 g of ionic liquid with 0.5 g of graphite was ground in a mortar for 10 min; Then it was placed in an ultrasonic bath (0.40 kW, water temperature 24 °C.) for 8 h, then centrifuged for 30 min at 4000 rpm; The gray to black liquid phase was recovered (sample 2).

2.3   Preparing Graphene Oxide Associated to Ionic Liquide Using Ultrasons (Second Methode)

Since the previous method does not give (GO) alone in an ionic liquid matrix. Preferably, (GO) is prepared by the method mentioned above Sect. 2.1. Then the obtained (GO) is mixed with the ionic liquid. The resulting mixture was treated using an ultrasonic source.

A gray to black liquid phase was recovered (sample 3). Then another long treatment to obtain a black liquid phase (sample 4).

3   Caracterisations

3.1   Spectroscopy DRX

Figure 2 shows three spectra, the first spectrum (GO-IL) corresponds to graphite and ionic liquid mixture (sample 2), the second corresponds to graphene oxide alone (GO) (sample 1) and the third spectrum to graphite alone (control sample). The first observation on these spectra is the absence of the graphite peak in the graphene oxide spectrum, which is in favor of the method used at the top for the preparation of the first (GO) sample alone. From the spectrum named (GO-IL) (in red in Fig. 2), graphene and graphene oxide are present in the sample, and graphite is also present, which means that the method (1st method) is interesting, but insufficient for the abundance of the (GO) in the complex.

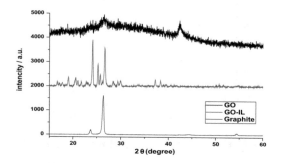

Fig. 2. DRX spectrum of graphite, graphene oxide prepared with hummer method and of ionic liquid mixture with graphene oxide prepared with the first method (Sample 2)

We try to obtain a stable graphene oxide (GO) alone in an ionic liquid medium, for which the first method of synthesis of the (GO) associated with the ionic liquid in

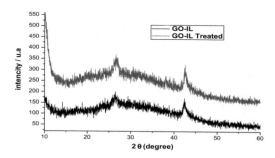

Fig. 3. DRX spectrum of ionic liquid mixture with graphene oxide prepared with the second method, before treatment with ultrasound (Sample 3) and after another long treatment with ultrasound (Sample 4)

addition to an ultrasonic irradiation, seems to be more efficient. Figure 3 shows the DRX spectrum of the ionic liquid mixture with graphene oxide obtained by this method (in black) (sample 3), and the sample treated with the ultrasound (sample 4) is represented in the same FIG. (Red spectrum).

3.2 Absorption UV-Vis

The spectrum of Fig. 4 representing the UV-vis absorption of graphene oxide (GO) is removed from the literature (Graphene Oxide Synthesized by using Modified Hummers Approach). This spectrum shows a maximum absorption peak at ~237 nm attributed to the transition $\pi-\pi^*$ of the CC atomic bonds, and the shoulder peak at ~300 nm which is due to the transition $n-\pi^*$ of the aromatic bonds CC.

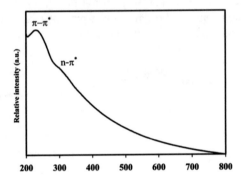

Fig. 4. UV-visabsorption of graphene oxide (GO) [11].

Three samples were prepared for UV-vis characterization; The first is a suspension of the graphene oxide associated with the ionic liquid and the second a suspension of the graphene oxide complex associated with the ionic liquid and (THF), the third is (THF) alone. Their UV-vis absorption spectra are shown in Fig. 5.

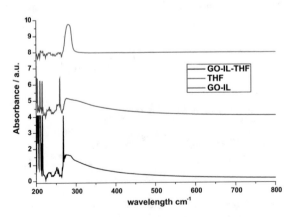

Fig. 5. GO-IL-THF UV absorption spectra.

4 Conclusion

In our work, we have synthesized the graphene oxide associated with the ionic liquid, and also the graphene oxide alone as a control sample, followed by a characterization by DRX and UV.

It develop requirement for performing an electrochemical process in order to generate nano-sensor constituted by graphene oxide, ionic liquids and tetrahydrofurane (THF).

References

1. Novoselov, K.S., Geim, A.K., Morozov, S.V., Jiang, D., Zhang, Y., Dubonos, S.V., Grigorieva, I.V., Firsov, A.A.: Electric field effect in atomically thin carbon films. Science **306**(5696), 666–669 (2004)
2. Balandin, A.A., Ghosh, S., Bao, W., Calizo, I., Teweldebrhan, D., Miao, F., Lau, C.N.: Superior thermal conductivity of single-layer graphene. Nano Lett. **8**(3), 902–907 (2008)
3. Kannappan, S., Kaliyappan, K., Manian, R.K., Pandian, A.S., Yang, H., Lee, Y.S., Jang, J.-H., Lu, W.: Graphene based supercapacitors with improved specific capacitance and fast charging time at high current density. ArXiv Preprint ArXiv:1311.1548. https://arxiv.org/abs/1311.1548 (2013)
4. Dikin, D.A., Stankovich, S., Zimney, E.J., Piner, R.D., Dommett, G.H., Evmenenko, G., Nguyen, S.T., Ruoff, R.S.: Preparation and characterization of graphene oxide paper. Nature **448**(7152), 457 (2007)
5. Hagiwara, R., Ito, Y.: Room temperature ionic liquids of alkylimidazolium cations and fluoroanions. J. Fluorine Chem. **105**(2), 221–227 (2000)
6. Kadari, M., Belarbi, E.H., Moumene, T., Bresson, S., Haddad, B., Abbas, O., Khelifa, B.: Comparative study between 1-Propyl-3-methylimidazolium bromide and trimethylene bis-methylimidazolium bromide ionic liquids by FTIR/ATR and FT-RAMAN spectroscopies. J. Mol. Struct. **1143**, 91–99 (2017)
7. Stankovich, S., Dikin, D.A., Piner, R.D., Kohlhaas, K.A., Kleinhammes, A., Jia, Y., Wu, Y., Nguyen, S.T., Ruoff, R.S.: Synthesis of graphene-based nanosheets via chemical reduction of exfoliated graphite oxide. Carbon **45**(7), 1558–1565 (2007)
8. Nuvoli, D., Valentini, L., Alzari, V., Scognamillo, S., Bon, S.B., Piccinini, M., Illescas, J., Mariani, A.: High concentration few-layer graphene sheets obtained by liquid phase exfoliation of graphite in ionic liquid. J. Mater. Chem. **21**(10), 3428–3431 (2011)
9. Kang, H.S.: Theoretical study of binding of metal-doped graphene sheet and carbon nanotubes with dioxin. J. Am. Chem. Soc. **127**(27), 9839–9843 (2005)
10. El Achaby, M.: Nanocomposites Graphene-Polymere Thermoplastique: Fabrication et Étude Des Propriétés Structurales, Thermiques, Rhéologiques et Mécaniques. Université Mohammed V-Agdal, Faculté des Sciences de Rabat, Faculté des sciences de Rabat (2012). https://tel.archives-ouvertes.fr/tel-00818644/
11. Shahriary, L., Athawale, A.A.: Graphene oxide synthesized by using modified hummers approach. Int. J. Renew. Energy Environ. Eng. **2**(01), 58–63 (2014)
12. Song, J., Wang, X., Chang, C.-T.: Preparation and characterization of graphene oxide. J. Nanomater. (2014). https://www.hindawi.com/journals/jnm/2014/276143/abs/
13. Kang, J.H., Kim, T., Choi, J., Park, J., Kim, Y.S., Chang, M.S., Jung, H., Park, K.T., Yang, S.J., Park, C.R.: Hidden second oxidation step of hummers method. Chem. Mater. **28**(3), 756–764 (2016)

Temperature Effect on InGaN/GaN Multiwell Quantum Solar Cells Performances

N. Harchouch[1], Abdelkader Aissat[1,2(✉)], A. Laidouci[1], and J. P. Vilcot[2]

[1] Laboratory LATSI, University of Blida. 1, 09000 Blida, Algeria
sakre23@yahoo.fr
[2] Institut d'Electronique, de Micro électronique et de Nanotechnologie (IEMN), UMR CNRS 8520, Université des Sciences et Technologies de Lille 1, Avenue Poincaré, BP 60069, 59652 Villeneuve d'Ascq, France

Abstract. In this paper, we are interested in modeling and simulations of InGaN/GaN Quantum Wells Solar Cell, like all other semiconductor devices solar cells are sensitive to temperature, the most parameter affected by the rise in temperature is the open circuit voltage V_{oc}. In this paper, the thermal effects on InGaN/GaN multiple quantum well solar cells (MQWSCs) with an Indium concentration of 0.28 are studied. A temperature of 280 K gives better results of open circuit voltage, maximum power output and conversion efficiency their values are 1.84 V, 24.86 mW/cm^2 and 24.87% respectively.

Keywords: Solar cell · Quantum wells · Temperature · Thermal effect

1 Introduction

The temperature rising is a negative factor which can affect the electrical properties of solar cells such as open circuit voltage, short circuit current, the fill factor and conversion efficiency [1]. Actually many study are concentrated on the thermal effect of III-V based solar cell among them $In_xGa_{1-x}N$ alloys they have been widely investigated [2], they present a very high optical properties and ultra high efficiency solar cells, which make them very widely used for space applications [3]. Their band gap energy tunable from 0.7 to 3.4 eV by changing the indium content, and thus covering almost the whole solar spectrum, superior photo voltaic characteristics, high optical absorption coefficient [4], high resistance to radiation. Combining the high optical properties of InGaN alloys with the use of quantum wells and applying this on solar cells offers a profound effect [5], the concept of quantum well consist of the insertion of material with a low band gap (which is InGaN in our work) by material with a higher band gap (which is GaN), this allow the absorption of low energy photons and therefore improves the conversion efficiency of solar cells [6, 7]. In this work, we are interested in modeling and simulation of InGaN/GaN Quantum Wells Solar Cell in goal to enhance the standard solar cell performances and to show the effect of the temperature on the characteristics parameters of the studied solar cell. The schematic diagram of the present $In_{0.28}Ga_{0.72}N$/GaN Quantum Wells Solar Cell structure is shown in Fig. 1. $In_{0.28}Ga_{0.72}N$ QWs are

M. Hatti (ed.), *Artificial Intelligence in Renewable Energetic Systems*, Lecture Notes in Networks and Systems 35, https://doi.org/10.1007/978-3-319-73192-6_52

sandwiched between GaN p- and n-type layers the thicknesses of the wells and the barriers are 3 nm and 7 nm respectively, we have used a sapphire as substrate.

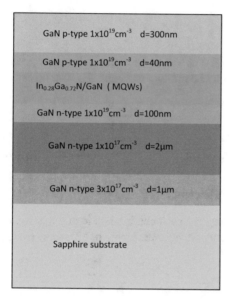

GaN p-type $1 \times 10^{19} cm^{-3}$ d=300nm

GaN p-type $1 \times 10^{19} cm^{-3}$ d=40nm

$In_{0.28}Ga_{0.72}N$/GaN (MQWs)

GaN n-type $1 \times 10^{19} cm^{-3}$ d=100nm

GaN n-type $1 \times 10^{17} cm^{-3}$ d=2μm

GaN n-type $3 \times 10^{17} cm^{-3}$ d=1μm

Sapphire substrate

Fig. 1. The structure of $In_{0.28}Ga_{0.72}N$/GaN quantum well solar cell.

2   Theoretical Model

In this section we will see the equations used in the modeling and simulation of this structure and the relation between the main parameters of solar cells and the temperature of an idealized single-junction PV cell.

2.1   The Open-Circuit Voltage

The open-circuit voltage is the maximum voltage available from a solar cell; Voc is strongly dependent on temperature. For an ideal p-n junction, Voc can be given as [8, 9]:

$$V_{oc} = \frac{KT}{q} + \ln\left[\left(\frac{I_{ph}}{I_s}\right) + 1\right] \tag{1}$$

Where, K is Boltzmann constant, T is the temperature, q is the electronic charge, I_{ph} is the photocurrent, and I_s, is the diode saturation current.

2.2   The Short Circuit Current

The short circuit current (I_{sc}) increases slightly with temperature, for an ideal p-n junction, I_{sc} can be given as [10]:

$$I_{sc} = I_{ph} \left[Exp\left(\frac{qV_{oc}}{KT} \right) - 1 \right] - I_{ph} \tag{2}$$

2.3 The Fill Factor

The Fill factor is defined as the ratio of the maximum power output (P_{max}) at the maximum power point to the product of I_{sc} and V_{oc}. The fill factor can be given as [11]:

$$FF = \frac{I_m \times V_m}{I_{sc} \times V_{oc}} \tag{3}$$

Where I_m is the maximum current and V_m is the maximum voltage:

2.4 The Conversion Efficiency

The conversion efficiency of a solar cell is the ratio of the power output corresponding to the maximum power point to the Power input, the efficiency (η) can be given as [11, 12]:

$$\eta = \frac{I_{sc} \times V_{oc} \times FF}{P_{in}} \tag{4}$$

Where Pin is the power input.

3 Results and Discussion

In this work, we are interested in modeling and simulation of $In_{0.28}Ga_{0.72}N/GaN$ quantum well solar cell in goal to enhance the standard solar cell performances. All simulations were performed at room temperature using AM1.5 of one sun. We have studied the effect of the temperature on the characteristics parameters of the $In_{0.28}Ga_{0.72}N/GaN$ solar cell with 15 quantum wells. The current density–voltage and power density-voltage characteristics of $In_{0.28}Ga_{0.72}N/GaN$ QWSC with 15- QWs are presented in Figs. 2 and 3.

It's obviously in Fig. 2 that both the J (V) and P (V) characteristics change with the increase of the temperature. As we can see, the solar cell provide short-circuit current density of 18, 64 mA/cm^2, an open voltage of 1.84 V approximately and a maximum output power of 24, 86 mW/cm^2 at room temperature. We have taken the values of the maximum power for each temperature; the results are shown in the Fig. 3. The Fig. 3 shows the variation of the maximum output power and conversion efficiency under different operating temperatures, as it is shown a decreasing of the P_{max} is happening when a temperature is increasing the same case for the conversion efficiency (η). Consequently, the temperature dependences of the short-circuit current, open circuit voltage, fill factor and conversion efficiency of the p–MQWs–n solar cell structure are calculated and shown in Table 1 and the simulation is shown in Figs. 4 and 5.

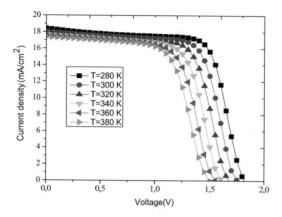

Fig. 2. The J(V) of 15 $In_{0.28}Ga_{0.72}N/GaN$ QWs quantum wells solar cell versus temperatures.

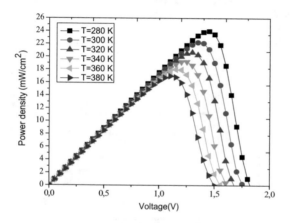

Fig. 3. The P(V) of 15 $In_{0.28}Ga_{0.72}N/GaN$ QWs quantum wells solar cell versus temperatures.

Table 1. The important parameters of $In_{0.28}Ga_{0.72}N/GaN$ quantum wells solar cell with different temperature.

Temperature (K)	Jsc (mA/cm^2)	V_{oc} (V)	FF (%)	η (%)
280	18.7	1.84	72	24.87
300	18,21	1.8	70.9	23.81
320	17.85	1.74	71	21.85
340	17.66	1.68	70.5	20.67
360	17.57	1.62	70	20.05
380	17.48	1.56	69	18.88

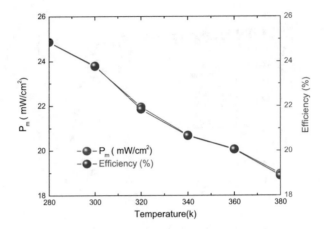

Fig. 4. Maximum output power (Pm) and conversion efficiency (η) versus temperature of InGaN/GaN MQWs Solar Cell.

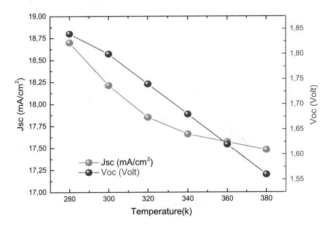

Fig. 5. Short-circuit current density (Jsc) and open circuit voltage (Voc) versus temperature of InGaN/GaN MQWs Solar Cell.

As depicted in the figures, the short circuit current, open-circuit voltage and the conversion efficiency decrease with increasing temperature which reach 18.64 mA/cm², 1.8 V, 24% respectively at room temperature. All these study were driven us to study the variation of External Quantum Efficiency (EQE) under different temperatures. EQE exhibits the ratio of the number of electrons collected by the solar cell to the number of incident photons.

The simulation was done these results as shown in Fig. 5. It is observed in this figure, that the increase of temperature leads a slightly decreasing of the EQE. In this temperature range from 300 K to 380 K the maximum value of EQE is not overly influenced by the temperature rise, for T = 300 K EQE = 31.4% and for T = 380 K EQE = 29.54%. These results confirm what we have obtained and presented previously in Fig. 2. This proves that InGaN/GaN QWs solar cells are resistive structure to high temperature, making it the most widely used for space applications [13–15] (Fig. 6).

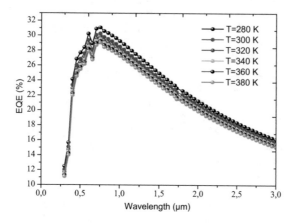

Fig. 6. External Quantum Efficiency Cell versus temperature of InGaN/GaN MQWs Solar.

4 Conclusion

In this work, the effects of temperature on photovoltaic modules are studied. The conversion efficiency of solar cells depends on the sun insulation reaching its surface. The present work indicates an increase in temperature leads to a degradation of the conversion efficiency of the PV module. The InGaN/GaN QWs solar cell represents an improvement regarding the open circuit voltage, short circuit current, maximum power output and the conversion efficiency due to the concept of quantum wells combined with a material which presents a high thermal conductivity and low saturation current, PV cells made from InGaN alloys are expected to have excellent potential to operate under high solar concentration.

References

1. Deng, Q., Wang, X., Xiao, H., Wang, C., Yin, H., Chen, H., Hou, Q., Lin, D., Li, J., Wang, Z., Hou, X.: An investigation on In x Ga $_{1-x}$N/GaN multiple quantum well solar cells. J. Phys. D Appl. Phys. **44**(26), 265103 (2011)
2. Ambacher, O.: Growth and applications of group III-nitrides. J. Phys. D Appl. Phys. **31**, 2653 (1998)
3. Namkoong, G., Boland, P., Bae, S.-Y., Shim, J.-P., Lee, D.-S., Jeon, S.-R., Foe, K., Latimer, K., Doolittle, W.A.: Effect of III- nitride polarization on Voc in p–i–n and MQW solar cells. Phys. Stat. Sol. RRL **5**(2), 86–88 (2011)
4. Feng, G.S.W., Lai, C.M., Chen, C.H., Sun, W.C., Tu, L.W.: Theoretical simulations of the effects of the indium content, thickness, and defect density of the i-layer on the performance of p-i-n InGaN single homojunction solar cells. J. Appl. Phys. **108**(9), 093118 (2010)

5. Shen, X., Lin, S., Li, F., Wei, Y., Zhong, S., Wan, H., Li, J.: Simulation of the InGaN-based tandem solar cells. In: von Roedern, B., Delahoy, A.E. (eds.) Proceedings of SPIE, Photovoltaic Cell and Module Technologies II, vol. 7045 (2008)
6. Mukhtarova, A.: InGaN/GaN Multiple Quantum Wells for Photovoltaics. Ph.D. thesis (2015)
7. Jani, O., Ferguson, I., Honsberg, C., Kurtz, S.: Design and characterization of GaN/InGaN solar cells. Appl. Phys. Lett. **91**(13), 132117 (2007)
8. Çakmak, H., Arslan, E., Rudziński, M., Demirel, P., Unalan, H.E., Strupiński, W., Turan, R., Öztürk, M., Özbay, E.: Indium rich InGaN solar cells grown by MOCVD. J. Mater. Sci. Mater. Electron. **25**(8), 3652–3658 (2014)
9. Skoplaki, E., Palyvos, J.A.: Operating temperature of photovoltaic modules: a survey of pertinent correlations. Renew Energy **34**, 23e9 (2009). Press
10. Mustapha, B.U., Dikwa, M.M.K., Abbagana, M.: Electrical parameters estimation of solar photovoltaic module. J. Eng. Appl. Sci. **4**, 28–37 (2012)
11. Landis, G., Rafaelle, R., Merritt, D.: High temperature solar cell development. In: 19th European Photovoltaic Science and Engineering Conference, Paris, France, 7–11 June 2004
12. Liang, Z.C., Chen, D.M., Liang, X.Q., Yang, Z.J., Shen, H., Shi, J.: Crystalline Si solar cells based on solar grade silicon materials. Renew. Energy **35**(10), 2297–2300 (2010)
13. Tsai, Y.-L., Wang, S.-W., Huang, J.-K., Hsu, L.-H., Chiu, C.-H., Lee, P.-T., Yu, P., Lin, C.-C., Kuo, H.-C.: Enhanced power conversion efficiency in InGaNbased solar cells via graded composition multiple quantum wells, vol. 23, No. 24, November 2015. https://doi.org/10.1364/oe.23.0a1434
14. Liu, C.Y., Lai, C.C., Liao, J.H., Cheng, L.C., Liu, H.H., Chang, C.C., Lee, G.Y., Chyi, J.-I., Yeh, L.K., He, J.H., Chung, T.Y., Huang, L.C., Lai, K.Y.: Nitride-based concentrator solar cells grown on Si substrates. Sol. Energy Mater. Sol. Cells **117**, 54–58 (2013)
15. Xiao-Bin, Z., Xiao-Liang, W., Cui-Bai, X.H.-L.Y., Qi-Feng, H., Hai-Bo, Y., Hong, C., Zhan-Guo, W.: InGaN/GaN multiple quantum well solar cells with an enhanced open-circuit voltage. Chin. Phys. B **20**(2), 028402 (2011)

Evaluation of Numerical Algorithms of a Single and Two Diodes Models

Kelthom Hammaoui[1(✉)], M. Hamouda[2], and Bouchra Benabdelkrim[3]

[1] Department of Sciences techniq Institute of Science and Technology,
University of Ahmed Draia, Adrar, Algeria
hammaouikelthoml@gmail.com
[2] Laboratoire de développement durable et informatique,
Institute of Science and Technology, University of Ahmed Draia,
Adrar, Algeria
jhamouda@yahoo.fr
[3] Department of Material Sciences, Institute of Science and Technology,
University of Ahmed Draia, Adrar, Algeria
benaekbouchra@gmail.com

Abstract. The photovoltaic module is typically represented by an equivalent circuit whose parameters are calculated using the experimental current voltage characteristic I–V. The description of photovoltaic cells current–voltage mathematical is usually defined by a coupled nonlinear equation, difficult to solve using analytical methods. This difficulty has led to the development of several algorithms for solving this equation using numerical techniques. This paper compares three different algorithms of a 5-parameter single-diode and algorithm of two diodes solar PV model using manufacturer's data sheet. This comparison is based on calculated the error relative between the obtained results and experimental values in the important points of the I–V curves, the peak power (Imax, Vmax), short-circuit current Isc and open circuit voltage Voc. We have compared via Matlab developments for multi-crystalline, mono-crystalline and thin-film modules.

Keywords: Numerical algorithms · Single-diode model
Two-diode model · PV irradiation variation · PV temperature variation

1 Introduction

The renewable energy sources will play an important role in the future of power systems, because of it increased use. The unpredicted nature of solar power lead to a need for energy storage as prevalence increases. The increase of research and development work in the area of photovoltaic (PV) systems has made the PV one of the fastest growing renewable energy technologies for electricity generation around the world. The PV systems are excellent for clean sustainable power generation. Different mathematical models have developed by researchers in order to understand the effect of these changing conditions on PV electrical output. The widely applied model in literature is lumped parameter electric circuit based models because it is the more successful. There are different lumped parameter models which are classified based on the

© Springer International Publishing AG 2018
M. Hatti (ed.), *Artificial Intelligence in Renewable Energetic Systems*, Lecture Notes
in Networks and Systems 35, https://doi.org/10.1007/978-3-319-73192-6_53

number of diodes. They are single diode model, double diode model and in recent times the three diode model. The single-diode, 5-parameter model offers the compromise, only if a number of equations are solved to extract the initial parameters that are applied to the model. There are a lot of methods to extract these parameters. Amongst these methods are the artificial neural network, genetic algorithm, the analytical based approach and the numerical methods, and these have shown different levels of accuracy. The researchers have compared the algorithm to extracting the parameter PV and they found out that Newton–Raphson method is the most favoured for the extraction the parameters. The modelling of a photovoltaic (PV) module is an indispensable step for the evaluation of the efficiency of photovoltaic energy production systems. Modelling allows presenting the I–V characteristics of a module depending on a set of parameters (as temperature and illumination of PV cells) and estimating the optimal PV module performances. In this work therefore evaluates the performances of three algorithms of one diode (T. Estram, Villalva and Vika algorithms) end algorithm of two diodes (Ishaque) The performance of these algorithms is evaluated by compares of the I–V curves at various temperatures, irradiance for multi-crystalline (Kyocera KG200GT), mono-crystalline (Shell SQ150-PC) and thin-film (Shell ST 40) types.

2 Modeling of Photovoltaic Modules

2.1 One Diode Model

Figure 1 shows the simplest model to represent a PV module which is the ideal diode model with the equivalent circuit model. The simplest model consists of a single diode connected in parallel with a light generated current source (IPV) model [1]. The basic equation from the theory of semiconductors that mathematically describes the I–V characteristic of the ideal Diode model is:

$$I = I_{pv} - I_0 \left[exp\left(\frac{V}{aV_T}\right) - 1 \right] \text{ where: } V_T = N_s \cdot kT_c / q \tag{1}$$

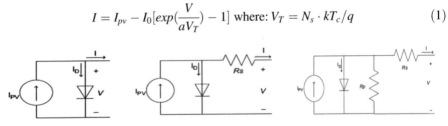

Fig. 1. Ideal diode model   **Fig. 2.** Single diode model with   **Fig. 3.** Single diode model with
Rs.   Rs and Rp.

where I_{PV} is the current generated by the incidence of light, I_0 is the reverse saturation current, V_T is the thermal voltage of the PV module having N_S cells connected in series, q is the electron charge $(1.60217646 \cdot 10^{-19}C)$, k is the Boltzmann constant $(1.3806503 \cdot 10^{-23}J/K)$, T_c is the temperature of the $p-n$ junction in K and a is the diode ideality factor. The single diode model which includes the series resistance, Rs is depicted in Fig. 2. The output current is:

$$I = I_{pv} - I_0[exp(\frac{V + IR_s}{aV_T}) - 1] \qquad (2)$$

Equation (2) does not adequately represent the behaviour of the cell when subjected to environmental variations, especially at low voltage [2]. Amore practical model can be seen in Fig. 3, it consists of a current source anti parallel with a diode, with a series and parallel resistor connected at the terminal of the diode.

Applying Kirchhoff's law and the Shockley's diode equation to the equivalent circuit in Fig. 3, then, the corresponding mathematical equations for the generated current in the PV cell at a particular operating voltage can be obtained as [3]:

$$I = I_{pv} - I_0 \left[exp\left(\frac{V + IR_s}{aV_T}\right) - 1 \right] - \left[V - \frac{IR_s}{R_p} \right] \qquad (3)$$

R_s and R_p represent the series and parallel resistances, respectively. Figure 3 depicted the electrical equivalent circuit of a 5-parameter single diode model [4–6]. This parameters are (I_0, I_{pv}, a, R_s and R_p), and are not provided by the manufacturer but rather only a few experimental values at the standard test condition (STC) $T_{STC} = 25\,°C$ et $G_{STC} = 1000\,w/m^2$ are provided. The values that are provided by the manufacturers are the nominal open-circuit voltage V_{oc} the nominal short-circuit current I_{sc}, the voltage at the maximum power point V_{mp}, the current at the maximum power point I_{mp} and the maximum experimental peak output power $P_{max,e}$. To get the unknown five parameter input to the mode we should always use procedures. There are three step approach to obtain the performance of photovoltaic modules: extracting the 5-parameters at the standard test condition (STC), adjusting the 5-parameters to the required environmental conditions and solving the I–V equation of the cell.

The 5 Parameters Extracted at STC

The modelling process begins with obtaining data that is available in the manufacturer's data sheet which depends on solving Eq. (3). Since there are five unknown in the I–V equations, then, five independent equations will be required. Equations (4)–(6) are derived from Eq. (3) by applying the short circuit, open circuit and maximum power point conditions. Similarly, Eqs. (10)–(17) are derived by differentiating the values of the power and the current of the photovoltaic cell with respect to voltage [3]. Atshort circuit point $(0, I_{SC})$, open circuit point $(V_{OC}, 0)$ and maximum power point (V_{mp}, I_{mp})

Then I_{sc} can be obtained from (3) as:

$$I_{sc} = I_{pv} - I_0 \left[exp\left(\frac{I_{sc}R_s}{aV_T}\right) - 1 \right] - \left[\frac{I_{sc}R_s}{R_P} \right] \qquad (4)$$

Therefore, Eq. (3) can be re-written as:

$$0 = I_{pv} - I_0 \left[exp\left(\frac{V_{oc}}{aV_T}\right) - 1 \right] - \left[\frac{V_{oc}}{R_p} \right] \qquad (5)$$

And Eq. (3) can be re-written as:

$$I_{mp} = I_{pv} - I_0[\exp(\frac{V_{mp} + I_{mp}R_s}{aV_T} - 1] - \left[\frac{V_{mp} + I_{mp}R_s}{R_p}\right] \qquad (6)$$

Generally, in photovoltaic models this assumption is used

$$I_{sc} = I_{PV} \qquad (7)$$

Because in the practical devices the series resistance is low and the parallel resistance is high [7].

Expression for Current saturation Io can be obtained from (5) as:

$$I_0 = \frac{I_{pv} - \frac{V_{oc}}{R_p}}{\exp\left(\frac{V_{oc}}{aV_T}\right) - 1} \qquad (8)$$

Through the assumption that $R_p < V_{oc}$ [3] Eq. (8) resolves to [8]:

$$I_0 = \frac{I_{pv}}{\exp\left(\frac{V_{oc}}{aV_T}\right) - 1} \qquad (9)$$

Also, differentiating (3) will result into:

$$\frac{dI}{dV} = -I_0\left[\frac{1}{aV_T}\left(1 + \frac{dI}{dV} \cdot R_s\right)\exp\left(\frac{V + I \cdot R_s}{aV_T}\right)\right] - \frac{1}{R_p}\left(1 + \frac{dI}{dV}R_s\right) \qquad (10)$$

At the short circuit condition to (10) we obtain:

$$\frac{dI}{dV}\Big|_{V=0} = -I_0\left[\frac{1}{aV_T}\left(1 + \frac{dI}{dV}\Big|_{V=0}R_s\right) \cdot \exp\left(\frac{I_{sc}R_s}{aV_T}\right)\right] - \frac{1}{R_p}\left(1 + \frac{dI}{dV}\Big|_{V=0}R_s\right) \qquad (11)$$

At the open circuit point Eq. (10) can also be written as:

$$\frac{dI}{dV}\Big|_{I=0} = -I_0\left[\frac{1}{aV_T}\left(1 + \frac{dI}{dV}\Big|_{II=0}R_s\right) \cdot \exp\left(\frac{V_{oc}}{aV_T}\right)\right] - \frac{1}{R_p}\left(1 + \frac{dI}{dV}\Big|_{I=0}R_s\right) \qquad (12)$$

At the maximum power point the derivative of the solar PV power with respect to the voltage is equal to zero.

$$\frac{dP}{dV}\Big|_{\substack{V = V_{mp} \\ I = I_{mp}}} = 0, \frac{dP}{dV} = \frac{dIV}{dV} = IV\frac{dI}{dV} \qquad (13)$$

$$\frac{dI}{dV} = -\frac{V_{mp}}{I_{mp}} \tag{14}$$

From (14) and (10) at maximum power point get:

$$-\frac{V_{mp}}{I_{mp}} = -I_0 \frac{1}{aV_T}\left(1 - \frac{V_{mp}}{I_{mp}}R_s\right) - \frac{1}{R_p}\left(1 - \frac{V_{mp}}{I_{mp}}R_s\right) \tag{15}$$

For notational convenience we express

$$\left.\frac{dI}{dV}\right|_{V=V_{oc}} = R_{s0} \tag{16}$$

$$\left.\frac{dI}{dV}\right|_{I=I_{sc}} = R_{p0} \tag{17}$$

2.1.1 Esram Algorithm

The Esram method is premised on the assumption that the slope of the I–V curve at V_{oc} and I_{sc} is controlled by the series and shunt resistance, respectively, hence R_{s0} and R_{p0} can then be approximated as R_s and R_p respectively. Esram further made additional simplifications by assuming I_{pv} to be short equivalent to the circuit current in Eq. (7), and also that I_0 can be obtained from Eq. (8). Subsequently R_s, R_p and a can be obtained by simultaneous solution of Eqs. (6), (10) and (15) [3, 7]. The simultaneous equation can be easily solved in MATLAB environment with Newton–Raphson technique using fsolve symbolic function.

2.1.2 Vika Algorithm

In this method the ideality factor is also approximated to be based on the PV technology. I_0 is initialised by using Eq. (7) with an initial assumption that (5) holds. The remaining parameters R_s, R_p and I_0 are mutually dependent as seen from (9), (10) and (13). Iterative techniques can then be applied such that a combination of R_s, R_p and I_{pv} that fits well with the datasheet points for a fixed number of iterations giving the least minimum error is taken as the value for R_s, R_p and I_{pv} To obtain the errors [3]. The saturation current I_0 dependence of the temperature and one of those is shown in this equation:

$$I_0(T) = I_0\left(\frac{T}{T_{STC}}\right)^3 \exp\left[\left(\frac{qE_g}{a \cdot K}\right)\left(\frac{T}{T_{STC}} - 1\right)\right] \tag{18}$$

E_g is the band gap energy of the semiconductor. Improvement on Eq. (18) by substituting in Eq. (19).

$$I_0(T) = \frac{I_{sc} + K_I.\Delta T}{\exp\left(\frac{V_{oc} + K_V \cdot \Delta T}{a \cdot V_T}\right) - 1} \text{ Where } \Delta T = T - T_{STC} \tag{19}$$

The photoelectric current I_{PV} depends on the temperature and solar radiation as shown in Eq. (20).

$$I_{PV} = \left[I_{PV,STC} + K_I \cdot \Delta T\right] \cdot G/G_{STC} \tag{20}$$

K_V is the open circuit voltage temperature coefficient and K_I is the short circuit current temperature coefficient.

The series resistance R_s is expressed as (21) can be calculated by the maximum power rating at STC.

$$\left.\frac{dP}{dV}\right|_{mpp} = 0 \rightarrow \left.\frac{dI \cdot V}{dV}\right|_{mpp} = I_{mp} + V_{mp} \cdot \left.\frac{dI}{dV}\right|_{mpp} = 0, R_S$$
$$= \frac{V_{mp}}{I_{mp}} - \frac{a \cdot V_T R_P}{I_0 R_P \exp\left(\frac{V_{mp} + I_{mp} \cdot R_s}{a \cdot V_T}\right) + a \cdot V_T} \tag{21}$$

Subsequently, the error value for Rs is obtained by expressing (21) as (22)

$$Err1 = \frac{V_{mp}}{I_{mp}} - \frac{a \cdot V_T R_P}{I_0 R_P \cdot \exp\left(\frac{V_{mp} + I_{mp} \cdot R_s}{a \cdot V_T}\right) + a \cdot V_T} - R_S \tag{22}$$

The parallel resistance at the maximum power point, Eq. (7) results in Eq. (23):

$$\frac{P_{mp}}{V_{mp}} = I_{mp} = I_{pv} - I_0 \cdot \left[\exp\left(\frac{V_{mp} + I_{mp} \cdot R_s}{a \cdot V_T}\right) - 1\right] - \frac{V_{mp} + I_{mp} \cdot R_s}{R_p} \tag{23}$$

This can be rearranged into Eq. (24):

$$R_P = \frac{V_{mp} + I_{mp} \cdot R_s}{I_{pv} - I_{mp} - I_0 \cdot \left[\exp\left(\frac{V_{mp} + I_{mp} \cdot R_s}{a \cdot V_T}\right) - 1\right]} \tag{24}$$

Error values for R_p is re-written as:

$$Err2 = \frac{V_{mp} + I_{mp} \cdot R_s}{I_{pv} - I_{mp} - I_0 \cdot \left[\exp\left(\frac{V_{mp} + I_{mp} \cdot R_s}{a \cdot V_T}\right) - 1\right]} - R_P \tag{25}$$

Photoelectric current is often assumed that the photoelectric current I_{pv} is equal to the short circuit current I_{sc} it as shown in Eq. (26).

$$I_{PV} = \frac{R_p + R_s}{R_p} I_{sc} \tag{26}$$

The error for I_{pv} can then be written as:

$$Err3 = \frac{R_p + R_s}{R_p} I_{sc} - I_{pv} \tag{27}$$

In [3] it is suggested to iterate the parameters and minimize the error values given in this equation:

$$Err = (Err1)^2 + (Err2)^2 + (Err3)^2 \tag{28}$$

2.1.3   Villalva Algorithm

This algorithm appears to be the most popularly used. This method assumes that $I_{sc} = I_{pv}$ in Eq. (5) holds, whilst the diode reverse saturation current I_0 can be calculated from (7). It further assumes that the ideality factor is based on the technology applied in the manufacturing of the PV modules. The remaining two parameters R_s and R_p can then be computed from the fact that only a unique pair of these two parameters gives the maximum power at every operating condition. This unique pair can be obtained using iterative technique by gradually adjusting the values of R_s and then calculating the corresponding value of R_p that gives a close approximation with the maximum experimental power obtained from the datasheet. The value of R_p and the calculated maximum power is computed using (29) and (30), [7] respectively.

$$P_{max,e} = P_{max,m} = V_{mp} \cdot I_{mp}$$

$$P_{max,m} = V_{mp} \left\{ I_{PV} - I_0 \left[exp \left(V_{mp} + \frac{I_{mp} \cdot R_s}{a \cdot V_T} \right) - 1 \right] - \left(V_{mp} - \frac{I_{mp} \cdot R_S}{R_P} \right) \right\} = P_{max,e} \tag{29}$$

$$R_P = \frac{V_{mp}[V_{mp} + I_{mp} \cdot R_S]}{\left(V_{mp} I_{PV} - V_{mp} I_0 \left[exp \left(\frac{V_{mp} + I_{mp} R_S}{V_T} \right) + 1 \right] - P_{max,e} \right)} \tag{30}$$

2.2   Tow Diode Model

The following equation describes the output current of the cell:

$$I = I_{pv} - I_{01} \left[exp \left(\frac{V + IR_s}{a_1 \cdot V_{T1}} \right) - 1 \right] - I_{02} \left[exp \left(\frac{V + IR_s}{a_2 \cdot V_{T2}} \right) - 1 \right] - \frac{V + IR_s}{R_p} \tag{31}$$

Attempt has been made to this equation to solve the saturation currents. To maintain where I_{01} and I_{02} are there verse saturation currents of diode1 and diode2, respectively, V_{T2} and V_{T2} are the thermal voltages of respective diodes. a_1 and a_2 represent the diode ideality constants. Although greater accuracy can be achieved using this model, it requires the computation of seven parameters, namely I_{pv}, I_{01}, I_{02}, R_p, R_s, a_1 and a_2. Furthermore I_{01}, I_{02}, R_p and R_s are obtained through iteration. For the simplification the majority of researches propose that $a_1 = 1$ and $a_2 = 2$.

The equation for PV current as a function of temperature and irradiance can be written as:

$$I_{pv} = [I_{pv,STC} + K_I \cdot \Delta T] \cdot G/G_{STC} \tag{32}$$

2.2.1   Algorithm Ishaque

This algorithm appears an improved modeling approach for the two-diode model of photovoltaic (PV) module. The main contribution of this algorithm is the simplification of the current equation. Furthermore the values of the series and parallel resistances are computed using a simple and fast iterative method. Ishaque, propose a modification of Eq. (19) and apply it to the two-diode model. No the equation in the same form as in Eq. (19), both reverse saturation currents I_{01}, I_{02} are set to be equal in magnitude.

$$I_{01} = I_{02} = I_0 = (I_{sc-STC} + K_I \cdot \Delta T)/[\exp[\frac{V_{oc-STC} + K_V \cdot \Delta T}{[(a_1 + a_2)/p] \cdot V_T}]-1] \tag{33}$$

Diode ideality factors a_1 and a_2 represent the diffusion and recombination current components, respectively. Ishaque put $\frac{a_1 + a_2}{p} = 1$ and $a_1 = 1$, it follows that variable p can be chosen to be $p \geq 2.2$. This generalization can eliminate the ambiguity in selecting the values of a_1 and a_2. Equation (33) can be simplified in terms of p as

$$I = I_{PV} - I_0[exp((V + I \cdot R_s)/V_T) + \exp((V + I \cdot R_s)/(p-1)V_T) + 2] - \left[V - I \cdot \frac{R_s}{R_p}\right] \tag{34}$$

The R_s and R_p are calculated simultaneously, similar to the procedure proposed in [9]. From Eq. (34) at maximum power point condition, the expression for R_p can be rearranged and rewritten as (35).

$$R_P = [V_{mp} + I_{mp} \\ \cdot R_S]/\left(I_{PV} - I_0\left[exp\left(\frac{V_{mp} + I_{mp}R_S}{V_T}\right) + \exp(\frac{V_{mp} + I_{mp}R_S}{(p-1)V_T}) + 2\right] - \frac{P_{max,e}}{V_{mp}}\right) \tag{35}$$

The initial conditions for both resistances are given by [10]:

$$R_{s0} = 0, \ R_{p0} = \frac{V_{mp}}{I_{sc,STC} - I_{mp}} - \frac{V_{oc,STC} - V_{mp}}{I_{mp}}$$

3   Results and Discussion

In this section compared the one diode and the two diode models for different photovoltaic, is validated by measured parameters of selected PV modules. The experimental (V, I) data are extracted from the manufacturer's datasheet and from. Three different modules of different brands/ models are utilized for verification; these include the multi-crystalline (Kyocera KG200GT), mono-crystalline (Shell SQ150-PC) as well as thin-film (Shell ST 40) types. The specifications of the modules are summarized in Table 1.

Table 1. Specifications for the used modules

Parameter	Multi-crystalline Kyocera KC200GT	Mono-crystalline Shell SQ150-PC	Mono-crystalline Shell ST 40
V_{oc} (V)	32.9	43.4	23.3
I_{sc} (A)	8.21	4.8	2.68
V_{mp} (V)	26.3	34	16.6
I_{mp} (A)	7.61	4.4	2.41
K_V (mV/°C)	−123	−128	−100
K_I (mA/°C)	3.18	1.4	0.35
N_s	54	60	36
P_{max} (W)	200	150	40

In this paper we calculated and extracted parameters for PV models parameters using three algorithms of one diode model (T. Esram, Villalva and Vika) end one algorithm of tow diodes model (Ishaque). Table 2 shows the parameters that are used for the module multi-crystalline Kyocera KC200GT for the different algorithms which mentioned before. This parameters are calculated namely, Io, I_{PV}, Rs, Rp, pmax. Table 3 represents the parameters for mono-crystalline Shell SQ150-PC. Table 4 shows the parameters for as thin-film (Shell ST 40).

Table 2. Calculated parameters for module multi-crystalline Kyocera KC200GT at STC

Parameters	Ishaque	Vika	T_Esram	Villalva
$I_{PV}(A)$	8.21	8.214	8.21187	8.21458
a	1.3	1.2	1.341	1.3
$R_S(\Omega)$	0.33	0.261	0.21715	0.23
$R_p(\Omega)$	174.1551	314	951.93	601.3368
$I_0(A)$	4.128e−10	2.151826e−8	1.71064e−7	9.8252e−8

Table 3. The parameters for mono-crystalline Shell SQ150-PC at STC

Parameters	Ishaque	Vika	T_Esram	Villalva
$I_{PV}(A)$	4.8	4.825	4.8024	4.815
a	1.3	1.3	1.7288	1.3
$R_S(\Omega)$	1.05	0.848	0.5907	0.84
$R_p(\Omega)$	210.736	267	1164.217	276.279
$I_0(A)$	2.847e−12	1.89088e−9	7.00698e−7	1.888e−9

Table 4. Shows the parameters for as thin-filM (Shell ST 40) at STC.

Parameters	Ishaque	Vika	T_Esram	Villalva
$I_{PV}(A)$	2.68	2.78	2.684	2.696
a	1.3	1.3	1.5041	1.3
$R_S(\Omega)$	1.71	1.15	1.4226	1.51
$R_p(\Omega)$	198.941	72	952.9119	266.5478
$I_0(A)$	3.075e−11	1.0307e−8	1.4066e−7	1.0292e−8

Figures 4, 5, 6, 7, 8 and 9 show the I–V curves for KC200GT and SQ150-PC and Shell ST 40 for different levels of irradiation and temperature. The calculated values from the two-diode (Ishaque) and one-diode models (T. Esram, Villalva and Vika) are evaluated against measured data from the manufacturer's datasheet.

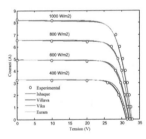 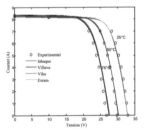 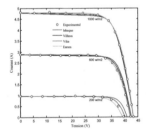

Fig. 4. I–V characteristic for Kyocera KC200GT multi-crystalline PV module using the different models and for various irradiances.

Fig. 5. I–V characteristic for Kyocera KC200GT multi-crystalline PV module using the different models and for various temperature.

Fig. 6. I–V characteristic for Shell SQ150-PC mono-crystalline PV module using the different models and for various irradiances

To further evaluate the predictive capability of the algorithms, they were utilised to determine P_{max}, V_{oc}, and I_{sc} for the three selected PV technologies. The effective the models for different technology, comparisons between SQ150-PC (mono-crystalline), KC200GT multi-crystalline) and ST40 (thin-film) can be concluded, more accurate results are obtained from the T. Esram model for the all three technologies because the

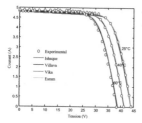

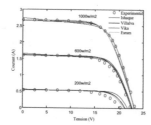

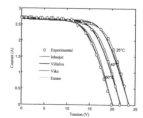

Fig. 7. I–V characteristic for Shell SQ150-PC mono-crystalline PV module using the different models and for various temperatures.

Fig. 8. I–V characteristic for Shell ST 40 thin-film PV module using the different models and for various irradiances

Fig. 9. I–V characteristic for Shell ST 40 thin-film PV module using the different models and for various temperatures.

five parameter model proposed by (Villalva, Vika) and the two diode model proposed by (Ishaque) focus on three remarkable points, the short circuit current Isc, the open source voltage Voc and the maximum power point MPP which are assured by forcing the choice of R_s and R_p, to fit them. The particular obvious weakness of these two models is the arbitrarily choice of the values of diodes ideality factors $(1 \leq a \leq 1.5)$. It is well known that the ideality factor affects the curvature of the I–V curves and decreases the I–V fitting accuracy of the model, bat T. Esram Calculate R_s, R_p, and ideality factor a in the three remarkable points.

4 Conclusion

In this paper, an inquired the three algorithms improved modeling approach for the one-diode model of photovoltaic (PV) module and algorithm of a two-diode solar PV cells. The inquiry of the four models is evaluated using practical data from the manufacturers of three PV modules of different types SQ150-PC (mono-crystalline), KC200GT (multi-crystalline) and ST40 (thin-film). Its performances are compared to the models of single diode (Esram.Villalva.Vika) and two diode model (Ishaque). The four models are approvable when subjected to irradiance and temperature variations. In particular, the two diode model exhibits excellent accuracy at lower irradiance conditions. And T. Esram algorithm performs best.

References

1. Ayodele, T.R., Ogunjuyigbe, A.S.O., Ekoh, E.E.: Evaluation of numerical algorithms used in extracting the parameters of a single-diode photovoltaic model. Sustain. Energy Technol. Assess. **13**, 51–59 (2016)
2. Esram, T.: Modeling and Control of an Alternating-Current Photovoltaic Module. Submitted in partial fulfillment of the requirements for the degree of Doctor of Philosophy in Electrical and Computer Engineering in the Graduate College of the University of Illinois at Urbana-Champaign (2010)

3. Vika, H.B.: Modelling of Photovoltaic Modules with Battery Energy Storage in Simulink/Matlab. Master of Energy and Environmental Engineering, Norwegian University of Science and Technology, Department of Electric Power Engineering, June 2014
4. Villalva, M.G., Gazoli, J.R., Filho, E.R.: Comprehensive approach to modeling and simulation of photovoltaic arrays. IEEE Trans. Power Electron. 24, 1198–1208 (2009)
5. Yuncong, J., Qahouq, J.A.A., Orabi, M.: Matlab/Pspice hybrid simulation modeling of solar PV cell/module. In: 2011 Twenty-Sixth Annual IEEE on Applied Power Electronics Conference and Exposition (APEC), pp. 1244–1250 (2011)
6. Chatterjee, A., Keyhani, A., Kapoor, D.: Identification of photovoltaic source models. IEEE Trans. Energy Convers. **PP**, 1–7 (2011)
7. Said, S., Massoud, A., Benammar, M., Ahmed, S.: A matlab/simulink - based photovoltaic array model employing SimPowerSystems toolbox. J. Energy Power Eng. **6**, 1965–1975 (2012)
8. Villalva, M.G., Gazoli, J.R., Ruppert Filho, E.: Modeling and circuit-based Simulation of photovoltaic arrays. 978-1-4244-3370-4/09/$25.00 © 2009 IEEE
9. Ishaque, K., Salam, Z., Taheri, H.: Simple, fast and accurate two-diode model for photovoltaic modules. Sol. Energy Mater. Sol. Cells **95**, 586–594 (2011)
10. Ishaque, K., Salam, Z., Taheri, H., Shamsudin, A.: A critical evaluation of EA computational methods for photovoltaic cell parameter extraction based on two diode model. Sol. Energy **85**, 1768–1779 (2011)

First Principles Calculations of Structural, Electronic and Optical Properties of Ternary ZnO Alloys: Te Doped

A. Zouaneb[1,2(✉)], F. Benhamied[2], and A. Rouabhia[3]

[1] Laboratory Energy, Environment and Information Systems,
Faculty of Science and Technology, University African Ahmed Draia Adrar,
Adrar, Algeria
Aichaaz@yahoo.fr

[2] Department of Materials Science, Faculty of Science and Technology,
University African Ahmad Draia Adrar, Adrar, Algeria

[3] Unité de Recherche en Energies Renouvelables en Milieu Saharien URERMS,
Centre de Développement des Energies Renouvelables, Adrar, Algeria

Abstract. The role of interactions is certainly one of the most difficult and important phenomena to be solved in the physics of condensed matter. Thus, we use the Full Potential Linearized Augmented Plane Wave (FP-LAPW) method, which is based on the resolution of the Kohn-Sham equations in two arbitrarily defined regions in the elementary mesh. The Generalized Gradient Approximation (GGA) is used to process binary alloys which are ZnO and ZnTe. We will compare these predictions with the experimental results and with the theoretical work devoted to this deviation. The Wien2k code based on density functional theory (DFT) is explored to determine the various structural and electronic properties of ternary $ZnTe_{1-x}O_x$ alloys for x = 0.25, 0.50, 0.75 as well as the curvature parameters by the ab-initio calculation means.

Keywords: FP-LAPW · GGA · DFT · Binary alloys · Ternary alloys

1 Introduction

The use of transparent conductive oxides (TCO) began with the discovery of indium oxide with tin (In_2O_3) in 1954 by Rupprecht [1]. In 1960, it was found that thin layers of binary compounds such as ZnO, SnO_2, In_2O_3 and their alloys are also good TCO [2]. In 1980, ternary alloys such as Cd_2SnO_4, $CdSnO_3$ and $CdIn_2O4$ were developed as well as multi-component oxides of ZnO, CdO, SnO_2 and In_2O_3 [3]. Most of these TCO materials are n-type semiconductors. The type p was observed in 1993 by Sato et al. for NiO [4]. The interest in ZnO was became more apparent in 1995 when obtaining UV laser emissions by ZnO thin films at room temperature [5]. Recently, zinc oxide (ZnO), non-doped and doped, are the most studied TCO, because they have interesting optical and electrical properties [6, 7]. In addition, ZnO is non toxic, cheap and abundant on earth. ZnO is an important II-VI semiconductor that crystallizes in the hexagonal wurtzite structure. Zinc Telluride is a binary alloy with the chemical formula ZnTe, it crystallizes in the zinc blende structure and can easily be doped. Therefore, it is one of

© Springer International Publishing AG 2018
M. Hatti (ed.), *Artificial Intelligence in Renewable Energetic Systems*, Lecture Notes in Networks and Systems 35, https://doi.org/10.1007/978-3-319-73192-6_54

the most common semiconductor materials used in optoelectronics, such as visual display, the high-density optical memories, photodetectors, transparent conductors, solar cells [8], high-efficiency tin films and light emitting diodes [9]. The objective of this work is to define the structural, optical and electrical properties of ZnO non dope and doped by Te.

2   Calculation Method

In our work, we studied the physical properties of $ZnTe_{1-x}O_x$ ternary alloys using the augmented and linearized plane wave method (FP_LAPW) [10] based on generalized gradient approximation (GGA) [11] which is implemented in the Wien2k code [12].

We calculate the energy as a function of volume to determine the equilibrium parameters for each binary ZnO, ZnTe and ternary $ZnTe_{1-x}O_x$ alloys.

In the FP-LAPW method, space is divided into two regions: spheres that do not overlap and are centered on each atom (called muffin-tin spheres) of RMT rays and the interstitial region between the spheres. The basic functions, the electronic densities and the potentials are developed on the one hand in a combination of spherical harmonics inside the muffin-tin spheres for a maximum value and on the other hand, in the Fourier series in the interstitial region with a cut-off radius RMT * K_{max} = 8 (RMT is the smallest radius of the muffin-tin spheres and K_{max} represents the largest wave vector in the extension of the plane waves of the interstitial region). The convergence of self-consistent cycles is only achieved when the total energy is stable at 0.1 mRy.

3   Results and Discussion

3.1   Structural Properties

In the structural optimization, we determine the equilibrium lattice parameters (a_0), the Bulk modulus (B_0) and its derivative (B_0') of the binary compounds and their ternary alloys. The empirical Murnaghan equation of state [13] is used to determine the equilibrium lattice parameters by minimizing the total energy depending on the volume of $ZnTe_{1-x}O_x$ alloys (x = 0, 0.25, 0.50, 0.75 and 1):

$$E(V) = \frac{B_0 V}{B_0'} \left[\frac{(V_0/V)^{B_0'}}{B_0' - 1} - 1 \right] + cste \tag{1}$$

E(V): Represents the total energy variation versus volume, V_0 and B_0 are respectively the volume at equilibrium of the elementary mesh and the equilibrium Bulk modulus. We have shown in (Fig. 1) the total energy obtained for the ZnO, ZnTe and ternary $ZnTe_{75}O_{25}$, $ZnTe_{50}O_{50}$, $ZnTe_{25}O_{75}$ alloys.

The numerical results obtained using the GGA approximation are given in Table 1. In order to assess the quality of our results, this table also contains experimental data and other results obtained by other theoretical methods. It should be noted that the experimental or theoretical results available in the literature concern only binary alloys.

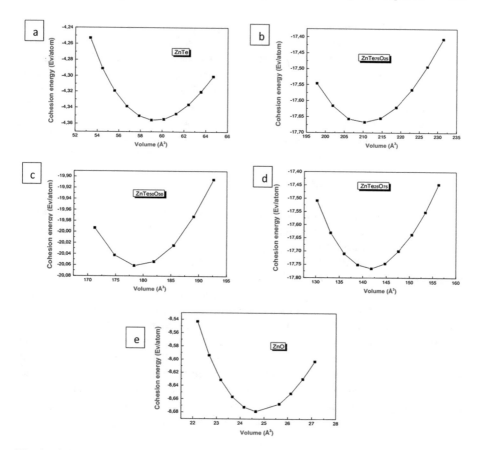

Fig. 1. Calculated cohesion energy optimization variation versus volume for $ZnTe_{1-x}O_x$ alloys at (a) x = 0, (b) x = 0.25, (c) x = 0.50, (d) x = 0.75 and (e) x = 1.

Table 1. Calculated lattice parameter a_0 and bulk modulus B_0 for the alloys ZnO, ZnTe, $ZnTe_{75}O_{25}$, $ZnTe_{50}O_{50}$, and $ZnTe_{25}O_{75}$.

X	$ZnTe_{1-x}O_x$	Lattice parameter a_0 (Å)			Bulk modulus B_0 (Gap)			Cohesion energy E_c (eV/atom)
		This work	Exp.	Cal.	This work	Exp.	Cal.	
0	ZnTe	6.192	6.103[a]	6.074[a]	43.5356	50.9[b]	51.75[c]	4.355
0.25	$ZnTe_{75}O_{25}$	5.945	-	-	46.883	-	-	17.667
0.50	$ZnTe_{50}O_{50}$	5.637	-	-	57.008	-	53.68[d]	20.061
0.75	$ZnTe_{25}O_{75}$	5.209	-	-	78.325	-	-	17.767
1	ZnO	4.683	4.463[e]	4.616[f]	130.058	-	157.86[d]	8.678
		-	-	-	-	-	125.345[f]	-

[a]Ref. [14], [b]Ref. [15], [c]Ref. [16], [d]Ref. [17], [e]Ref. [18], [f]Ref. [19].

For binary compounds, the equilibrium lattice parameters are in good agreement with other works available in the experimental one. The values obtained using the PBE-GGA approximation overestimate the lattice parameter and underestimate the Bulk modulus with respect to the experiment. For the ternary alloys $ZnTe_{1-x}O_x$; no experimental data or theoretical calculations are available in the literature, so our results are predictive.

Concerning the composition dependence of the lattice parameter, Vegard's law [20] indicates that for substitution solid solutions, the lattice parameter varies linearly with the atomic composition. For $AB_{1-x}C_x$ alloys, the lattice parameter is written:

$$a\left(AB_{1-x}C_x\right) = x\,a_{AC} + (1-x)a_{AB} \qquad (2)$$

Where a_{AC} and a_{AB} are the lattice parameters of the binary alloys AC and AB.

The calculated lattice parameter optimization variation versus the concentration X for $ZnTe_{1-x}O_x$ alloys is shown in (Fig. 2). The lattice constant varies almost linearly with the composition, thus obeys the Vegard law. The bowing deviations are small and equal to 0.00571 Å. These parameters were determined by an adjustment of the curves giving the total energy versus volume, by a polynomial function of the second degree.

The bulk modulus versus composition is shown in (Fig. 2), it increases with the composition(x) of the alloys $ZnTe_{1-x}O_x$. We observe an important deviation of the variation with respect to the law of the linear concentration dependence (LCD), an important value has been found for the bowing, which is equal to 99.23 GPa for $ZnTe_{1-x}O_x$. This big value of "bowing" is attributed to the big differences of the bulk modulus of the parent binary compounds. It is clear from the figure that when the concentration of oxygen increases, the bulk modulus increases, this suggests that when the concentration (x) increases, materials become less compressible.

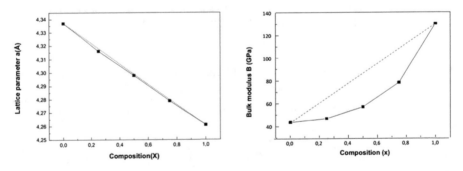

Fig. 2. The lattice parameter (a) and bulk modulus variation versus the concentration for ZnTe1-xOx alloys using Vegard's law.

4 Electronic Properties

4.1 Band Structure

The ZnO, ZnTe and $ZnTe_{1-x}O_x$ alloys were calculated using the FP-LAPW method and using the PBE-GGA approximation. This study focuses on the dependence of the band structures with the composition of the oxygen. The band structure energies of ternary $ZnTe_{1-x}O_x$ alloys were calculated using the PBE-GGA scheme at their equilibrium lattice parameters along the high symmetry directions of the first Brillouin zone are depicted in (Fig. 3), the results obtained are listed in (Table 2).

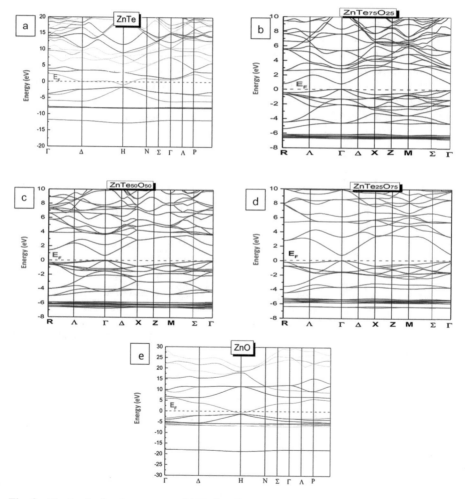

Fig. 3. Electronic band structure of ZB for $ZnTe_{1-x}O_x$ alloys at (a) x = 0, (b) x = 0.25, (c) x = 0.50, (d) x = 0.75 and (e) x = 1.

Table 2. Calculated results of the different energy gaps of $ZnTe_{1-x}O_x$ alloys.

X	Alloy	E_g (eV)		
		This work	Exp.	Cal.
0	ZnTe	0.702	–	–
0.25	$ZnTe_{75}O_{25}$	0.724	–	–
0.50	$ZnTe_{50}O_{50}$	0.730	–	–
0.75	$ZnTe_{25}O_{75}$	0.773	–	–
1	ZnO	0.753	–	–

4.2   Density of States

Electronic state density (DOS) is one of the most important electronic properties that informs us about the behavior and electronic nature of the system. It also allows us to know the nature of the chemical bonds between the atoms of crystal or of molecule. From the partial state density curves (PDOS), it is possible to determine the predominant character for each region. The calculations of the total and partial density of states plots of $ZnTe_{1-x}O_x$ alloy obtained by the GGA approximation are presented in (Figs. 4 and 5), where the Fermi level is taken as the origin of energies.

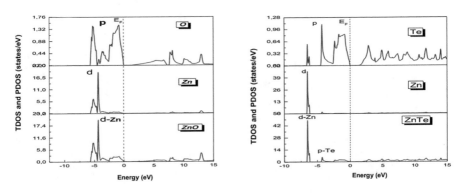

Fig. 4. Total density of states (TDOS) and partial (PDOS) for ZnO and ZnTe alloys using (GGA) approximation.

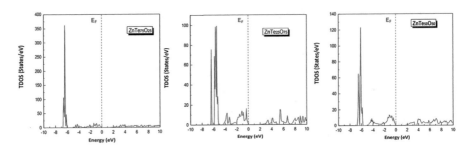

Fig. 5. Total density of states (TDOS) for $ZnTe_{1-x}O_x$ alloys using (GGA) approximation.

5   Optic Properties

5.1   Imaginary and Real Part of the Dielectric Function

The variations of the imaginary and real part of the dielectric function as a function of the energy for $ZnTe_{1-x}O_x$ alloys are presented in (Fig. 6). The peaks that appear in the graphs giving its variation as a function of the energy are connected to optical transitions. The absorption thresholds correspond to the optical gaps. We also show in (Fig. 6) the variations of $\varepsilon_2(\omega)$ and $\varepsilon_1(\omega)$ for the studied alloys. Our analyses of the curves show that the energy threshold corresponds to the main peaks of the spectra.

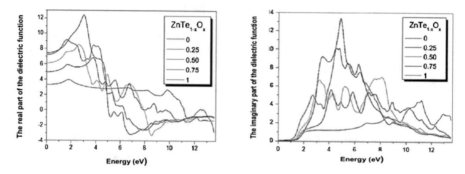

Fig. 6. The real part and the imaginary part of the dielectric function as a function of the energy for $ZnTe_{1-x}O_x$.

6   Conclusion

In this work, we studied the structural, electronic and optical properties of the ternary alloys $ZnTe_{1-x}O_x$ which consists of binary compounds and ternary alloys. The calculations were carried out by the ab-initio method called augmented plane waves (FP-LAPW) in the framework of the functional theory of density (DFT). Our work concerns a detailed study of the structural, electronic, optical properties of $ZnTe_{1-x}O_x$ alloys.

We presented an ab-initio simulation and optimization study of electronic properties of Zinc family semiconductor II-VI materials (ZnO and ZnTe) and ternary alloys $ZnTe_{1-x}O_x$ for the different concentrations; x = 25%, x = 50%, and x = 75%. For the ternary alloys, we have shown that the crystalline parameter varies almost linearly, which is in agreement with the Vegard's law. On the other hand the gap varies in a non-linear manner, which is translated by a disorder factor, for the latter our results are consistent with those of the other published works.

References

1. Rupprecht, G.: Untersuchungen der elektrischen und lichtelektrischen Leitfiihigkeit diinner Indiumoxydsehiehten. Zeitschrift für Medizinische Physik **139**, 504–517 (1954)
2. Haacke, G.: Transparent conducting coatings. Ann. Rev. Mater. Sci. **7**, 73–93 (1977)
3. Minami, T.: New n-type transparent conducting oxides. MRS Bull. **25**, 38–44 (2000)
4. Sato, H., Minami, T., Takata, S., Yamada, T.: Transparent conducting p-type NiO thin films prepared by magnetron sputtering. Thin Solid Films **236**, 27–31 (1993)
5. Klingshirn, C.: ZnO: material, physics and applications. ChemPhysChem **8**, 782 (2007)
6. Suzuki, S., Miyata, T., Ishii, M., Minami, T.: Transparent conducting V-co-doped AZO thin films prepared by magnetron sputtering. Thin Solid Films **434**, 14–19 (2003)
7. Breivik, T.H., Diplas, S., Ulyashin, A.G., Gunnaes, A.E., Olaisen, B.R., Wright, D.N., Holt, A., Olsen, A.: Nano-structural properties of ZnO films for Si based heterojunction solar cells. Thin Solid Films **515**, 8479–8483 (2007)
8. Karazhanov, S.Z., Ravindran, P., Kjekshus, A., Fjellvag, H., Svensson, B.G.: Electronic structure and optical properties of ZnX (X = O, S, Se, Te): a density functional study. Phys. Rev. B **75**, 155104 (2007)
9. Sato, K., Hanafusa, M., Noda, A., Arakawa, A., Uchida, M., Asahi, T., Oda, O., Cryst, J.: Development of pure green LEDs based on ZnTe. phys. stat. sol. (a) **180**(1), 214–215 (2000)
10. Anderson, O.K.: prescription for the density of states of liquid iron. Phys. Rev. B **12**(8), 2869 (1975)
11. Cottenier, S.: Density Functional Theory and the Family of (L) APW-Methods: A Step-by-Step Introduction. Instituut voor Kern- en Stralingsfysica, K. U. Leuven, Belgium (2002)
12. Blaha, P., Schwarz, K., Madsen, G.K.H., Kvasnicka, D., Luitz, J.: WIEN2k, An Augmented Plane Wave Plus Local Orbitals Program for Calculating Crystal Properties. Vienna University of Technology, Vienna (2008)
13. Murnaghan, F.D.: The Compressibility of Media under Extreme Pressures. Proc. Natl. Acad. Sci. USA **30**, 5390 (1944)
14. Okuyama, H., Kishita, Y., Ishibashi, A.: Quaternary alloy Zn1–xMgxSySe1–y. Phys. Rev. B **57**, 2257 (1998)
15. Biering, S., Schwerdtfeger, P.: A comparative density functional study of the low pressure phases of solid ZnX, CdX, and HgX: trends and relativistic effects. J. Chem. Phys. **136**, 034504 (2012)
16. Ameri, M., Rached, D., Rabah, M., El Haj Hassan, F., Khenata, R., Doui-Aici, M.: First principles study of structural and electronic properties of Be$_x$Zn$_{1-x}$S and Be$_x$Zn$_{1-x}$Te alloys. phys. state sol. (b) **245**(1), 106–113 (2008)
17. Saib, S., Bouarissa, N.: Structural parameters and transition pressures of ZnO: ab-initio calculations. phys. stat. sol. (b) **244**, 1063–1069 (2007)
18. Özgur, Ü., Aliov, Y.I., Liu, C., Teke, A., Rechicov, M.A., Dogan, S., Avrutin, V., Cho, S.-J., Morkoç, H.: Appl. Phys. Rev. **98**, 041301 (2005)
19. Mohammadi, A.S., Baizaee, S.M., Salehi, H.: Density functional approach to study electronic structure of ZnO single crystal. World Appl. Sci. J. **14**(10), 1530–1536 (2011)
20. Vegard, L.: Die Konstitution der Mischkristalle und die Raumfüllung der Atome. Zeitschrift für Physik **5**, 17–26 (1921)

Study and Simulation of a New Structures Containing GaInAsSb/GaInSb for Photovoltaic

A. Aissat[1,2(✉)], H. Guesmi[1], and J. P. Vilcot[2]

[1] Laboratory LATSI, University of Blida 1, 09000 Blida, Algeria
sakre23@yahoo.fr
[2] Institut d'Electronique, de Microe électronique et de Nanotechnologie (IEMN),
UMR CNRS 8520, Université des Sciences et Technologies de Lille 1,
Avenue Poincaré, BP 60069, 59652 Villeneuve d'Ascq, France

Abstract. This article aims to modeling and to simulate new solar cells for a photovoltaic application with semiconductor materials III-V, more exactly based of $InAs_xSb_{1-x}/GaAs$. We initially studied the influence of the indium concentration on the various alloy parameters GaInAsSb epitaxed on GaInSb substratum. Indeed the increase of the density of indium decreases the band gap of the structure, in order to absorb the maximum of solar spectrum. The study includes results giving the variations of the various factors influencing the efficiency on conversion according to the indium concentration in the structure.

Keywords: Semiconductor materials · Solar cell · Photovoltaic

1 Introduction

During last year, the researchers aim at improving the efficiency of solar cells, to override limit of conversion estimated at 30% by Schockley-Quisser [1, 2]. The development of these cells passes essentially by the control of materials used in the conception of components. The technologies based on the silicon are promising for the microelectronics field, but the major challenges live in the improvement of their electronic properties, in particular the properties of transport. Recently, most of the laboratories are directed to the research for new materials. Semiconductor III-V presents mostly a structure of band in direct gap, the maximum of the valence band and the minimum of the conduction band are in the same points of the Brillouin zone, what gives us a high detection efficiency and a high light emission. The practical relevance of these materials improves the possibility of making a ternary and quaternary substitution of the partial substitution of one of the elements by other one of the same column in the periodic table because of these good physical and optical properties [3], who helps to create an optoelectronics device. But the challenge is to find a system of materials with a coupled network parameter and lower band gap energy. The InAsSb appeared as an attractive material during recent years, due to its high mobility and low band gap energy [4, 5]. The study presents in this work, concerns the modeling, the simulation and the characterization of a photovoltaic cell, for that we study the influence of the indium concentration on the various physical parameters of alloy

© Springer International Publishing AG 2018
M. Hatti (ed.), *Artificial Intelligence in Renewable Energetic Systems*, Lecture Notes in Networks and Systems 35, https://doi.org/10.1007/978-3-319-73192-6_55

$Ga_{1-x}In_xSb$, as well as the simulation and the study of the effect of thickness of the insulating layer and the fraction of indium on the efficiency on conversion of the cell.

2   Theoretical Model

The following section describes the physical models used for this simulation.

Vegard's law is used to calculate the ternary and quaternary mesh parameters [6].

$$a_{GaInSb}(x, y) = xa_{InSb} + (1 - x)a_{GaAs} \tag{1}$$

$$a_{(GaInAsSb)}(x, y) = xya_{(InSb)} + (1 - x)ya_{GaSb} + x(1 - y)a_{InAs} + (1 - x)(1 - y)a_{GaSb} \tag{2}$$

In these equations, x and y are the compositions of materials used.

The value of the critical thickness in the case of a material of sphalerite structure epitaxial on the surface 001 is given by the following equation [7].

$$h_c = \frac{a_{epit}}{k \cdot \sqrt{2} \cdot \pi \cdot \Delta_a} \times \frac{1 - (0.25 \cdot Y)}{(1 + Y)} \times \ln\left(\frac{h_c\sqrt{2}}{a_{epit}} + 1\right) \tag{3}$$

Or a_{epit} is the lattice parameter of the relaxed layer Δ_a is the parametric detuning and Y is the Poisson's coefficient given by:

$$Y = \frac{C_{12}}{c_{11} + c_{12}}$$

With C_{ij} are the elastic coefficients (cm^{-2}).

K is a coefficient equal to 1 in the case of a super lattice, to 2 for a quantum well and to 4 in the case of a single layer. The strain bandgap equation is given by [8]:

$$E_g^{cont} = E_c - E_v = E_g + \Delta E_c^{hyd} - \Delta E_{v,moy}^{hyd} - max\left(\Delta E_{hh}^{cisa}, \Delta E_{lh}^{cisa}\right) \tag{4}$$

Or ΔE_c^{hyd} and $\Delta E_{v,moy}^{hyd}$ are the energetic shifts of the center of gravity of the valence band and the conduction band at k = 0 induced by the hydrostatic stress, ΔE_{hh}^{cisa}, ΔE_{lh}^{cisa} are the energetic shift induced by the shear stress in each of the bands constituting the valence band [9].

3   Electric Characteristics of a Photovoltaic Cell

In a photovoltaic cell, two currents oppose, the current of illumination (photocurrent I_{ph}) and the diode current called current of darkness I_{obs} which results from the polarization of the component. The resultant current I (V) is [10, 11].

$$I(V) = I_{obs}(V) - I_{ph} \tag{5}$$

$$I_{obs}(V) = I_s\left(e^{\left(\frac{qV}{nkT}\right)} - 1\right) \tag{6}$$

$$I_{ph} = q \int_{\lambda_2}^{\lambda_1} F(\lambda)EQE_{Total}(\lambda)d\lambda \tag{7}$$

Or V represent the tension to the borders of the junction, q elementary load, k is the constant of Boltzmann ($k = 1.38 \times 10^{-23}$ J $\cdot$ K^{-1}), T is the temperature (K), F (λ) is the solar spectrum, λ is the wavelength (m), and EQE (λ) is the external quantum efficiency (%). I_s is the diode saturation current, n the diode ideality factor according to the quality of junction (equal to 1 if the diode is ideal and equal to 2 if the diode is real). From the current-voltage characteristic of a photovoltaic cell, the parameters specific to the photovoltaic cell can be deduced:

I_{sc}: Current of short circuit (obtained for V = 0).
V_{OC}: Tension in open circuit (obtained for I = 0).
I_M: Current to the maximal power of functioning of the PV cell.
V_M: Tension in the maximal power of functioning of the PV cell.
η: Conversion efficiency.
FF: Form factor.

$$\eta = \frac{V_M I_M}{P_i \cdot A} = \frac{FF \cdot V_{OC} \cdot I_{CC}}{P_i \cdot A} \tag{8}$$

P_i is the power of illumination received by unit area.

4 Results and Discussion

This part is devoted to the simulation and the determination of the characteristics of the structure $Ga_{1-x}In_xAs_ySb_{1-y}/Ga_{1-x}In_xSb$. First of all we are going to vary the concentration of indium on the alloy $Ga_{1-x}In_xSb$, in continuation the concentrations of indium and of antimony on the interval [0, 1] in the quaternary, in order to have the appropriate percentages of the elements that fit into the component. According to this simulation, we fixed the concentration of the antimony to 70% which gives us a better agreement, and the choice of 30% of indium returns to the high cost of this element. After the choice of the adequate concentrations, the second simulation of the disagreement was realized in 2D while varying the concentration of indium in the ternary on the entire interval [0, 1]. Figure 1 shows the variation of the disagreement according to the indium in the ternary, we note a increase of 7% also best values of disagreement are obtained for concentrations lower than 30%, we also note that there is a single constraint (forced in compression) whatever the indium concentration. Figure 2 shows the variation of the critical thickness according to the indium concentration in the ternary. We can noted that more the concentration increases more we have a decrease of the critical thickness; also better results are obtained for x < 30%.

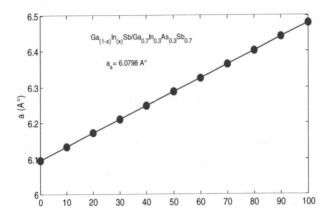

Fig. 1. Variation of the disagreement according to the indium concentration.

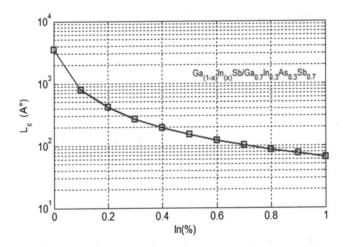

Fig. 2. The critical thickness according to the composition x of indium.

Figure 3 shows the variation of the gap of alloy $Ga_{1-x}In_xSb$ forces on the quaternary $Ga_{1-x}In_xAs_ySb_{1-y}$. We notice a reduction of the width in the gap band with an increase of the indium concentration, and notice also; because we have a single constraint (forced in compression); that the gap of the heavy holes (E_{ghh}) is always upper to the light ones (E_{gLh}). Figure 4 shows the variation of coefficient of absorption according to the energy of the incidental photon on the material for several concentrations of indium. We notice that for the energies less than 0.6 eV the coefficient of absorption is zero whatever the value of indium concentration, because this energy is insufficient so that electrons can cross the gap band of the ternary alloy, as soon as the energy exceeds this value the coefficient of absorption increases. Figure 5 shows the variation of absorption coefficient in function of the indium concentration for fixed incidental photon energies. Seen the decrease of the gap according to the concentration of indium, an increase of the coefficient of absorption is obtained.

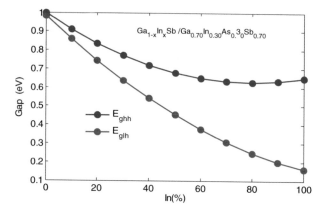

Fig. 3. Variation of the band gap according to the indium concentration.

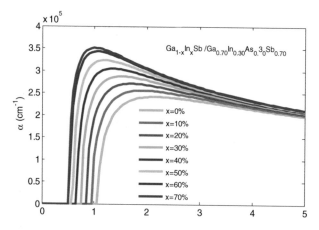

Fig. 4. Variation of the absorption coefficient according to the incidental energy.

We also notice a saturation of the latter with the increase of the energy of the incidental photon; this saturation can be explained that with a strong incidental energy is a wider excitement of electrons, i.e. all the electrons of the band of valence have the necessary energy to immigrate towards the band of conduction. Figure 6 shows the variation of the photo-current under the spectra AMG1.5 which takes into account the direct and indirect radiations of the solar radiation. We can note an increase of the photo-current with the increase of the concentration of indium until 50% and values between 80 and 80.5 mA/cm² for concentrations superior in 50%. This saturation is due to the strong shortening of the band gap for concentrations of indium interesting.

Figure 7 shows the current-voltage characteristic of the structure studied for several concentrations of indium. An improvement of amplitude of current was observed with the increase of the concentration of indium, this improvement pulled an increase of the power freed by this structure (Fig. 8). Figure 9 shows the variation of the tension of the

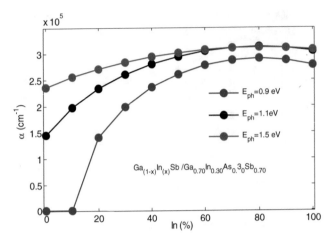

Fig. 5. Variation of the coefficient of absorption according to the indium concentration

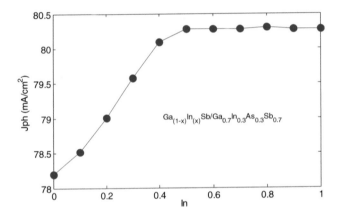

Fig. 6. Variation of the photo-current according to the concentration of indium

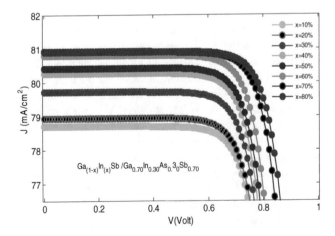

Fig. 7. Current-Voltage characteristic for several indium concentrations.

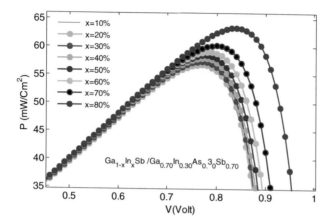

Fig. 8. Variation of the power for several concentrations of indium

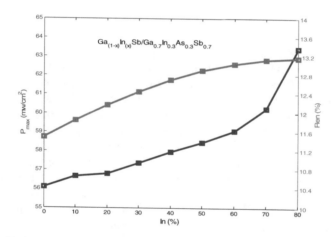

Fig. 9. Variation of the Form Factor and V_{OC} according to the indium concentration

open circuit and the Form Factor according to the concentration of indium. We notice that the tension decreases with the increase of the concentration of indium, and as the Form Factor is a report between the power delivered by the cell unit and the tension of the open circuit, we observe an increase which varies from 49% to 63% of form factor according to the concentration of indium. Figure 10 shows the variation of the maximal power and the efficiency according to the indium concentration, we notice that the power increases with the growth of the indium concentration, and also we have an efficiency which varies between 11.5 and 13.2% is obtained. This efficiency increases with the indium concentration but the adequate choice of the latter coincides with several constraints, among them the high cost of indium and the considerable cost of the technology used for the manufacturing of this type of structure.

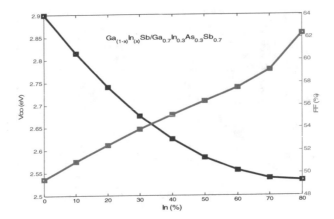

Fig. 10. Variation of the efficiency and the maximal according to the indium concentration

5   Conclusion

In this article, we brought back a study of modeling of the heterostructure GaInAsSb/GaInSb which is mainly dedicated to discern the effects of the levels of in and as concentration on the various parameters of the cell. This study shows that an enormous increase of the Form Factor and the efficiency is obtained with the increase of the concentrations of In. For a 70 and 30% of concentration of As and In respectively on the quaternary, we notice a increase of 51.01% until 62.36% of the Form Factor and from 11.84 to 13.6% of the efficiency for various of the indium concentrations in the ternary, for the same As and In concentrations in quaternary, also a decrease of the width of the band gap band about 0.29 eV on the interval [0, 0.8]. This decrease of the energy of the band gap allows these materials to arrange several wavelengths in applications of telecommunications and can establish an attractive solution to the structures of multi-junctions solar cells.

References

1. Shockley, W., Queisser, H.: Detailed balance limit of efficiency of *p-n* junction solar cells. J. Appl. Phys. **32**, 510 (1961)
2. Green, M.A.: Third generation photovoltaics: ultra-high conversion efficiency at low cost. Prog. Photovolt. Res. Appl. **9**, 123–135 (2001)
3. Sealy, B.J.: J. Inst. Electron. RadioEng. **57**(1S), S2–S12 (1987)
4. Tsukamoto, S., Battacharya, P., Chen, Y.C., Kim, J.H.: J. Appl. Phys. **67**, 6819 (1990)
5. Kudo, M., Mishma, T., Tanaka, T.: J. Vac. Sci. Technol. B 18, **746** (2000)
6. Vurgaftman, I., Meyer, J.R.: Appl. Phys. **94**, 3675 (2003)
7. Koksal, K., Gönül, B.: Critical layer thickness of GaIn (N) As (Sb) QWs on GaAs and InP substrates for (001) and (111) orientations. Eur. Phys. J. **69**, 211–218 (2009)
8. Cuminal, Y.: Réalisation et étude de diodes lasers à base de GaSb émettant vers2.3 mm pour application à l'analyse de gaz (Ph.D.), Montpellier II University, France (1997)

9. Ghione, G.: Semiconductor Devices for High-Speed Optoelectronics. Cambridge University Press, Cambridge (2009)
10. Abdo, F.: Croissance des Couches minces de Silicium par Epitaxie en Phase Liquide a basse Température pour Applications Photovoltaïques, Institut National des Science Appliquées de Lyon (2007)
11. Rimada, J.C., Hernàndez, L., Connolly, J.P., Barnham, K.W.J.: Conversion efficiency enhancement of AlGaAs quantum well solar cells. Microelectron. J. **38**, 513–518 (2007)

Author Index

© Springer International Publishing AG 2018
M. Hatti (ed.), *Artificial Intelligence in Renewable Energetic Systems*, Lecture Notes in Networks and Systems 35, https://doi.org/10.1007/978-3-319-73192-6

Printed in the United States
By Bookmasters